Sustainable Corrosion Inhibitors I: Fundamentals, Methodologies, and Industrial Applications

ACS SYMPOSIUM SERIES **1403**

Sustainable Corrosion Inhibitors I: Fundamentals, Methodologies, and Industrial Applications

Chaudhery Mustansar Hussain, Editor
Department of Chemistry and Environmental Science, New Jersey Institute of Technology
Newark, New Jersey, United States

Chandrabhan Verma, Editor
Interdisciplinary Research Center for Advanced Materials, King Fahd University of Petroleum and Minerals
Dhahran, Saudi Arabia

American Chemical Society, Washington, DC

Library of Congress Cataloging-in-Publication Data

Names: Hussain, Chaudhery Mustansar, editor. | Verma, Chandrabhan, editor.
Title: Sustainable corrosion inhibitors I : fundamentals, methodologies, and industrial applications / Chaudhery Mustansar Hussain, Department of Chemistry and Environmental Science, New Jersey Institute of Technology, Newark, New Jersey, United States, Chandrabhan Verma, Interdisciplinary Research Center for Advanced Materials, King Fahd University of Petroleum and Minerals, Dhahran, Saudi Arabia, editors.
Description: Washington, DC : American Chemical Society, [2021] | Series: ACS symposium series ; 1403 | Includes bibliographical references and index.
Identifiers: LCCN 2021054463 (print) | LCCN 2021054464 (ebook) | ISBN 9780841297906 (hardcover) | ISBN 9780841297890 (ebook other)
Subjects: LCSH: Corrosion and anti-corrosives.
Classification: LCC TA462 .S793 2021 (print) | LCC TA462 (ebook) | DDC 620.1/1223--dc23/eng/20211118
LC record available at https://lccn.loc.gov/2021054463
LC ebook record available at https://lccn.loc.gov/2021054464

The paper used in this publication meets the minimum requirements of American National Standard for Information Sciences—Permanence of Paper for Printed Library Materials, ANSI Z39.48n1984.

Foreword

The ACS Symposium Series is an established program that publishes high-quality volumes of thematic manuscripts. For over 40 years, the ACS Symposium Series has been delivering essential research from world leading scientists, including 36 Chemistry Nobel Laureates, to audiences spanning disciplines and applications.

Books are developed from successful symposia sponsored by the ACS or other organizations. Topics span the entirety of chemistry, including applications, basic research, and interdisciplinary reviews.

Before agreeing to publish a book, prospective editors submit a proposal, including a table of contents. The proposal is reviewed for originality, coverage, and interest to the audience. Some manuscripts may be excluded to better focus the book; others may be added to aid comprehensiveness. All chapters are peer reviewed prior to final acceptance or rejection.

As a rule, only original research papers and original review papers are included in the volumes. Verbatim reproductions of previous published papers are not accepted.

ACS Books

Contents

Indexes

Preface

Corrosion is a highly challenging and damaging phenomenon that causes huge safety and economic losses. The use of traditional organic compounds, especially heterocyclic compounds, is associated with several disadvantages including being non-environmentally friendly. The growing demands of green chemistry and sustainable developments do not permit application of these toxic chemicals for corrosion protection. In view of this, several relatively newer and eco-friendly alternatives have been developed and are being implemented. Many researchers have recently focused on design and application of green corrosion inhibitors. This book collects the new developments on sustainable corrosion inhibitors.

A book compiling the developments in recent advances in sustainable corrosion inhibitors including design, performance, and industrial scale applications, is long overdue. Drs. Hussain and Verma have addressed these recent advances in a book that attends to fundamental characteristics of corrosion, corrosion protection, chronological growths on sustainable corrosion inhibitors and the industrial scale applications. The book offers a collection of natural and synthetic corrosion inhibitors. The corrosion inhibition application of sustainable alternatives is broad ranging.

The book is divided into two volumes. Chapters 1–3 of this volume describe the fundamentals of corrosion including cost of corrosion, corrosion protection, and factors affecting corrosion rate. Chapters 4–6 describe experimental and computational methods of corrosion monitoring. Advantages and disadvantages of these metals along with the instrumentation of these techniques have also been discussed. Chapters 7–9 aim to describe the mechanism of corrosion and corrosion protection in various types of electrolytes including acidic, basic, neutral, and petroleum based electrolytes. Chapter 10 discusses the recent advancements on sustainable corrosion inhibitors for crude oil refinery units, and Chapter 11 talks about sustainable corrosion inhibitors for high temperature corrosion in the oil and gas industry.

Overall, this book is written for scholars in academia and industry, working corrosion engineers, and materials science students focused on applied and engineering chemistry. The editors and contributors are well-known researchers, scientists and true professionals from academia and industry. On behalf of the American Chemical Society (ACS), we are very thankful to the authors of all chapters for their amazing and passionate efforts in the making of this book. Special thanks to Amanda Koenig (acquisitions editor) and Hayes Simpson and Hayley Nestor (Technical Editorial at American Chemical Society [ACS]), for their dedicated support and help during this project. In the end all thanks to American Chemical Society (ACS) for publishing the book.

Chaudhery Mustansar Hussain, Ph.D.
Adjunct Professor & Director of Laboratories
Department of Chemistry and Environmental Science
New Jersey Institute of Technology
Newark, New Jersey 07102, United States

Chandrabhan Verma, Ph.D.
Postdoctoral Fellow (PDF)
Interdisciplinary Research Center for Advanced Materials
King Fahd University of Petroleum and Minerals
Dhahran 31261, Saudi Arabia

Chapter 1

An Overview of Corrosion

Brahim El Ibrahimi,[1,2,*] Jéssica Verger Nardeli,[3] and Lei Guo[4]

[1]Faculty of Applied Sciences, Ibn Zohr University, 86153 Aït Melloul, Morocco

[2]Team of Physical Chemistry and Environment, Faculty of Sciences, Ibn Zohr University, 80000 Agadir, Morocco

[3]Universidade Estadual Paulista-UNESP, Instituto de Química, 14800-060 Araraquara, SP, Brazil

[4]School of Material and Chemical Engineering, Tongren University, Tongren 554300, China

[*]Email: brahimmhm@gmail.com.

In its simplest definition, corrosion is the deterioration of desired physicochemical properties of materials by the environment. This natural process can affect both metallic and nonmetallic materials (e.g., plastic, linings, and paint). Metallic corrosion, the main subject of this chapter, is a spontaneous phenomenon by which a metallic material returns to its stable thermodynamic state (i.e., metal oxides or metal sulfides). Most corrosion processes are electrochemical in nature, and their rate is greatly affected by the operating conditions. Based on the corrosion mechanism or the nature of the surrounding environment, metallic corrosion can be classified into different forms. However, corrosion is costly and can lead to serious damage. In order to eliminate or reduce the undesirable effects of corrosion, it is imperative to monitor this process and apply appropriate prevention methods. In this regard, the compromise between a less expensive method and a more effective one to protect against corrosion, as well as engineering fusibility, requires innovative methods. Given this view, this chapter provides an overview of the fundamentals of metallic corrosion through its mechanisms, costs, forms, study and monitoring, and preventative methods.

Corrosion and Affecting Factors

Definition

Corrosion is the deterioration of a material and its desired properties by the environment. This natural process can impact numerous sectors, such as transportation, infrastructure, production, utilities, and manufacturing. It can affect various materials such as engineering materials (metals, alloys, plastics, paint, and rubber), covalent and ionic substances, aggregates (composite materials and concrete), and wood (*1*). For instance, corrosive environments include acidic (hydrochloride acid, sulfuric acid, and nitric acid), alkaline (sodium hydroxide and potassium hydroxide) media, seawater, and saline solutions (*2*).

In terms of metallic materials, according to Fontana's description, corrosion is a spontaneous phenomenon by which a metal material returns to its stable thermodynamic state. Most corrosion processes are electrochemical (i.e., oxidation-reduction reactions) in nature, and they are greatly affected by operating conditions (*3*, *4*).

The field of corrosion has been more widely studied for various reasons. For example, scientists need to have a deeper understanding of the corrosion behavior of newly developed materials (such as alloys). Furthermore, the aqueous and atmospheric phases become more corrosive because of pollution and acidification, as well industrial activities (*5*).

Factors Affecting Corrosion

The corrosion process of metals and alloys in a solution is significantly affected by several factors, as shown in Figure 1 (*6*). Here, we limit this section to highlighting factors on the metallic corrosion. In this context, an elevation of temperature results in an increase in the corrosion rates of metals as the kinetics of electrochemical reactions augment.

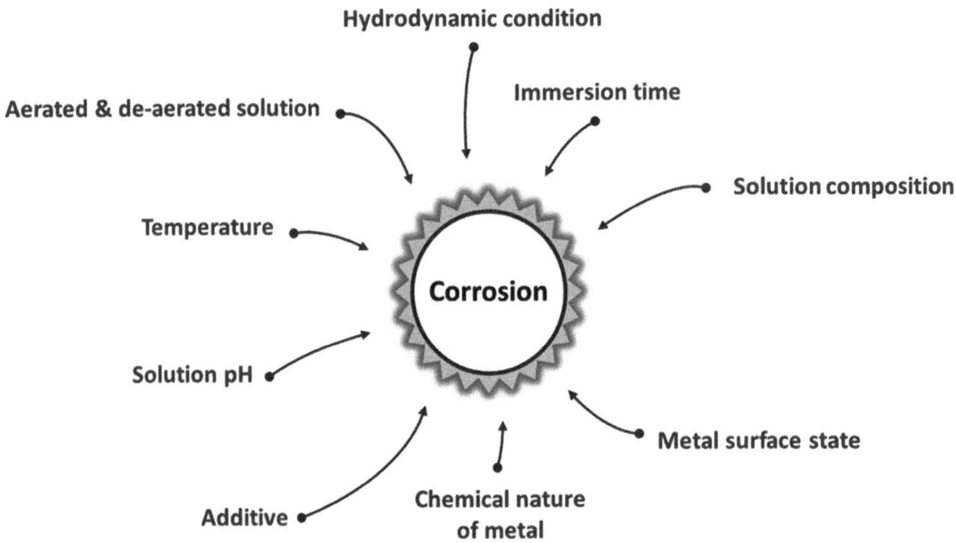

Figure 1. Some factors affecting the rate of aqueous corrosion of metallic materials. Reproduced with permission from reference (6). Copyright 2020 Elsevier.

The flow or movement of corrosive fluids can have an important influence on the corrosion speed of metallic components. In some cases, the increase of flow of corrodent media over the metal surface leads to an amplification of its corrosion rate. In other cases, such as when passivate materials are concerned, an increase in velocity can actually conduct to reduce the damage to the metal.

The state of the metal surface also has a considerable impact on the corrosion severity and what form it takes. Deposits or films can act as initiation sites of attacks. Furthermore, biological growths can modify the chemistry of the underlying surface to stimulate metal dissolution. Tight gaps or rough surfaces can engender the formation of concentrations of cells that promote corrosion. Microstructure also affects the surface state in metallic corrosion. In this instance, the grain orientation and size or the presence of secondary phases can impact how corrosion occurs.

The immersion (or contact) time is a determining factor on the rate of a corrosion attack on equipment. In this regard, some corrosion processes are rapid and severe, whereas other ones are mostly slow—almost indiscernible on a day-to-day basis.

Costs of Corrosion

Corrosion is costly and can lead to major damage. The three key causes for concern with corrosion are economics, safety, and environmental damage. In this context, failure of operating equipment or structures, such as bridges, caused by corrosion can lead to disastrous results (i.e., human injury or even death).

Economic Cost

The economic cost of corrosion includes direct and indirect losses, which can be subdivided into four main categories, namely capital (e.g., replacement of equipment), control (e.g., corrosion control, repair, and maintenance), design (e.g., corrosion allowance and materials of construction), and associated (e.g., insurance and technical support) costs. Table 1 shows the estimated costs of corrosion in some countries around the world. It is well known that corrosion costs have significant consequences on the national economies of countries, which can reach about 2–4% of gross national product, according to an investigation by the National Association of Corrosion Engineers. To minimize such economic losses and combat corrosion, it is necessary for management to implement effective control measures; in this case, about 25% of corrosion costs are avoidable.

Table 1. Corrosion Costs in Select Countries (in Billions of Dollars)

Country	*Corrosion costs (year)*
United States	549 (2010)
Canada	32.8 (2010)
Japan	9.2 (1976–1977)
Russia	6.7 (1969)
United Kingdom	3.2 (1969–1970)

Human Safety

Along with economic losses, corrosion can also affect human safety, so industry personnel particularly need to pay attention to this phenomenon. Failure of steam pipes in nuclear power plants, bridge collapses, gas pipeline bursts, and airline accidents can all cause death. Furthermore, corrosion of iron-based hulls in ships poses a threat to the crew. For many years, lead-based pipes were extensively employed in water transportation until it was found that because of corrosion, lead can enter into water, causing lead poisoning in humans. Fatal accidents related to corrosion are more likely to arise in industrial plants in which corrosive compounds are applied (e.g., cyclohexene in Flixborough in 1974 and hydrogen cyanide in Bhopal in 1984).

Impact on the Environment

Corrosion can further contaminate and affect the ecosystem, destroying irreplaceable fauna or flora. Failures of oil pipelines, tanks, or gas pipelines caused by corrosion can cause severe pollution in the water and air. In addition, the corrosion of equipment employed to limit air pollution can reduce their effectiveness. Such unwanted situations cause volatile contaminants from manufacturing processes to pass into the air. However, corrosion can also damage the environment via extensive use of limited resources of metallic materials, as well as the associated industrial processing (7).

Forms of Metallic Corrosion

Corrosion of metallic materials can manifest in many forms. According to Fontana, there are mainly eight forms of corrosion: general (or uniform) corrosion, pitting corrosion, galvanic corrosion, intergranular corrosion, crevice corrosion, parting, stress corrosion cracking, and erosion corrosion. Additional corrosion forms have also been defined, such as biological corrosion, hydrogen damage, and filiform corrosion. This section will highlight some of these forms.

Uniform Corrosion

Uniform corrosion is the most common, simplest form of metal corrosion. It is characterized by a homogenous attack on the exposed surface by the corrosive environment. Metal loss occurs through a chemical attack or dissolution of the metallic material into metallic ions, which corresponds to oxidation reactions. At higher temperatures, the uniform corrosion of metal arises by combining with other species of corrosive media such as oxygen to produce oxides. One example of this is the dissolution of zinc in a hydrochloric acid solution.

Galvanic Corrosion

Galvanic corrosion, or dissimilar metal corrosion, occurs when two different metallic materials are electrically connected and immersed in an electrolyte. The existence of this difference of potential in the metal couple forces the dissolution of the less noble material acting as an anode. This means that more noble (cathode) metals (platinum and gold) exhibit a low tendency to be dissolved in such a situation. Nevertheless, these cathodic metals could suffer from hydrogen damage. As examples of galvanic systems, the combination of magnesium and copper or magnesium alloys and aluminum shows severe galvanic corrosion behaviors. Furthermore, intense galvanic corrosion can also occur when an anode with a small surface area is coupled with a cathode with a large surface area.

Pitting Corrosion

Pitting corrosion is a form of localized attack that is characterized by a local penetration of the metal leading to a localized loss of metal. This form of corrosion is one of the most insidious and destructive forms, and it can cause components to fail. Pits usually require a long initiation period (from months to years); they are commonly developed and grow in the gravitational direction (i.e., downwards from horizontal surfaces). The initiation of those pits is combined with a local failure of the protective film, which was previously formed on the metal surface. This initial breakdown of the film is related to irregularities in the chemical or physical structure of the target metal surface. Such localized corrosion can be accelerated when dissimilar metals are connected or if there is a concentration of cells.

Crevice Corrosion

In addition to pitting corrosion, crevice corrosion is another localized corrosion that occurs within narrow gaps or openings formed by metal–metal or nonmetal–metal conjunctions. This type of corrosion can occur if there is a local difference in concentration of oxygen, lap joints, gaskets, or when small amounts of liquids are collected and become stagnant in crevices under a bolt or around rivet heads. It is interesting to note that any metallic materials can be susceptible to crevice corrosion. However, metals with the ability to form an oxide film, such in the case of aluminum and stainless steels, are prone to crevice corrosion—particularly in saline media (e.g., seawater).

Erosion Corrosion

Erosion corrosion is a mechanically assisted corrosion type, which also includes corrosion fatigue, fretting corrosion, and cavitation damage. The rapid movement or flow of aggressive media within equipment results in increased mechanical wear on the surface. Whether it is in the corrosion of solid products or soluble ions, the metal is detached from the surface by the movement of fluid. Erosion corrosion is generally revealed in piping systems in which the equipment is exposed to moving fluids such as centrifuges, turbine blades, and pumps. Many factors affect the rate of erosion corrosion such as impingements, hardness, chemical composition, and velocity of the moving fluid. For instance, an increase in velocity generally results in a stronger attack against the metal. Some metallic materials exhibit an outstanding resistance against this type of corrosion, such as all grades of stainless steels, while other ones are more susceptible to erosion corrosion (aluminum and copper).

Stress Corrosion Cracking

In certain aggressive media in which there is tensile stress on the metal, stress corrosion cracking can occur. In such an attack, the metal's surface may appear to be undamaged, but fine cracks can develop throughout it, which can have serious consequences. The major factors that affect this form of corrosion are the chemical composition of the metallic materials and their structure, the nature of the corrosive solution, temperature, and stress.

Lab Studies on Corrosion

The common electrochemical techniques used in corrosion lab testing include destructive and nondestructive techniques. These techniques are widely used to quantitatively and qualitatively examine the corrosion behavior of metallic materials in various corrosive environments.

Destructive Techniques

Weight Loss Method

The weight loss method (or gravimetric method) is a destructive method by which the corrosion rate (C_R) is determined using Equation 1 (*8*). It considers the weight of the specimen of a determined surface (S, in cm^2) before (W_i, in g) and after (W_f, in g) corrosion for a given immersion time (t, in h) in a corrosive medium. Prior to and after the specimen is immersed, its surface should be treated according to the known standards, such as the American Society for Testing and Materials (ASTM) standard procedures (*9*). Therefore, this method is very fast and simple; nevertheless, it does not provide a precise analysis.

$$C_R = \frac{W_i - W_f}{S \times t} \quad (1)$$

Volume of Liberated Hydrogen Gas

Another way to study corrosion is by monitoring the volume of hydrogen gas that has evolved (or gasometric method) with time through corrosion. Such a method is applied when the corrosion process is associated with a degaging of gas, mainly dihydrogen gas (H_2). Generally, it is used to investigate the corrosion of metallic materials in concentrated acidic solutions. Figure 2 shows the evolution of the volume of released hydrogen gas for hot-dip galvanized (HDG) steel in a simulated fresh concrete pore solution with the addition of NaCl in the absence and presence of Na_2MoO_4 (*10*).

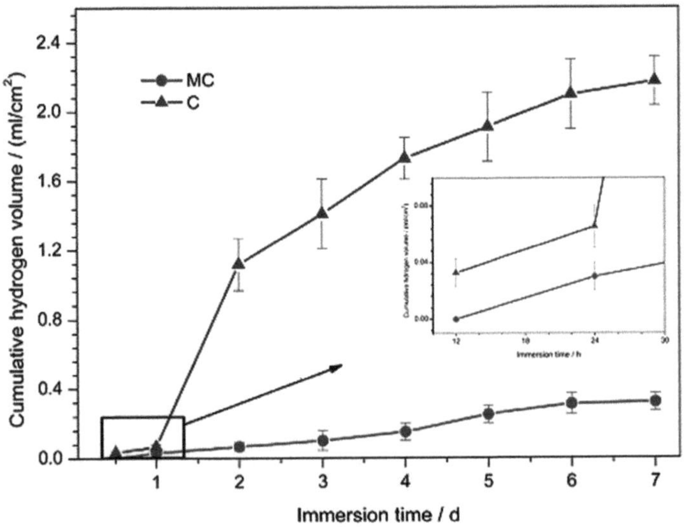

Figure 2. The cumulative volume of liberated hydrogen of HDG steel in saturated $Ca(OH)_2$ solution + 0.2 M KOH + 0.25 M NaCl with (MC) and without (C) Na_2MoO_4 at 10 mM. Reproduced with permission from reference (10). Copyright 2018 Elsevier.

Thermometric Method

The thermometric method is another destructive technique to determine the corrosion behavior of metals and alloys. The evolution of the corrosion process is monitored by the determination of changes in temperature of the corrosion system (metal or corrosive solution) as a function of time employing a calibrated thermometer. After that, the reaction number (R_N) is calculated using Equation 2. In the first stage, the initial temperature of system components before the experiment should be kept at the same value (*11*). Figure 3 shows changes in system temperature because of the corrosion reaction of aluminum in a 2 M HCl medium both with and without the addition of an inhibitor at different concentrations (*12*).

$$R_N = \frac{T_m - T_i}{t} \tag{2}$$

where T_m and T_i denote the maximum temperature (in °C) attained by the corrosion system and its initial temperature (in °C), respectively, and t denotes the time elapsed to reach T_m (in min).

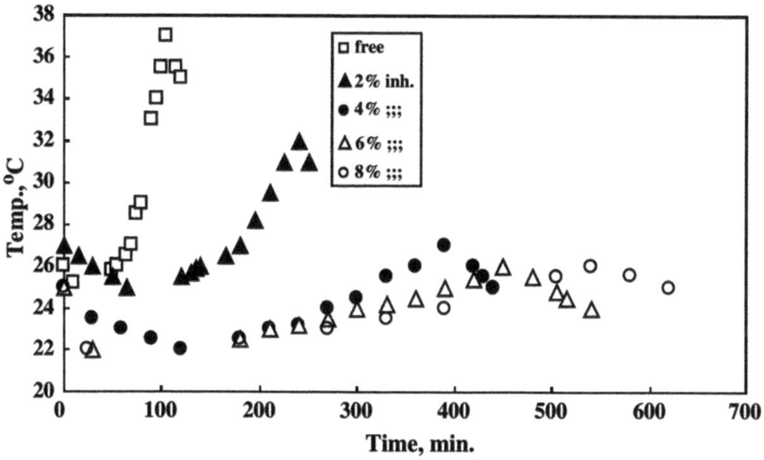

Figure 3. Thermometric curves of aluminum metal in free and inhibited 2M HCl solutions. Reproduced with permission from reference (12). Copyright 2003 Elsevier.

Potentiodynamic Polarization

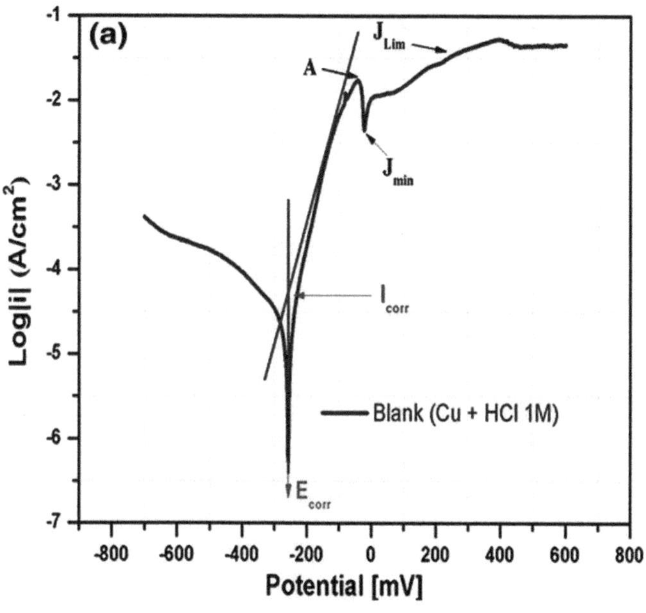

Figure 4. Potentiodynamic polarization curve, Log|i| = f(E), of copper electrode in natural aerated hydrochloric acid solution 25° C. Reproduced with permission from reference (15). Copyright 2017 Springer Nature.

Potentiodynamic polarization corrosion testing provides information on, for example, the difference in the results between certain alloys using the same electrolyte (*13*, *14*). However, this technique is applied only to metallic surfaces or metallic coatings. In this destructive technique, it is not recommended to use the same sample for other analysis sequentially. Figure 4 shows an example of a polarization curve of a copper electrode in an acidic solution with an anodic Tafel line (*15*). Based on this technique, some relevant electrochemical corrosion parameters can be determined, especially corrosion current density (I_{corr}) and corrosion potential (E_{corr}), in addition to cathodic and anodic

Tafel slopes. Corrosion current density is directly related to the corrosion rate, in which an increase of the current corresponds to an increase of corrosion rate and vice versa (*16, 17*).

Nondestructive Techniques

Open Circuit Potential

The open circuit potential (E_{OCP}) is a common electrochemical technique used in corrosion and coating analysis. Monitoring these values over time can provide information about when the system has reached a steady state and when transitions between different states, such as passive and active behaviors, occur (*18–20*). For example, E_{OCP} (also referred to as the free corrosion potential) is the electrical potential difference between two conductors in a specific electrolyte with zero current flow between them (*21, 22*). It was used to evaluate the behavior of tannin corrosion inhibitors in a solution when it comes into contact with an aluminum alloy substrate. Through an analysis of the behavior of the E_{OCP}, it was suggested that, under the experimental conditions, the tannin functioned predominantly as an anodic inhibitor.

Electrochemical Impedances Spectroscopy

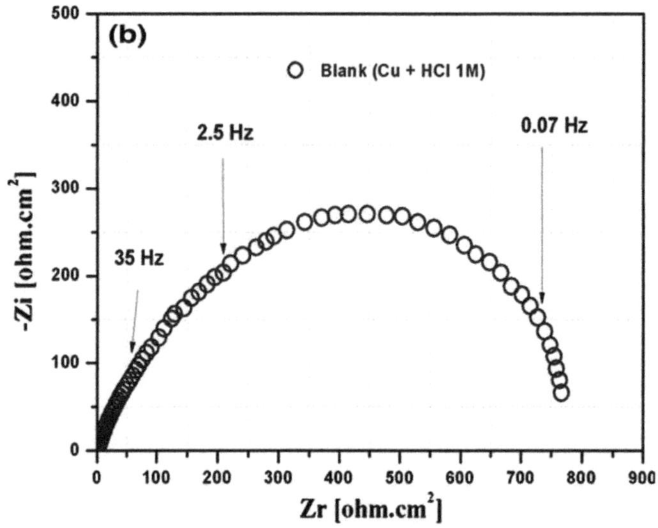

Figure 5. EIS spectrum (present under Nyquist form) of copper electrode in a natural aerated hydrochloric acid solution at 25° C. Reproduced with permission from reference (15). Copyright 2017 Springer Nature.

Electrochemical impedance spectroscopy (EIS) is a powerful electrochemical technique used in corrosion and coating analysis, especially for monitoring corrosion in long-term immersion to achieve higher accuracy. Figure 5 illustrates the EIS spectrum of a copper electrode immersed in a hydrochloric acid solution (*15*). EIS data can be presented under two different presentations, namely Nyquist and Bode presentations (*23, 24*). It is further used to estimate the durability of the materials and coatings. An electric current flows past a circuit with various components (resistors, inductors, or capacitors), and that combination provides measurement of impedance. In general, it is a transient technique in which excitation is applied to the system and the response (as a function of frequency) is observed. Many electrical parameters of the system can be determined, and studies can be expanded to combinations of equivalent electrical circuits. Many studies use EIS to examine the stability and protection of coatings for corrosion protection of metallic substrates. Some experiments take place for a short or long period of immersion (*25–28*); however, they allow for a record of a complete study of the system, and they are thus able to predict the durability according to the exposed solution. Nardeli et al. (*29*) used EIS to study different coating formulations intended for corrosion protection of aluminum alloy AA1200. They observed that some formulations failed to protect against corrosion, while others protected longer (approximately one year). Based on this, they proposed another system (also using EIS) in which an inhibitor was added in solution (*30*), and the EIS technique allowed them to observe that the addition of a corrosion inhibitor increased the protection period.

Localized Electrochemical Impedance

A localized EIS is carried out using a smaller sample than in EIS studies, and it allows for the use of conditions aimed at accelerating corrosion since the analysis time is usually shorter. Local electrochemical impedance mapping is used to investigate delamination phenomena on the metal or coating. One of the ways to speed up the study of corrosion is to apply an artificial defect in the sample and analyze the region of interest. Some studies (*31*) use different artificial defects to meet this goal. When a scratch is used as an artificial defect, it is possible to clearly separate the delaminated region from the intact one (*32*).

Electrochemical Frequency Modulation

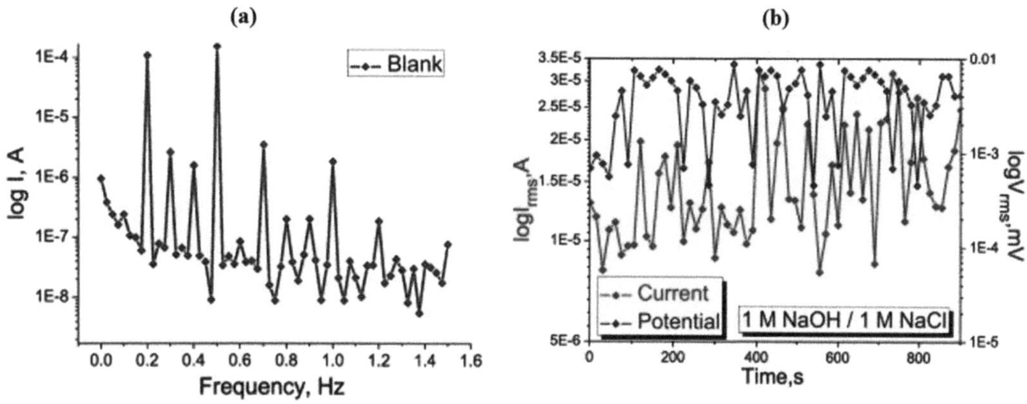

Figure 6. (a) EFM intermodulation spectra and (b) time versus current or potential noise for the carbon steel in 1M NaOH + 1M NaCl mixture solutions. Reproduced with permission from reference (34). Copyright 2021 Elsevier.

Electrochemical frequency modulation (EFM) is another nondestructive corrosion measurement technique, that can directly provide the value of the corrosion current (I_{corr}) without prior knowledge of anodic and cathodic Tafel constants (βa and βc, respectively). Therefore, EFM can determine the kinetic parameters of corrosion (i.e., I_{corr}, βa, and βc) in which small alternative current signals are used, much like in EIS. Nevertheless, unlike EIS, two sine waves at different frequencies are applied simultaneously (*33*). To judge the accuracy of the obtained EFM outcomes, two factors called (causality factors) are calculated in Equations 3 and 4, respectively. Figure 6(a) shows an EFM intermodulation spectrum of a carbon steel sample in a corrosive solution. An analysis of this spectrum demonstrates that the value of I_{corr}, βa, and βc of the investigated metal are 619.1 µA, 254.1, and 399.3 mV dec$-^{1}$, respectively (*34*).

$$CF\text{–}2 = \frac{I_{\omega_1 \pm \omega_2}}{I_{2\omega_1}} \quad (3)$$

$$CF\text{–}3 = \frac{I_{2\omega_1 \pm \omega_2}}{I_{3\omega_1}} \quad (4)$$

where $I_{\omega_1 \pm \omega_2}$ and $I_{2\omega_1 \pm \omega_2}$ are intermodulation current peaks at angular frequencies $\omega_1 \pm \omega_2$ and $2\omega_1 \pm \omega_2$, respectively. $I_{2\omega 1}$ and $I_{3\omega 1}$ are current peaks at harmonic frequency $2\omega_1$ and $3\omega_1$, respectively.

Electrochemical Noise Analysis

Electrochemical noise (EN) is a nondestructive technique that can monitor the current (I_{rms}) and potential (E_{rms}) fluctuations that naturally result for a given metal (electrode) as a function of time. It can be also used to determine the current or potential noises of the tested component by applying potential or current on this electrode, respectively. These fluctuations are measured between two identical electrodes or between a reference and the working electrodes. In comparison to other electrochemical techniques, EN is faster and is capable of providing highly accurate information about the kinetics and mechanisms of metallic corrosion (*35*). Figure 6(b) presents the plots of statistical time versus potential and current changes of carbon steel immersed in a 1M NaOH + 1M NaCl solution. According to these plots, the average potential ($\bar{E}_{rms}$) and current ($\bar{I}_{rms}$) noise fluctuated are around 15.52 µV and 115.3 µA, respectively. These values can be used to calculate the average noise resistance ($\bar{R}_{rms}$), as shown in Equation 5.

$$\bar{R}_{rms} = \frac{\bar{E}_{rms}}{\bar{I}_{rms}} \quad (5)$$

Surface Morphology Analysis

Visual samples observation by optical microscopy or scanning electron microscopy (SEM) shows the existence or absence of corrosion independently of the time of immersion (*36*). It is necessary to note that such observations can help to detect potential corrosion and auxiliaries in the proposed protection mechanism (*37, 38*). For instance, Figure 7(a) presents SEM images of a mild steel coupon after immersion in acidic solution (*39*). It clearly shows severe corrosion. In many cases, SEM can also provide information about when corrosion propagates under a coating (*22*). Energy-dispersive X-ray spectroscopy (EDS) is another analytical technique largely employed for the elemental analysis or chemical characterization of a sample during the corrosion process (*40, 41*). This technique is generally coupled with SEM. An additional technique is used for surface analysis is atomic force microscopy (AFM) (*39, 42, 43*). AFM micrographs provide information about morphology and roughness of the surface at nanoscale with grant of distribution of corrosion inhibitors or corrosion products, as shown in Figure 7(b). In order to identify the chemical composition of the corroded surface in depth, X-ray photoelectron spectroscopy is used (*44, 45*). It is a surface-sensitive quantitative and qualitative spectroscopic technique that can provide an elemental analysis of the surface and its chemical state and electronic structure.

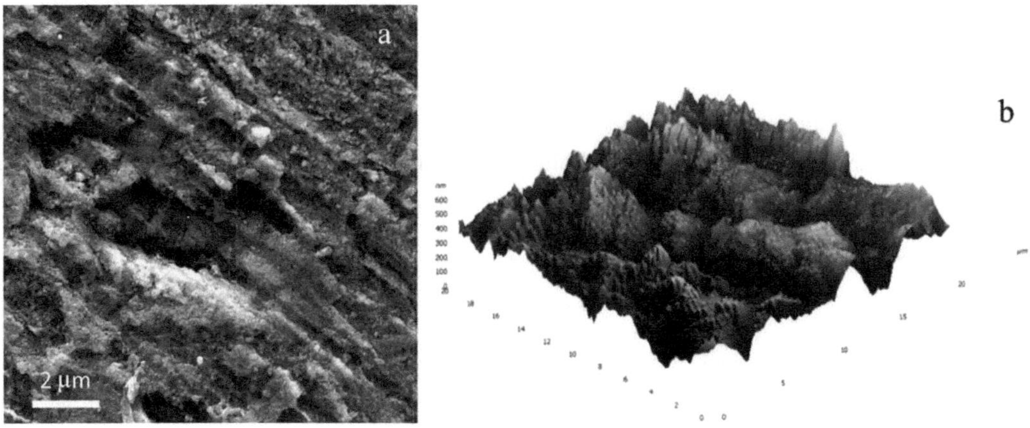

Figure 7. (a) SEM and (b) AFM micrographs of metal surface after immersion in hydrochloric acid solution. Reproduced with permission from reference (39). Copyright 2021 Elsevier.

Corrosion Monitoring

To avoid corrosion-related failures and subsequent severe consequences of corrosion in industrial plants, it is important to control corrosion processes. This can be achieved via continuous monitoring of corrosion of these plants. Corrosion monitoring helps with industrial maintenance and inspection periodically, and it lets observers know if injecting anticorrosion agents (i.e., inhibitors) or applying coatings or other corrosion-resistant measures are needed.

In the industrial sector, the use of various (two or three) monitoring techniques together is mandatory to gain useful insights into the corrosion processes over time. A combination of several techniques is requested because there are many forms of corrosion (i.e., corrosion of equipment can be either localized or general) and each technique has its own strengths and weaknesses. Response time and sensitivity are the two key properties of a given system of corrosion monitoring, and they have an inverse relationship (*46*).

Table 2. Direct and Indirect Corrosion Monitoring Techniques

Direct techniques	*Indirect techniques*
• Intrusive techniques: - Physical techniques (mass-loss methods, electrical resistance, and visual inspection) - Electrochemical direct current techniques (linear polarization resistance and electrochemical noise) - Electrochemical alternative current techniques (harmonic distortion analysis and EIS) • Nonintrusive techniques: - Physical techniques for metal loss (ultrasonics, magnetic flux leakage, electrical field mapping, and radiographic testing) - Physical techniques for crack detection and propagation (acoustic emission)	• Online techniques: - Corrosion products (H_2 monitoring) - Electrochemical techniques (E_{corr}) - Water chemistry parameters (pH, conductivity, and dissolved O_2) - Fluid detection (flow regime and velocity) - Process parameters (pressure and temperature) • Offline techniques: - Water chemistry parameters (alkalinity, metal ion analysis, concentration of dissolved solids, and gas analysis) - Residual inhibitor (inhibitor film) - Chemical analysis of process samples (sulfur, nitrogen, and salt contents)

There are two main kinds of corrosion monitoring techniques: direct and indirect techniques. Direct ones monitor parameters affected by corrosion, while indirect ones aim to get information on key parameters that either affect or are affected by the corrosion or the aggressive media. Moreover, whether these techniques can be intrusive or nonintrusive depends on the necessity of a vessel or a pipe for measuring the rate of corrosion. Table 2 summarizes monitoring techniques of corrosion processes, some which will be highlighted in this section. More details on those techniques can be found in the literature (47).

Ultrasonic Testing

Ultrasonic testing is a direct nonintrusive technique that is used to measure the thickness of the wall of corroded components over a long period of time. This technique is based on the determination of the time that is required for a wave (of known velocity) to travel via the target metallic equipment; such information lets researchers know its thickness or pit depth (consequently, the pitting rate). The latest measure is achieved by coupling a transducer to the metal surface. Ultrasonic testing precision was improved through implementing microcomputers. Development of transducers can also improve precision. This technique is largely employed for inspection of, for example, the integrity of a vessel, or it can be used as an ongoing corrosion monitor if severe metal loss has occurred.

Radiographic Testing

This monitoring technique is based on the utilization of radiographic images to measure the thickness of corroded equipment. The radiographic method has been employed in harsh oil fields and can be used online. It is nevertheless insensitive to deliver real-time information. Further, by employing manual or automated real-time radiographic structures, this technique can inexpensively determine the piping integrity and equipment over large areas.

Flow Velocity

As mentioned in subsection dealing with factors on corrosion process, the flow velocity has a strong influence on this interfacial process. Determining the velocity of a single-phase flow can be concluded from the mass flow rate, taking into consideration the effects of temperature and pressure. In the case of a multiphase flow, the velocity cannot be easily determined. This indirect technique is used online for real-time measurement.

Monitoring Hydrogen Penetration

In acidic corrosive environments, hydrogen gas is generated at the cathode region of the corrodent. In its atomic state, hydrogen can diffuse via the pipe or vessel walls and then recombine to produce hydrogen molecules on the external surface. In this regard, the formation of atomic hydrogen can be used for corrosion monitoring, in which the monitoring devices are placed on the outside walls of the plant components. Such a monitoring technique is largely applied in the petrochemical industry and in oil refining.

Concentration of Dissolved Solids

The concentration of the total dissolved solids can also monitor corrosion processes. This technique aims to determine the sum of all minerals in the medium, which is achieved by the evaporation of the solution. It is particularly valuable in determining the scaling index.

Prevention of Corrosion

The adoption of preventive measures is strictly dependent on many factors, such as the type of corrosion, type of metal, surface state, pH of the solution, presence of ions, temperature, and hydrodynamic condition of the corrosive environment. Therefore, choosing a method to hinder corrosion is not an easy task, and the previously listed factors must be considered before finalizing this choice. According to Vernon, prevention methods against corrosion can be classified into four mains categories: modification of the environment, modification of the metal, modification of the procedures, and protective coating (*48*). For instance, the adopted protection methods in the industrial sector comprise the selection of the appropriate materials for the particular environment and design of equipment (i.e., evade particular geometrical structures that help corrosion to promote, as well as avoid the undesirable conjunction between metal–metal or metal–nonmetal that facilitate corrosion). Also, it is important to minimize mechanical effects, including stresses that can cause fretting, fatigue corrosion, or stress corrosion cracking.

Additionally, one of the precautions to protect the metal surface against corrosion is the application of coatings (i.e., modification of metal surface) or by the addition of inhibitor (i.e., modification of environment) compounds into the aggressive media. There are numerous coating compositions with different applications and durability. In this context, different approaches have been pursued to minimize corrosion problems in metallic substrates, and the most effective ones use organic coatings, namely primers modified with different anticorrosion pigments (*49–52*). In general, numerous studies (*53–55*) have explored the idea of applying environmentally friendly coatings as anticorrosion barriers for different metallic surfaces. For example, an environmentally friendly tannic acid/Zn conversion film was used to protect against corrosion of iron, which exhibited excellent corrosion protection with a prevention effectiveness around of 92.39% (*55*). Polyurethane bio-based coatings prepared from vegetable oils with incorporation of zinc microflakes showed better anticorrosion performance for AA7475 aluminum alloy. This was attributed to its self-healing effect, efficient anodic protection, and cathodic inhibition by corrosion products (*56*). Effective corrosion protection and self-repair effects depend on the response of the system.

Another way of protecting metallic substrates against corrosion is the incorporation of inhibitors (i.e., anticorrosion compounds) in the solution. Inhibitors of corrosion are chemical substances that, when added in small amounts into a corrosive medium in which a metallic material would corrode, will decrease the corrosion rate of the material. Wide ranges of organic and inorganic substances have been employed as potent, cost-effective corrosion inhibitors (*57*). Concerning the toxicity of extensively applied corrosion inhibitors such as hexavalent chromium and 2-mercaptobenzothiazole, it is acknowledged that these compounds are highly toxic and carcinogenic. Furthermore, it is documented that these inhibitors are very poisonous to aquatic life and have long-term effects (*58*). Accordingly, discovering and developing effective and eco-friendly replacements of toxic corrosion inhibitors have become a trend in this field. In this regard, many studies have explored corrosion inhibitors derived from plant extracts as additives in aggressive solutions (*59–63*). Such extracts contain a variety of natural organic products and substances that rich in functional groups, and they have been reported as being effective, eco-friendly corrosion inhibitors (*64, 65*). The general mechanism by which an organic-based inhibitor compound can prevent metal corrosion is based on its adsorption into the surface, which leads to the formation of protective films. The inhibitor can be absorbed by direct interactions with the metal surface or by indirect adsorption via pre-adsorbed species such as chlorine ions (*66–68*). Inorganic inhibitors, such as molybdate anions, mitigate the corrosion by either stabilizing or repairing the deficiencies in the formed passive film on the metal's surface by establishing insoluble compounds and preventing the adsorption of corrosive anions through competitive adsorption (*69, 70*).

Innovative methods are needed to find a compromise between a less expensive method and a more effective one to protect against corrosion, as well as engineering fusibility. In general, there is a tendency to search for novel environmentally friendly alternatives based on nontoxic, biodegradable inhibitors (*71*). There is also a need to introduce self-healing abilities in the protective coatings (*72*).

Conclusion

Rather than being just an inevitable natural phenomenon, corrosion is very important in terms of economics, ecosystem pollutions, and even accidents affecting human safety, which can lead to death. The major goal to understand this spontaneous process is choosing appropriate industrial processes, equipment, materials, or controls to optimize the exploitation of financial and our limited natural resources. The choice of corrosion prevention methods and analysis techniques must be planned to close a corrosion protection mechanism. It also depends on reaching a path in order to improve the formulation of the protection system. Ideally, a complete investigation should contain a combination of different techniques. However, it is more important to understand phenomena in each case and attribute the response to mechanism protection. In the industrial sector, constantly monitoring corrosion by adopting appropriate testing techniques can help a corrosion engineer identify damage on equipment over time and take effective remedial action before it gets out of control. One of the most widely used preventive measures to combat corrosion in the industrial field is the addition of corrosion inhibitors, which involves the elimination or a reduction of corrosion rate to an acceptable level. Because of strict regulations imposed by environmental agencies on the use of inhibitors comprising their synthesis, usage, and discharge, the field of corrosion inhibitors is undertaking dramatic developments. Therefore, inhibitors of corrosion are required to be eco-friendly and safe—or, as they are often referred, sustainable corrosion inhibitors.

References

1. Sastri, V. S. *Green Corrosion Inhibitors: Theory and Practice*, First ed., John Wiley & Sons Ltd, Hoboken, N.J., 2011.
2. El Ibrahimi, B.; Guo, L. Azole-Based Compounds as Corrosion Inhibitors for Metallic Materials, in *Azoles - Synthesis, Properties, Applications and Perspectives*, Kuznetsov, A. ed., IntechOpen, U.K., 2020.
3. Marinescu, M. Recent Advances in the Use of Benzimidazoles as Corrosion Inhibitors. *BMC Chemistry* **2019**, *13*, 136.
4. M.E. Mashuga, M. E.; Olasunkanmi, L. O.; Ebenso, E. E. Experimental and Theoretical Investigation of the Inhibitory Effect of New Pyridazine Derivatives for the Corrosion of Mild Steel in 1 M HCl. *J. Mol. Struct* **2017**, *1136*, 127–139.

5. Cicek, V.; Al-Numan, B. *Corrosion Chemistry*, Scrivener Publishing LLC, Co-published by John Wiley & Sons, Inc. Hoboken, New Jersey, and Scrivener Publishing LLC, Salem, Massachusetts., USA, 2011.
6. El Ibrahimi, B.; Jmiai, A.; Bazzi, L.; El Issami, S. Amino Acids and Their Derivatives as Corrosion Inhibitors for Metals and Alloys. *Arab. J. Chem.* **2020**, *13*, 740–771.
7. Sastri, V. S.; Ghali, E.; Elboujdaini, M. *Corrosion Prevention and Protection Practical Solutions*, First ed., John Wiley & Sons Ltd, USA, 2007.
8. Liu, L.; Cao, T.-T.; Zhang, Q.-W.; Cui, C.-W. Organic Phosphorus Compounds as Inhibitors of Corrosion of Carbon Steel in Circulating Cooling Water: Weight Loss Method and Thermodynamic and Quantum Chemical Studies. *Advances in Materials Science and Engineering* **2018**, *9*, 1653484.
9. ASTM G31-72(2004), *Standard Practice for Laboratory Immersion Corrosion Testing of Metals*, ASTM, West Conshohocken, PA.
10. Wang, Y.-Q.; Kong, G.; Che, C.-S.; Zhang, B. Inhibitive Effect of Sodium Molybdate on the Corrosion Behavior of Galvanized Steel in Simulated Concrete Pore Solution. *Construction and Building Materials* **2018**, *162*, 383–392.
11. Ebenso, E. E.; Eddy, N. O.; Odiongenyi, A. O. Corrosion Inhibitive Properties and Adsorption Behaviour of Ethanol Extract of Piper Guinensis as a Green Corrosion Inhibitor for Mild Steel in H_2SO_4. *African Journal of Pure and Applied Chemistry* **2008**, *2*, 107–115.
12. El-Etre, A. Y. Inhibition of Aluminum Corrosion Using Opuntia Extract. *Corrosion Science* **2003**, *45*, 2485–2495.
13. Holland, R. I. Corrosion Testing by Potentiodynamic Polarization in Various Electrolytes. *Dental Materials* **1992**, *8*, 241–245.
14. Berdimurodov, E.; Kholikov, A.; Akbarov, K.; Guo, L.; Abdullah, A. M.; Elik, M. A Gossypol Derivative as an Efficient Corrosion Inhibitor for St2 Steel in 1 M HCl + 1 M KCl: An Experimental and Theoretical Investigation. *Journal of Molecular Liquids* **2021**, *328*, 115475.
15. Jmiai, A.; El Ibrahimi, B.; Tara, A.; Oukhrib, R.; El Issami, S.; Jbara, O.; Bazzi, L.; Hilali, M. Chitosan as an Eco-Friendly Inhibitor for Copper Corrosion in Acidic Medium: Protocol and Characterization. *Cellulose* **2017**, *24*, 3843–3867.
16. Ait Addi, B.; El Ibrahimi, B.; Ait Addi, A.; Shaban, A.; Addi, E.-H. A.; Hamdani, M. Assessment of Corrosion Inhibition Performance and Adsorption Thermodynamics of Hydrogen Phosphate (HPO_4^{2-}) and Molybdate (MoO_4^{2-}) Oxyanions on Tin in Maleic Acid. *Electroanalysis* **2021**, *33*, 804–819.
17. Chadili, M.; Rguiti, M. M.; El Ibrahimi, B.; Oukhrib, R.; Jmiai, A.; Beelkhaouda, M.; Bammou, L.; Hilali, M.; Bazzi, L. Corrosion Inhibition of 3003 Aluminum Alloy in Molar Hydrochloric Acid Solution by Olive Oil Mill Liquid By-Product. *International Journal of Corrosion* **2021**, *2021*, 6662395.
18. El Ibrahimi, B.; El Mouaden, K.; Jmiai, A.; Baddouh, A.; El Issami, S.; Bazzi, L.; Hilali, M. Understanding the Influence of Solution's pH on the Corrosion of Tin in Saline Solution Containing Functional Amino Acids Using Electrochemical Techniques and Molecular Modeling. *Surf. Interfaces* **2019**, *17*, 100343.
19. El Ibrahimi, B.; Jmiai, A.; El Mouaden, K.; Baddouh, A.; El Issami, S.; Bazzi, L.; Hilali, M. Effect of Solution's pH and Molecular Structure of Three Linear α-Amino Acids on the

Corrosion of Tin in Salt Solution: A Combined Experimental and Theoretical Approach. *J. Mol. Struct.* **2019**, *1196*, 105–118.

20. El Ibrahimi, B.; Jmiai, A.; Somoue, A.; Oukhrib, R.; Chadili, M.; El Issami, S.; Bazzi, L. Cysteine Duality Effect on the Corrosion Inhibition and Acceleration of 3003 Aluminium Alloy in a 2% NaCl Solution. *Port. Electrochimica Acta* **2018**, *36* (6), 403–422.
21. Aramaki, K.; Shimizu, K.; Sakakibara, M.; Nishihara, H. Corrosion of Iron in Anhydrous Acetonitrile Solutions of Some Carboxylic Acids. *Journal of the Electrochemical Society* **1995**, *140*, 1561–1567.
22. Nardeli, J. V.; Fugivara, C. S.; Taryba, M.; Montemor, M. F.; Ribeiro, S. J. L.; Benedetti, A. V. Novel Healing Coatings Based on Natural Derived Polyurethane Modified With Tannins for Corrosion Protection of AA2024-T3. *Corrosion Science* **2020**, *162*, 108213.
23. El Ibrahimi, B.; Baddouh, A.; Oukhrib, R.; El Issami, S.; Hafidi, Z.; Bazzi, L. Electrochemical and in Silico Investigations Into the Corrosion Inhibition of Cyclic Amino Acids on Tin Metal in the Saline Environment. *Surfaces and Interfaces* **2021**, *23*, 100966.
24. El Ibrahimi, B.; Bazzi, L.; El Issam, S. The Role of PH in Corrosion Inhibition of Tin Using the Proline Amino Acid: Theoretical and Experimental Investigations. *RSC Adv.* **2020**, *10*, 29696.
25. Bierwagen, G. P.; He, L.; Li, J.; Ellingson, L.; Tallman, D. E. Studies of a New Accelerated Evaluation Method for Coating Corrosion Resistance - Thermal Cycling Testing. *Progress in Organic Coatings* **2000**, *39*, 67–78.
26. Liu, C.; Wu, H.; Qiang, Y.; Zhao, H.; Wang, L. Design of Smart Protective Coatings With Autonomous Self-Healing and Early Corrosion Reporting Properties. *Corrosion Science* **2021**, *184*, 109355.
27. Raps, D.; Hack, T.; Wehr, J.; Zheludkevich, M. L.; Bastos, A. C.; Ferreira, M. G. S.; Nuyken, O. Electrochemical Study of Inhibitor-Containing Organic–Inorganic Hybrid Coatings on AA2024. *Corrosion Science* **2009**, *51*, 1012–1021.
28. Recloux, I.; Andreatta, F.; Druart, M.-E.; Coelho, L. B.; Cepek, C.; Cossement, D.; Fedrizzi, L.; Olivier, M.-G. Stability of Benzotriazole-Based Films Against AA2024 Aluminium Alloy Corrosion Process in Neutral Chloride Electrolyte. *Journal of Alloys and Compounds* **2018**, *735*, 2512–2522.
29. Nardeli, J. V.; Snihirova, D. V.; Fugivara, C. S.; Montemor, M. F.; Pinto, E. R. P.; Messaddecq, Y.; Benedetti, A. V. Localised Corrosion Assessement of Crambe-Oil-Based Polyurethane Coatings Applied on the ASTM 1200 Aluminum Alloy. *Corrosion Science* **2016**, *111*, 422–435.
30. Nardeli, J. V.; Fugivara, C. S.; Taryba, M.; Pinto, E. R. P.; Montemor, M. F.; Benedetti, A. V. Tannin: A Natural Corrosion Inhibitor for Bare and Coated Aluminum Alloys. *Progress in Organic Coatings* **2019**, *135*, 368–381.
31. Jorcin, J.-B.; Aragon, E.; Merlatti, C.; Pébère, N. Delaminated Areas Beneath Organic Coating: A Local Electrochemical Impedance Approach. *Corrosion Science* **2006**, *48*, 1779–1790.
32. Nardeli, J. V.; Fugivara, C. S.; Taryba, M.; Montemor, M. F.; Benedetti, A. V. Self-Healing Ability Based on Hydrogen Bonds in Organic Coatings for Corrosion Protection of AA1200. *Corrosion Science* **2020**, *177*, 108984.
33. Obot, I. B.; Onyeachu, I. B. Electrochemical Frequency Modulation (EFM) Technique: Theory and Recent Practical Applications in Corrosion Research. *Journal of Molecular Liquids* **2018**, *249*, 83–96.

34. Berdimurodov, E.; Kholikov, A.; Akbarov, K.; Guo, L. Inhibition Properties of 4,5-Dihydroxy-4,5-di-P-Tolylimidazolidine-2-Thione for Use on Carbon Steel in an Aggressive Alkaline Medium With Chloride Ions: Thermodynamic, Electrochemical, Surface and Theoretical Analyses. *Journal of Molecular Liquids* **2021**, *327*, 114813.
35. Obot, I. B.; Onyeachu, I. B.; Zeino, A.; Umoren, S. A. Electrochemical Noise (EN) Technique: Review of Recent Practical Applications to Corrosion Electrochemistry Research. *Journal of Adhesion Science and Technology* **2019**, *33*, 1453–1496.
36. Berdimurodov, E.; Kholikov, A.; Akbarov, K.; Xu, G.; Abdullah, A. M.; Hosseini, M. New Anti-Corrosion Inhibitor (3ar,6ar)-3a,6a-di-p-Tolyltetrahydroimidazo[4,5-D]Imidazole-2,5(1H,3H)-Dithione for Carbon Steel in 1 M HCl Medium: Gravimetric, Electrochemical, Surface and Quantum Chemical Analyses. *Arabian Journal of Chemistry* **2020**, *13*, 7504–7523.
37. Verma, D. K.; Kazi, M.; Alqahtani, M. S.; Syed, R.; Berdimurodov, E.; Kaya, S.; Salim, R.; Asatkar, A.; Haldhar, R. N–Hydroxybenzothioamide Derivatives as Green and Efficient Corrosion Inhibitors for Mild Steel: Experimental, DFT and MC Simulation Approach. *Journal of Molecular Structure* **2021**, *1241*, 130648.
38. Jmiai, A.; El Ibrahimi, B.; Tara, A.; El Issami, S.; Jbara, O.; Bazzi, L. Alginate Biopolymer as Green Corrosion Inhibitor for Copper in 1 M Hydrochloric Acid: Experimental and Theoretical Approaches. *J. Mol. Struct.* **2018**, *1157*, 408–417.
39. Shahmoradi, A. R.; Ranjbarghanei, M.; Javidparvar, A. A.; Guo, L.; Berdimurodov, E.; Ramezanzadeh, B. Theoretical and Surface/Electrochemical Investigations of Walnut Fruit Green Husk Extract as Effective Inhibitor for Mild-Steel Corrosion in 1M HCl Electrolyte. *Journal of Molecular Liquids* **2021**, *338*, 116550.
40. Berdimurodov, E.; Kholikov, A.; Akbarov, K.; Obot, I. B.; Guo, L. Thioglycoluril Derivative as a New and Effective Corrosion Inhibitor for Low Carbon Steel in a 1 M HCl Medium: Experimental and Theoretical Investigation. *Journal of Molecular Structure* **2021**, *1234*, 130165.
41. Berdimurodov, E.; Kholikov, A.; Akbarov, K.; Guo, L. Experimental and Theoretical Assessment of New and Eco–Friendly Thioglycoluril Derivative as an Effective Corrosion Inhibitor of St2 Steel in the Aggressive Hydrochloric Acid With Sulfate Ions. *Journal of Molecular Liquids* **2021**, *335*, 116168.
42. Yi, P.; Dong, C.; Xiao, K.; Li, X. Study on Corrosion Behavior of β-Sn and Intermetallic Compounds Phases in SAC305 Alloy by in-Situ EC-AFM and First-Principles Calculation. *Corrosion Science* **2021**, *181*, 109244.
43. Sebastian, D.; Yao, C.-W.; Nipa, L.; Lian, I.; Twu, G. Corrosion Behavior and Mechanical Properties of a Nanocomposite Superhydrophobic Coating. *Coatings* **2021**, *11*, 652.
44. Boonruang, C.; Sanumang, W. Effect of Nano-Grain Carbide Formation on Electrochemical Behavior of 316L Stainless Steel. *Scientific Reports* **2021**, *11*, 12602.
45. Chen, T.; Gan, H.; Chen, Z.; Chen, M.; Fu, C. Eco-Friendly Approach to Corrosion Inhibition of AA5083 Aluminum Alloy in HCl Solution by the Expired Vitamin B1 Drugs. *Journal of Molecular Structure* **2021**, *1244*, 130881.
46. Dean, S. W. Corrosion: Fundamentals, Testing and Protection, in *Metals Park, OH, ASM International*, B. Covino ed., 2003, pp. 533-541.
47. Technical Committee Report: *Techniques for Monitoring Corrosion and Related Parameters in Field Applications*, NACE International, Houston, 1999.

48. Vernon, W. H. J. Metallic Corrosion and Conservation. *The Conservation of Natural Resources* **1957**, 105–133.
49. Buchheit, R. G.; Guan, H.; Mahajanam, S.; Wong, F. Active Corrosion Protection and Corrosion Sensing in Chromate-Free Organic Coatings. *Prog. Org. Coat.* **2003**, *47*, 174–182.
50. Mrad, M.; Amor, Y. B.; Dhouibi, L.; Montemor, M. F. Corrosion Prevention of AA2024-T3 Aluminum Alloy With a Polyaniline/Poly(γ-Glycidoxypropyltrimethoxysilane) Bilayer Coating: Comparative Study With Polyaniline Mono-Layer Feature. *Surf. Coat. Technol.* **2018**, *337*, 1–11.
51. Visser, P.; Terryn, H.; Mol, J. M. C. On the Importance of Irreversibility of Corrosion Inhibitors for Active Coating Protection of AA2024-t3. *Corros. Sci.* **2018**, *140*, 272–285.
52. Visser, P.; Liu, Y.; Zhou, X.; Hashimoto, T.; Thompson, G. E.; Lyon, S. B.; Ven, L. G. J. V. D.; Mol, A. J. M. C.; Terryn, H. A. The Corrosion Protection of AA2024-t3 Aluminium Alloy by Leaching of Lithium-Containing Salts From Organic Coatings. *Faraday Dicussions* **2015**, *180*, 511–526.
53. Verma, C.; Ebenso, E. E.; Quraishi, M. A.; Hussain, C. M. Recent Developments in Sustainable Corrosion Inhibitors: Design, Performance and Industrial Scale Applications. *Materials Advances* **2021**, *2*, 3806–3850.
54. Ni, L.; Li, S.; Liu, Y.; Jiang, X.; Cai, P.; Feng, L.; Zhang, S.; Gao, X. Fabrication of Active Corrosion Protection Waterborne Polyurethane Coatings Using Cerium Modified Palygorskite Nanocontainers. *Journal of Applied Polymer Science* **2021**, *138*, 50899.
55. Wu, Y.; Yan, R.; Duan, Y.; Qiu, J.; Chen, T.; Ma, H. An Environmental-Friendly Tannic Acid/Zn Conversion Film With a Good Corrosion Protection for Iron. *Surfaces and Interfaces* **2021**, *24*, 101078.
56. Nardeli, J. V.; Fugivara, C. S.; Taryba, M.; Montemor, M. F.; Benedetti, A. V. Biobased Self-Healing Polyurethane Coating With Zn Micro-Flakes for Corrosion Protection of AA7475. *Chemical Engineering Journal* **2021**, *404*, 126478.
57. El Ibrahimi, B. Sustainable Corrosion Inhibitors for Copper and its Alloys, in *Sustainable Corrosion Inhibitors*, Inamuddin ed., Materials Research Forum LLC, Materials Research Foundations, U.S.A, 2021, pp. 175-203.
58. El Ibrahimi, B.; Guo, L.; Nardeli, J. V.; Oukhrib, R. The Application of Chitosan-Based Compounds against Metallic Corrosion, in *Chitin and Chitosan - Physicochemical Properties and Industrial Applications*, IntechOpen, U.K., 2021.
59. Martinez, S.; Stem, I. Inhibitory Mechanism of Low-Carbon Steel Corrosion by Mimosa Tannin in Sulphuric Acid Solutions. *J. Appl. Electrochem.* **2001**, *31*, 973–978.
60. Martinez, S.; Stern, I. The Thermodynamic Characterization of Metal Dissolution and Inhibitor Adsorption Processes in the Low Carbon Steel/Mimosa Tannin/Sulfuric Acid System. *App. Surf. Sci.* **2002**, *199*, 83–89.
61. James, A. O.; Atela, A. O. Aloe Vera: An Inhibitor of Aluminium Corrosion in Hydrochloric Acid Solution. *Int. J. Pure Appl. Chem.* **2008**, 3, 159–163.
62. Gerengi, H.; Sahin, H. I. Schinopsis Lorentzii Extract as a Green Corrosion Inhibitor for Low Carbon Steel in 1 M HCl Solution. *Ind. Eng. Chem. Res.* **2012**, *51*, 780–787.

63. Krishnegowda, P. M.; Venkatesha, V. T.; Krishnegowda, P. K. M.; Shivayogiraju, S. B. Acalypha Torta Leaf Extract as Green Corrosion Inhibitor for Mild Steel in Hydrochloric Acid Solution. *Ind. Eng. Chem. Res.* **2013**, *52*, 722–728.
64. Palou, R. M.; Xomelt, O. O.; Likhanova, N. V. Environmentally Friendly Corrosion Inhibitors, in *Developments in Corrosion Protection*, M. Aliofkhazraei ed., IntechOpen, U.K., 2014.
65. Aliofkhazraei, M. *Developments in Corrosion Protection*, IntechOpen, U.K., 2014.
66. Heragh, M. F.; Tavakoli, H. Synergetic Effect of the Combination of Prosopis Farcta Extract With Sodium Dodecyl Sulfate on Corrosion Inhibition of St37 Steel in 1M HCl Medium. *Journal of Molecular Structure* **2021**, *1245*, 131086.
67. Pessu, F.; Barker, R.; Chang, F.; Chen, T.; Neville, A. Iron Sulphide Formation and Interaction With Corrosion Inhibitor in H_2S-Containing Environments. *Journal of Petroleum Science and Engineering* **2021**, *207*, 109152.
68. Natarajan, R.; Zahir Said Al Shibli, F. S. Synthesis of Biomass Derived Product From Ziziphus Spina-Christi and Application for Surface Protection of Metal Under Acidic Environment-Performance Evaluation and Thermodynamic Studies. *Chemosphere* **2021**, *284*, 131375.
69. Wilcox, G. D.; Gab, D. R.; Warwick, M. E. The role of molybdates in corrosion prevention. *Corros. Rev.* **1986**, *6*, 327–365.
70. J. Roti, and K. Saeder, *A Comprehensive Evaluation of Molybdate Based Cooling Water Treatment Technology*, Paper TP-88-03, Cooling Tower Institute, 1988.
71. Devarayan, K.; Mayakrishnan, G.; Sulochana, N. Green Inhibitors for Corrosion of Metals: A Review. *Chem. Sci. Rev. Lett.* **2021**, *1*, 1–18.
72. Zadeh, M. A.; Tedim, J.; Zheludkevich, M.; Zwaag, S. V. D.; Garcia, S. J. Synergetic Active Corrosion Protection of AA2024-T3 by 2d- Anionic and 3d-Cationic Nanocontainers Loaded With Ce and Mercaptobenzothiazole. *Corros. Sci.* **2018**, *135*, 35–45.

Books From California
51 W. Easy St
Simi Valley, CA 93065
UNITED STATES
customerservice@booksfromca.com

Books From California
51 W. Easy St
Simi Valley, CA 93065
UNITED STATES

To: Alibris APEX DC 76926368-71 - APEX
800 Avondale Ave.

Grandview Heights, OH 43212-3473
UNITED STATES

If this order is shipping from the US to an international buyer, remember to attach proper documentation. Complete and attach USPS Form CN-22 for all Priority Mail International Envelopes or First-Class Mail International shipments (available at your post office).

Order Number:	34033171
Ship Method:	Standard
Customer Name:	Alibris APEX DC 76926368-71 - APEX
Order Date:	9/15/2026
Alibris Order #:	76926368-71
Email:	

Items:

Qty	Item	Locator	Item ID	Condition
1	Sustainable Corrosion Inhibitors I: Fundamentals, Methodolog...ions (ACS SYMPOSIUM SERIES) Hussain, Chaudhery Mustansar,Verma, Chandrabhan SKU: mon0003925425 ISBN: 0841297908 - Books	D -1-01-328-001-1077	76926368-71	Very Good

Notes:
Very good Very Clean Copy-Over 500, 000 Internet Orders Filled. hardcover

alibris

Alibris guarantees the condition of every item as it is described on our Web site. If you are dissatisfied for any reason, return your purchase within 60 days of receipt for a refund of the item price. We'll also refund shipping costs if the return was a result of our error. All return requests must be submitted via the Alibris Web site. Any item returned without accompanying paperwork and/or more than 60 days after its shipment date will be discarded and you will not be eligible to receive a refund.

If you have any questions or concerns regarding this order, please contact us at customerservice@booksfromca.com
Thanks for your order!

Chapter 2

Fundamentals of Corrosion Chemistry

Sheetal,[1] Bhawna Chugh,[1] Sanjeeve Thakur,[1] Balaram Pani,[2] and Ashish Kumar Singh[3,*]

[1]Department of Chemistry, Netaji Subhas University of Technology, Sector 3, Dwarka, New Delhi-110078, India

[2]Department of Chemistry, Bhaskaracharya College of Applied Sciences, University of Delhi, Sector 2, Dwarka, New Delhi-110075, India

[3]Department of Applied Sciences, Bharati Vidyapeeth's College of Engineering, Paschim Vihar, New Delhi-110063, India

[*]Email: ashish.singh.rs.apc@itbhu.ac.in.

Corrosion can be defined as the destructive attack of a metal by chemical or electrochemical reactions with its environment. In other words, it is an irreversible interfacial reaction of metal with its environment, which leads to the dissolution of the metal. Electrochemical, kinetic, and thermodynamic aspects are mainly responsible for corrosion. In view of the importance of corrosion, this chapter deals with factors that primarily contribute to corrosion, basic principles of corrosion, and the mechanism of corrosion. This chapter focuses on all the basics required to get an understanding of corrosion.

Introduction

There are three main reasons the study of corrosion is so important: safety, economy, and preservation. Such things as buildings, structures, bridges, and operational instruments are made up of metals and their respective alloys. Structures and operating equipment can collapse because of corrosion, and that can cause significant harm to human life. To reduce the economic impact of corrosion, scientists and engineers in this field aim to reduce the material and economic losses caused by corrosion. The loss of metals alone is not of concern but rather the loss of energy and the efforts necessary to fabricate that piece of metal. Economic losses have been categorized into direct and indirect losses (*1*). Direct losses include the replacement of metal or a particular structural material plus the cost of labor, material, and maintenance. This category broadly consists of the cost of processes such as painting, applying protection coatings on the metals, and regular inspection with corrosion-monitoring instruments. It is rather difficult to estimate the exact cost of indirect losses. Typical examples include:

1. Shutdown of machinery: failure of working parts because of corrosion often costs more than does replacing that part.
2. Contamination: Copper salts accelerate the rancidity of soaps and decrease their storage time. Similarly, the traces of some metals change the color of dyes.
3. Loss of products: various losses such as oil and gas leakage have been reported from corroded pipes.
4. Loss of efficiency: it has been reported that corrosion slows down an engine's efficiency and thus represents losses of several billion dollars.

Because corrosion involves chemical change, it is necessary to know the basic chemistry principles to understand corrosion and its reactions. Because corrosion reactions are mostly electrochemical in nature, it is quite important to have a deep insight into electrochemistry. In addition to this, it is presumed that the structure and composition of metals often determine corrosion behavior. Still, practically no specific criterion can help in selecting a particular material for a particular purpose. A rational decision is made by keeping several factors in mind, such as the ease of fabrication and the cost and availability of material. Frequently, these decisions are made by keeping the economic aspect in mind. In other words, the material that is cheaper is ideally selected. Also, wherever the metals are being utilized, their physical and mechanical properties must be considered.

Along with these considerations, the behavior of metals in the environment, their reactions, and their chemical and electrochemical aspects essentially constitute the fundamentals. This chapter provides an insight into all the fundamentals involved in the process of corrosion (*2*, *3*).

Definition of Corrosion

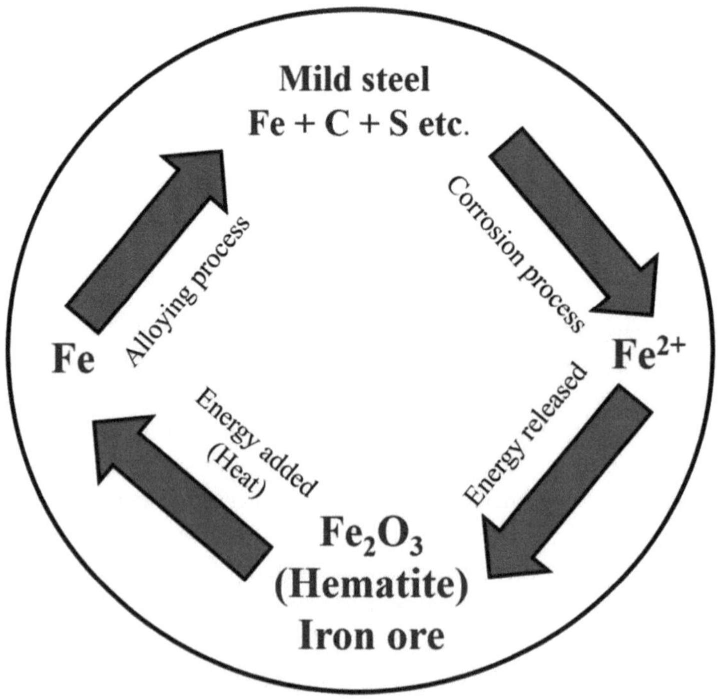

Figure 1. Stages of corrosion in mild steel.

The word corrosion is derived from the Latin word *corrodere*, which means "to attack or gnaw away." The gradual disintegration of metals caused by electrochemical, chemical, or biochemical interactions with their nearby environment is called corrosion. Corrosion occurs because of metals or materials that are thermodynamically unstable in their free state. Generally, metals are observed to be stable in the form of certain species. For example, pure metals are extracted from ore (thermodynamically steady-state) via the metallurgical process. However, the obtained pure metals are thermodynamically unstable in their free state. Therefore, these pure metals undergo corrosion to form corrosive products that are thermodynamically more stable than their pure form (*4–6*). Figure 1 explains various stages of corrosion for an alloy of iron (i.e., mild steel).

Because of corrosion, metals lose their weight, negatively impacting their unique physicochemical and mechanical properties. Corrosion causes waste of valuable materials, plant shutdowns, loss of product, immoderate maintenance, depletion in efficiency, and costly overdesign. The corrosion process can be explained by chemical or electrochemical theory; according to the electrochemical theory, metal undergoes corrosion in an aqueous environment mainly caused by both reduction and oxidation reactions occurring simultaneously on a metal surface. Oxidation occurs at an active anodic site and releases electrons toward the cathodic area via an electronic conductor. The reduction takes place at an active cathodic site and consumes electrons. It can be represented by a chemical reaction given as follows:

At anodic sites: solid metal (M) undergoes oxidation to form ions M^{n+} in an aqueous solution,

$$M\ (s) \rightarrow Mn^{+}(aq) + ne- \quad (1)$$

At cathodic sites: in the cathodic area, multiple reduction processes occur, which generally depend on aggressiveness and the nature of the corrosive system.

Evolution of hydrogen (acidic system):

$$2H^{+} + 2e^{-} \rightarrow H_2\ (g) \quad (2)$$

Reduction of oxygen (acidic system):

$$O_2\ (g) + 4H^{+} + 4e^{-} \rightarrow 2H_2O\ (l) \quad (3)$$

Reduction of oxygen (neutral or alkali system):

$$O_2\ (g) + 2H_2O\ (l) + 4e^{-} \rightarrow 4OH^{-}\ (aq) \quad (4)$$

Reduction of water (neutral or alkali system):

$$2H_2O\ (l) + 2e- \rightarrow 2OH^{-}(aq) + H_2(g) \quad (5)$$

Factors Affecting Corrosion

The degree of corrosion relies primarily on the metals as well as their surroundings. Conditions such as the structural characteristics of the metal, the type of environment, and the types of reaction at the metallic surface in contact with the environment play a crucial role in the corrosion process. The rate and degree of corrosion rely predominantly on the factors shown in Figure 2 (*7*).

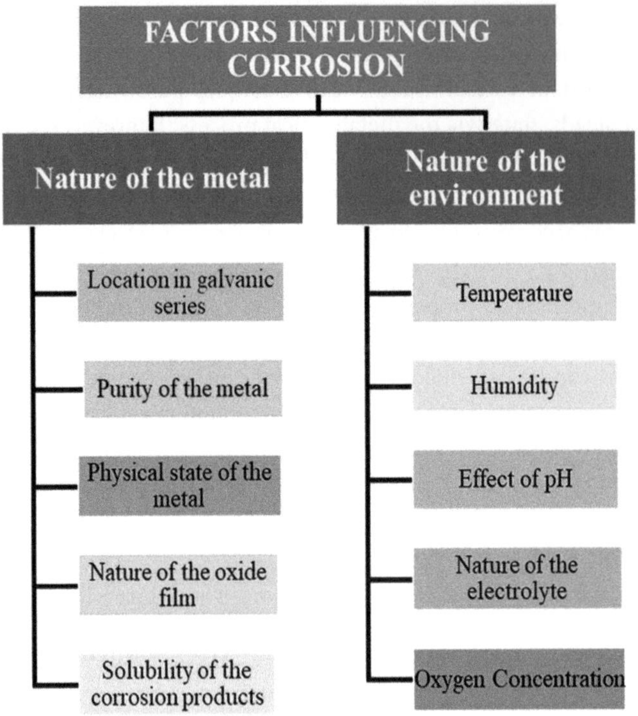

Figure 2. Schematic flow chart illustrating the factors influencing corrosion.

Nature of the Metal

Location in Galvanic Series

The degree of corrosion depends on the position of the metal in the galvanic series. The higher active metal (or the one above in the series) endures more corrosion. The extent of corrosiveness depends on the differences in their positions in the galvanic series. The larger the variation, the quicker the corrosion process of the anodic system (8).

Purity of the Metal

The impurities in any metal promote heterogeneity; consequently, galvanic cells are built up with different anodic and cathodic regions in the metal. The larger the number of impurities, the quicker the extent of corrosion at anodic sites. Therefore, the corrosion obstruction of metal might be enhanced by raising its purity.

Physical State of the Metal

Metal segments exposed to irregularly dispersed stresses are effectively destroyed. For pure metal, the regions under stress will usually be anodic and subsequently experience corrosion (9).

Nature of the Oxide Film

Some metals (e.g., Mg, Ca, and Ba) develop oxides with a volume less than the volume of the metal. The oxide layer they develop is porous, and oxygen quickly diffuses, consequently accelerating corrosion. Other metals (e.g., Al, Cr, and Ni) produce oxides with a volume more than that of the metal. The nonporous oxide layer they develop shields the metal from more corrosion.

Solubility of the Corrosion Products

Solubility of the corrosion products obtained is a significant aspect of corrosion. When the corrosive product is soluble in an acidic solution, the metallic corrosion process occurs at faster rate. But when the corrosion product is insoluble, the defensive covering inhibits corrosion (*10*).

Nature of the Environment

Temperature

The degree of chemical reactivity and the extent of dissemination of ions are enhanced with increasing temperature. The corrosion process is enhanced at a higher temperature because an inactive metal might turn dynamic at an elevated temperature.

Humidity

Atmospheric corrosion of metal is gradual in dry air yet enhances quickly in the presence of moisture. The moisture behaves as a solvent for the oxygen present in the atmosphere to sustain the electrolyte needed for establishing a corrosion cell.

Effect of pH

Metals such as iron dissolve in acids. The degree of corrosion for intermediate pH values is constrained by the transfer of oxidizer (typically solvable O_2) to the metallic surface. The pH may not influence noble metals such as gold and platinum.

Nature of the Electrolyte

The behavior of an electrolyte affects the extent of corrosion. If an electrolyte comprises silicate ions, they generate insoluble silicates and prohibit more corrosion. But if chloride ions exist, they demolish the protective layer, and the surface is subjected to more corrosion. If the conductance of the electrolyte is larger, the corrosion current is effectively conducted, and subsequently, the extent of corrosion is enhanced.

Oxygen Concentration

Oxygen is considered an aggressive gas or an oxidizing agent. If its amount rises, the corrosion rate will increase until the surface's diffusion rate attains a limit. Similar standards are applicable to many other oxidizing agents such as Cl_2, H^+, and Br_2.

Principles of Corrosion

Electrochemical Principle of Corrosion

Corrosion is the process in which a liquid or solid metal reacts with the nonmetallic environment. A reaction involving solid metal and the aqueous environment is termed wet or electrochemical corrosion. This term implies that the metal is converted into cations during the process and is also called anodic dissolution. This is simultaneously accompanied by cathodic reduction. The corrosion rate is faradaically equal to the amount of current moving between cathode and anode. It has been demonstrated that whether a metal surface is a few cm apart or few inches long, the two cathodic and anodic surfaces are created consecutively. The amount of net positive charge from the metal to the environment at the anode is equal to the net negative charge from the metal to the environment at the cathode. In cases in which the cathode and anode are on the same piece of metal, they tend to polarize each other, and this way, the more negative anode tries to pull the cathode in the negative direction and vice versa (*11*).

Polarization occurs, and the difference between the electrode potentials of anode and cathode resulting from this is responsible for the movement of the current (I) through the whole circuit of resistance (R). In instances in which the resistance is too small and $E_c = E_a$ (i.e., within 1 mV), these values of potentials may be identified with the value E_{corr}, which is considered to be a particular case of settlement or mixed potential, reached at the electrodes carrying two different reactions with zero ohmic drop between them.

The overview of this electrochemical mechanism of corrosion is illustrated with an example. The process of corrosion in the aqueous environment leads to electrochemical reactions, which is the most common type of corrosion. The moisture in air or various water resources such as rivers or seawater tends to serve as the aqueous medium for the metals so that the electrochemical reactions take place. There are significantly fewer chances when there is the possibility of pure water, and as it is known that there is a specific composition of gases dissolved in the water, and these gases render a property of dissociation to water and make it behave like an electrolyte. And the chemical nature of this electrolyte can be acidic, basic, or neutral (*12*). The most common reactions involve iron when exposed to an acid such as hydrochloric acid in which it experiences oxidation and loses electrons, and the reaction is characterized by the evolution of hydrogen gas from the medium. The general reaction can be given as

$$Fe\ (s) + 2HCl\ (aq) \rightarrow FeCl_2\ (aq) + H_2\ (g) \quad (6)$$

But from a mechanistic view, it involves an electron transfer from the metal to hydrogen gas.

$$Fe\ (s) + 2H^+(aq) + 2Cl^-\ (aq) \rightarrow Fe^{2+}(aq) + 2Cl^-(aq) + H_2\ (g) \quad (7)$$

Iron is transformed into a ferrous ion by losing two electrons, which are collected by the hydrogen. This way, hydrogen is evolved from the reaction medium and gives a characteristic feature of this reaction. Locations where electrons are donated are called anode, whereas the locations where electrons are being taken are called cathode. There is a difference between the two electrode's potential that is cathode and anode; this way, a circuit is completed. Electrons flow from their origin to the cathode, and hydrogen ions move toward the cathode to complete the circuit. The faster the dissolution of metal takes place or the faster the oxidation of metal occurs, the faster will be the rate of electron flow, and thus more current will flow through the circuit. Oxidation reactions are quite simple reactions in which the metal is just upgraded to a higher valence state. These reactions are responsible for the dissolution of metal, but cathodic reactions are also equally important. Metal atoms after corrosion have no accumulation of charges, so the electrons evolved by oxidation are consumed during reduction. This way, the electroneutrality of the metal is maintained. Cathodic reactions have been categorized as:

Metal deposition:

$$M^{3+}\ (aq) + 3e^- \rightarrow M\ (s) \tag{8}$$

Reduction of metal ions:

$$M^{3+}\ (aq) + e^- \rightarrow M^{2+}\ (aq) \tag{9}$$

Oxygen reduction in acidic environment:

$$O_2\ (g) + 4H^+ + 4e^- \rightarrow 2H_2O\ (aq) \tag{10}$$

Evolution of hydrogen in acidic media has been depicted earlier as eq 2 and reduction of oxygen as eq 4. Three types of cell formation take place, which can lead to further corrosion:

Dissimilar Electrode Cell

This type of cell is formed when there are two different metals or when two other electrodes set up on the same surface because of the heterogeneity of the surface. A Daniel cell is an example of this.

Concentration Cell

This type of cell is formed when there are two similar electrodes, but they are in contact with the solution of different compositions. An example of this is a salt concentration cell, which is formed when one electrode is in a dilute environment (called an anode) whereas the other is in a concentrated environment (called the cathode).

Differential Temperature Cell

This type of cell is formed when the same metal electrodes are dipped in the solution of the same initial composition at a varying temperature. This type of cell formation may take place in heat exchangers, boilers, and the like (*11*).

Thermodynamic Principle of Corrosion

Corrosion in the case of metals and their alloys follows electrochemical nature. Oxidation and reduction are the two main processes that take place in an electrochemical cell. Oxidation is coupled with the reduction of species in the surrounding electrolyte.

Uniform corrosion occurs when these anodic and cathodic reactions are distributed thermodynamically across the surface. On the other hand, when heterogeneities exist on the surface, it gives rise to a localized corrosion. In order to obey the rule of electroneutrality, it is expected that both the oxidation and reduction reactions take place simultaneously and with an equal rate. Both the anodic and cathodic sites must be connected electrically. The principle of thermodynamics is to be followed while these reactions are being carried out.

The first principle is that the value of cell potential decides the direction in which the oxidation or reduction will be allowed. When the potential of the electrode is higher than the equilibrium potential, oxidation occurs. But on the other hand, if the electrode potential is less than that of the equilibrium potential, the reaction will take place by reduction. If the value of electrode potential is equal to the equilibrium potential, no oxidation or reduction will occur.

Under the standard conditions, $E_0{}^+ = E$, no oxidation or reduction take place. If E_0 is less than E, then there are some chances of oxidation. And if E_0 is greater than E, then there will be some probability of reduction. Basically, these three conditions set a criterion for any reaction to take place in an electrochemical environment.

The second principle involves that according to De Donder (*13*): there is a relation established between the direction of the reaction and its chemical affinity denoted by A and velocity v in any reaction that has almost the same values as that of affinity or equals to zero.

$$A_v \geq 0 \tag{11}$$

And according to Pourbaix, it has been given

$$(E - E_0)\, I \geq 0 \tag{12}$$

This can be stated as the second thermodynamic principle of corrosion. Further, a concept of current is given as a positive sign in the process of oxidation, whereas it is negative in the process of reduction. Equation 12 establishes that the direction of electrochemical reaction (defined by the sign of the intensity of reaction current) is given by the difference ($E - E_0$) (i.e., between the electrode potential and the equilibrium electrode potential).

If at standard conditions $E > E_0$ and $i > 0$, then either the process of oxidation takes place, or no reaction will take place. On the other hand, if $E < E_0$ and $i < 0$, then there is the possibility of either reduction or no reaction. The main requirement involved in this thermodynamic principle requires zero velocity or the same sign as that of over potential. Equation 12 plays a vital role in the thermodynamic interpretation of the corrosion studies (*14*).

Potential pH Diagram

Potential pH diagrams are also termed Pourbaix diagrams. These are the graphical representations describing the stability of metals and their respective corrosion products as a function of the potential and pH (acidity and basicity) of the respective environment. The potential is plotted on the vertical axis and the pH on the horizontal axis. Pourbaix diagrams are constructed by the involvement of the Nernst equation and also the solubility data of metal compounds (*12*). For instance, the potential pH diagram has been illustrated as follows:

Because iron is metal and has good oxidation potential value, the following reactions are possible:

$$Fe\ (s) \rightarrow Fe^{2+}(aq) + 2e^- \tag{13}$$

$$Fe^{2+}\ (aq) \rightarrow Fe^{3+}(aq) + e^- \tag{14}$$

In Figure 3, the horizontal lines represent the pure electron loss process based on the electrode potential. Again, it should be noted that pH dependability plays no role in this regard.

The vertical lines represent the formation of hydroxides or oxides as corrosion products, reducing the concentration of Fe^{2+} and Fe^{3+}. These horizontal lines extend throughout the diagram until the pH is sufficiently high to promote hydroxide formation.

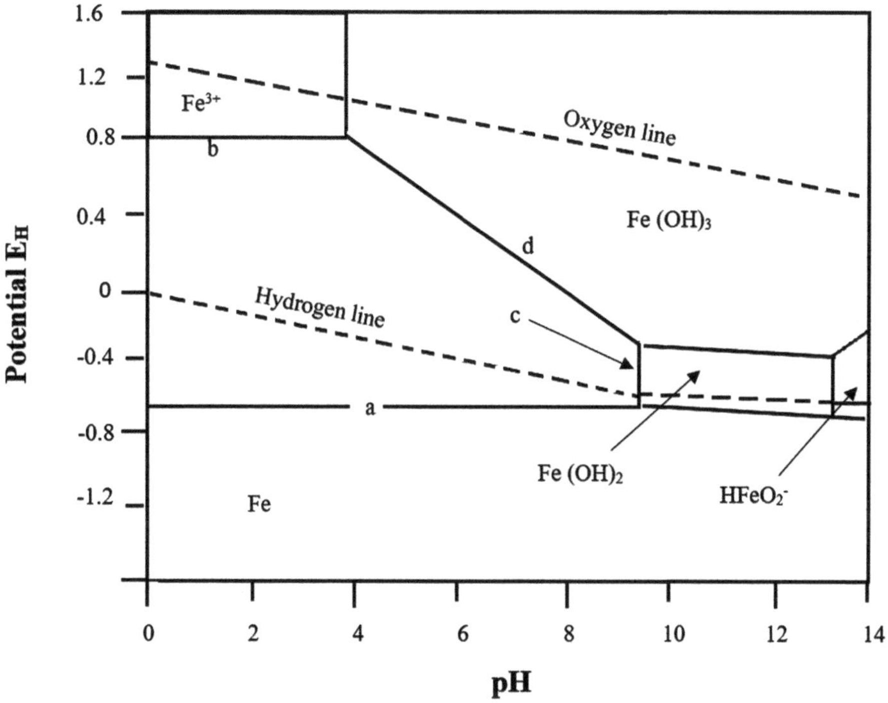

Figure 3. Pourbaix diagram for metal with water (Fe + H_2O). Reproduced with permission from reference (15). Copyright 2019, IOP Publishing Ltd. Content has been used under the terms of the Creative Commons Attribution 3.0 license.

These vertical lines can be further elaborated as

$$Fe^{2+}(aq) + 2H_2O\ (aq) \rightarrow Fe(OH)_2(s) + 2H^+\ (aq) \quad (15)$$

$$Fe^{3+}(aq) + 3H_2O\ (aq) \rightarrow Fe(OH)_3(s) \quad (16)$$

These vertical lines do not involve any electron movement but only affect a pH change throughout. Further, the sloping lines as indicated in Figure 3 involve the two processes simultaneously (i.e., electron transfer and pH change).

$$Fe^{2+}(aq) + 3H_2O\ (aq) \rightarrow Fe(OH)_3\ (s) + 3H^+(aq) + e^- \quad (17)$$

These two processes, namely oxidation and reduction, utilize two gases, oxygen and hydrogen, indicated by dotted lines on the graph.

In environments having pH less than 7, two reactions, namely eq 2 and eq 5, take place.

We can state that these two reactions are identical, and their pH reliance over the electrode potential is expressed as

$$E_{H^+/H_2} = E^0 - 0.0591\text{pH}$$

(18)

For the pH equal to zero, this concentration of $H^+ = 1$ and $E_0 = 0$, so following the straight-line equation slope is obtained as −0.0591.

The same reactions for oxygen in equilibrium with water at high and low pH take place, which have been mentioned earlier as eq 3 and eq 4, respectively.

The pH dependence of single electrode potential is represented by the following equation:

$$E_{O_2/H_2O} = E^0{}_{O_2/H_2O} - 0.0591\text{pH}$$

(19)

At pH = 0, $E^0{}_{O_2/H_2O} = 1.226\text{V}$, and at the value of pH = 1, $E^0{}_{O_2/H_2O} = 0.401\text{V}$. The slope of the line is obtained as −0.059 V. Water is found to be stable in the areas allotted by these two lines. Below the hydrogen line, water is reduced to hydrogen, and above the oxygen line, it is oxidized to oxygen. The potential pH plot is divided into three regions:

1. Immunity zone: in this zone and specifically in this potential and pH range, the metal iron remains in its metallic form.
2. Corrosion zone: in this region, iron, being a metal, gets corroded slowly with the formation of Fe^{2+} and Fe^{3+} ions.
3. Passive zone: when the specific potential and pH conditions prevail, the metal forms an anticorrosive or protective surface layer of ferric hydroxide, $Fe(OH)_3$, preventing iron from further corrosion.

Potential pH diagram plays a very significant role in different areas such as:

1. It helps predict the feasibility of the reaction.
2. It also helps evaluate the stability of products formed after corrosion such as hydroxides and oxides.
3. It also plays a significant role in determining the environmental conditions that could enhance or prevent the process of corrosion.

However, there are some unavoidable limitations in using these diagrams. The most noticeable limitation is that they depict the equilibrium conditions and cannot be relied upon for predicting the reaction rate. Furthermore, the assumption that the products formed after corrosion such as hydroxides and oxides lead to the prevention of further corrosion cannot be completely trusted because they may not precipitate over the surface of the metal and offer them protection. Also, the pH at the metal surface can vary based on several factors such as side reaction products, and predicting corrosion as a measure of bulk pH of the solution cannot be considered trustworthy (*12*, *16*).

Kinetic Aspects of Corrosion

The concept of corrosion is related to thermodynamics, and corrosion rates are determined primarily by corrosion kinetics. Generally, the potential change caused by the net flow of current to and from the electrode because of the electrons is termed polarization and is measured in volts. Figure 4 elucidates a general mechanism of polarization of metal in aqueous acid.

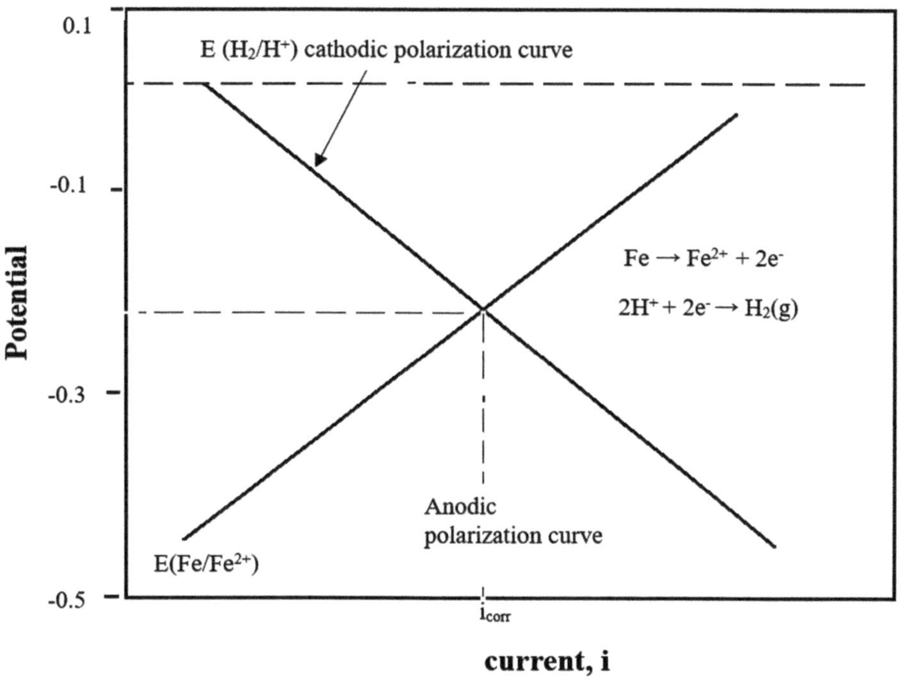

Figure 4. Polarization of iron in the presence of an acid.

Resistance in the circuit is determined by factors such as surface films, the medium's resistivity, and the metal itself. The intersection of two polarization curves gives a close approximation of corrosion current. Once the corrosion current is identified, the corrosion density can also be easily approximated. However, the polarization data are quite helpful in estimating corrosion rates because as the polarization increases, the corrosion decreases. Polarization can be either anodic or cathodic. There are three main causes of polarization, namely activation, concentration, and potential drop (*12*).

Activation Polarization

This arises because of the slow step of the electrode reaction for which the activation energy in the form of an increased potential is utilized to carry forward this reaction. For instance, H_2 evolution (eq 2) follows a specific sequence.

It takes place in a series of steps. First, hydrogen takes one electron from the metal to get adsorbed over the surface.

$$H^+ + e^- \rightarrow H_{ads} \quad (20)$$

Second, this H_{ads} atom reacts with another atom of its type to form a molecule.

$$H_{ads} + H_{ads} \rightarrow H_{2ads} \tag{21}$$

And in the third step, H_{ads} molecules combine and nucleate to form a bubble of hydrogen on the surface. Overall, from this sequence, the slowest step determines its speed and describes the rate and the overall rate. To have a higher rate of corrosion, the increase in potential should be affected. The relation between the reaction rate and change in potential is expressed by the Tafel equation given as

$$\eta_a = \pm\beta \log \frac{i}{i_o} \tag{22}$$

where η_a is the overvoltage polarization, and β is the Tafel constant and usually has the value of the order of 0.1V. The dissolution reactions are controlled by activation polarization, where usually the solution of ions is the rate-controlling step. The cathodic reactions are controlled by activation polarization when the concentration of hydrogen gas is high.

Concentration Polarization

If the reaction proceeds faster or slows down, then there are changes in the concentration of ions in the solution. As a result, the difference in the value of electrode potential can be observed with the help of the Nernst equation. This can be evidenced by the fact that while the zinc electrode is being corroded, the zinc concentration will subsequently increase. As a result, the value of a_{oxid} increases and shifts the potential toward a more positive direction. Simultaneously, hydrogen evolution at a faster rate brings the value of a_{oxid} down and shifts the electrode potential to a more negative side. A mathematical relation that relates concentration polarization is given by

$$\eta_c = \frac{2.3RT}{F} \log\left(1 - \frac{i}{i_l}\right) \tag{23}$$

η_c is the overvoltage in volts due to concentration polarization, R stands for universal gas constant, T is the temperature, F stands for Faraday, and i and i_l stand for current. When i approaches to i_l then the value of η_c becomes infinite.

Potential Drop

Potential drop is defined as the change in potential accompanied by change in the surrounding environment and the circuit between the cathodic and anodic sites. It also involves the effect of resistivity of surface films, media, and corrosion products (*17*).

Mechanism of Corrosion

Corrosion on the basis of a logical and scientific approach has been classified into two main types: wet corrosion and dry corrosion. The term wet includes all those reactions taking place in aqueous environments, whereas dry corrosion reactions involve those that occur in the absence of an aqueous medium. Both terms are quite ambiguous as it is not always understood whether wet means aqueous media. Both wet and dry with their respective mechanisms are elaborated as follows:

Dry Corrosion

This type of corrosion is also called chemical corrosion and involves the attack of nonmetals including gases such as oxygen and sulfide or vapors such as sulfur vapors over the surface of metals. In this type of corrosion, the environmental gases directly attack the surface of metals, for example, the chemical corrosion of silver metal in the presence of hydrogen sulfide gas. During these reactions, metal oxidation with reduction of nonmetal occurs, involving the formation of compounds that act as a protective barrier at the metal–nonmetal interface. If the compound formed is continuous, then it may provide a barrier for the reactants. Any interaction of the reactants allows them to diffuse through the film by processes such as diffusion of the nonmetal or diffusion of the respective ions of the reactants. Dry corrosion can be categorized into different categories such as oxidation, liquid metal, and other gases (*18*).

Oxidation Corrosion

This type of corrosion occurs when the metal comes in contact with the oxygen directly. No specific temperature conditions are required for this type of reaction to happen. Alkali and alkali metals form compounds such as oxides even at room temperature with oxygen.

Metals that are alkali, being electropositive in nature, lose electrons, and oxygen, being nonmetal, gain the electrons. As a result, the formation of oxide takes place on the metal–nonmetal interface. If the ions of metals or oxygen can enter via this layer with the help of diffusion, it creates more corrosion. This layer of metal oxide plays a principal role in oxidation corrosion. On the other hand, metals such as Pb and Al can form stable oxides and are not involved in further corrosion. However, in some instances where there is the possibility that oxygen may pass through the porous oxide film, the process of corrosion in the case of such metals continues until the whole metal gets converted to stable oxide. The porous oxide film is formed in alkali and alkaline earth metals whereas metals such as molybdenum form volatile oxide layers, which accelerates corrosion.

Liquid Metal Corrosion

Industries use many metals and their respective alloys such as iron in pipes or other types of equipment. Liquid or molten metals or substances flow through the pipes to meet the final product requirement. So when molten substances pass through pipes made of certain metals, that may lead to the dissolution of internal corrosion. For instance, mercury, which has a tendency to form amalgams with most metals, pushes those metals toward corrosion.

Corrosion by Other Gases

The environment contains gases such as oxygen, nitrogen, and sulfur dioxide. And corrosion caused by such gases can also occur and simultaneously lead to the deterioration of certain metals and their alloys. When reacting with metals, these gases form corrosion products that are protective or nonprotective in some instances. For example, dry chlorine gas reacts with metals such as silver and forms a protective covering of AgCl. In contrast, metals such as Sn react with the chlorine gas and form $SnCl_4$, which is volatile, so it is not protective.

Wet Corrosion

This type of corrosion requires a liquid medium. So the oxidation and the reduction processes take place in the solution phase. The oxidation and reduction take place at different metal surface areas with consequent electron transport from an electron-producing site (anode) to an electron-acceptor site (cathode). In other words, this can be defined as corrosion that take place in a liquid or aqueous environment or when two dissimilar electrodes partially share a solution. The corrosion product is formed at the interface of a metal. It has been reported that processes such as convection or diffusion might lead to the movement of cations and anions from the interface. Under such situations, the reactants will not be separated, and thus the rate law will tend to be linear. Wet corrosion involves two types of cathodic reactions, hydrogen evolution and oxygen absorption, based on the surrounding environment. These two reactions typically occur at the cathode, mainly because hydrogen and water molecules are invariably present in the aqueous medium. Most of the aqueous solutions are open, so they involve the addition of oxygen as well. In a more generalized way, the contribution made by hydrogen evolution reactions will increase in importance and decrease the pH. Simultaneously, it is also dependent on the nature of the metal and metal oxides.

Evolution of Hydrogen

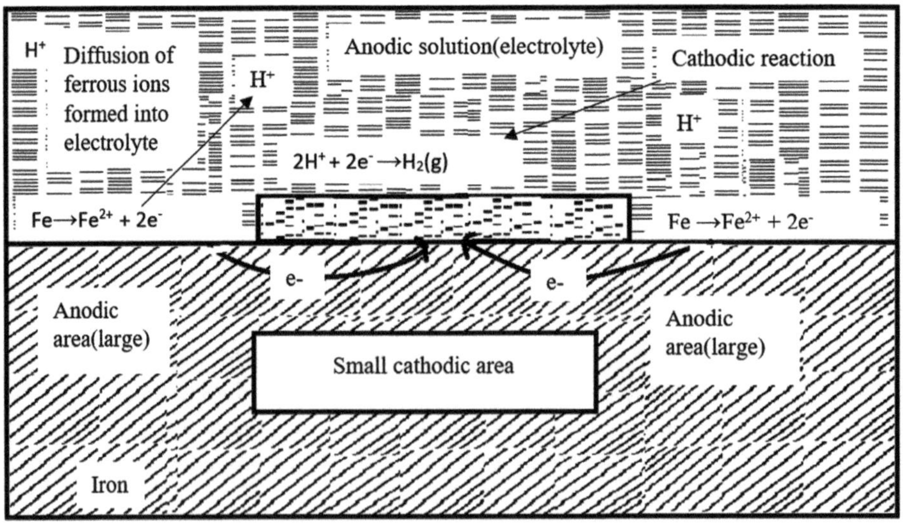

Figure 5. Mechanistic view of hydrogen evolution.

During wet corrosion, the formation of an electrochemical cell involves two different electrodes. This type of reaction occurs when the nearby environment of the electrochemical cell is acidic. There is an absence of oxygen in that environment, and the anode surface should be comparatively larger than the cathode. Examples include the rusting of iron. Here, the metal acts as an anode, and oxidation preferably takes place over there. Figure 5 depicts the evolvement of hydrogen at the cathode electrode.

Iron prefers to lose electrons and becomes Fe^{2+}. The main reactions included are eq 15 at the anode and eq 2 at the cathode.

Corrosion Reaction Caused by Oxygen Absorption

Here, the reaction takes place in the basic or neutral environment, and there should be the availability of oxygen. The other condition that needs to be followed is that the anode should be smaller in surface than the cathode. Involving again the instance of rusting of iron:

At anode: Iron loses two of its electrons to become Fe^{2+} (eq 15).

At cathode:

$$1/2O_2(g) + 2e^- + H_2O(aq) \rightarrow 2OH^-(aq) \tag{24}$$

The final reactions involve the formation of iron hydroxide.

All this largely depends on the availability of oxygen in the environment. If oxygen is present in a limited amount, then the formation of rust that is black in color takes place. On the other hand, if the availability of oxygen is sufficient or in excess, this leads to rust formation that is yellow in color (*19, 20*). Figure 6 provides a possible layout of the oxygen adsorption mechanism.

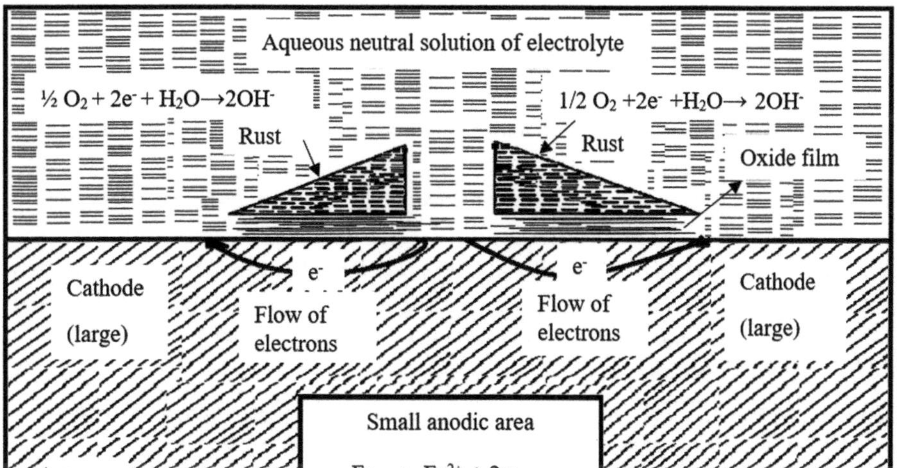

Figure 6. Mechanistic view of oxygen absorption.

Conclusion

Corrosion is simply comprised of electrochemical reactions of metals and their respective alloys with their environment and has become a calamity of significant concern. To combat this process, it is essential to have a deeper knowledge of the basic principles of electrochemistry, the rules of kinetics and thermodynamics that promote corrosion, and several other factors that describe the spontaneity of this process. Keeping in mind the understanding of these necessary factors, it will be quite easy for corrosion scientists to come up with new advances to tackle this corrosion process, such as cathodic protection and corrosion inhibitors among others. These could be available once the electrochemical procedure is known. However, modern, active corrosion protection systems are preferred for research, and a deep understanding of these fundamentals is required for the next generation to work in this field.

References

1. Arya, E. K.; Dhanya, B. S. Corrosion Control of Reinforced Concrete Structures in Construction Industry: A Review. *IOP Conf. Ser. Mater. Sci. Eng.* **2021**, *1114* (1), 01200.
2. Winston Revie, R. *Corrosion and Corrosion Control: An Introduction to Corrosion Science and Engineering*, 4th ed.; Wiley-Blackwell: Chichester, England, 2008, 1−4.
3. *Corrosion: Understanding the Basics*; Davis, J. R., Ed.; ASM International, USA, 2000, 1−17.
4. Uhlig, H. H. *Corrosion and Corrosion Control: An Introduction to Corrosion Science and Engineering*, 3rd ed.; John Wiley & Sons: Nashville, TN, 1985, 10−12.
5. McCafferty, E. *Introduction to Corrosion Science*; Springer, UK, 2010, 13−31.
6. Abbasov, V. M.; El-Lateef, H. M. A.; Aliyeva, L. I.; Ismayilov, T. A.; Ismayilov, I. T.; Mamedxanova, S. A. Evaluation of New Complex Surfactants Based on Vegetable Oils as Corrosion Inhibitors for Mild Steel in CO_2-Saturated 1.0% NaCl Solutions. *Int. J. Electrochem. Sci.* **2013**, *1*, 19–26.
7. *Corrosion Engineering: Principles and Practice*; McGraw-Hill Companies, New York, 2008, 23−38.
8. Cruz, J.; Martı̀nez, R.; Genesca, J.; Garcı̀a-Ochoa, E. Experimental and Theoretical Study of 1-(2-Ethylamino)-2-Methylimidazoline as an Inhibitor of Carbon Steel Corrosion in Acid Media. *J. Electroanal. Chem. (Lausanne Switz)* **2004**, *566*, 111–121.
9. Shukla, S. K.; Singh, A. K.; Ahamad, I.; Quraishi, M. A. Streptomycin: A Commercially Available Drug as Corrosion Inhibitor for Mild Steel in Hydrochloric Acid Solution. *Mater. Lett.* **2009**, *63*, 819–822.
10. Winston Revie, R.; Uhlig, H. H. *Corrosion and Corrosion Control: An Introduction to Corrosion Science and Engineering*; John Wiley & Sons, 2008, 115−138.
11. Hoar, T. P. Electrochemical Principles of the Corrosion and Protection of Metals. *J. Appl. Chem.* **2007**, *11*, 121–130.
12. Schweitzer, P. A. *Fundamentals of Corrosion: Mechanisms, Causes, and Preventative Methods*; CRC Press: Boca Raton, FL, 2010, 5−13.
13. Oonder, T. H. D. E. *L'Affinit ((R~Sdaction Nouvellc Par Pierre Van Ryssclberghc.* Gauthier Villars, Paris, 1936; pp 23–97.
14. Van Muylder, J. Thermodynamics of Corrosion. In *Electrochemical Materials Science*; Springer US: Boston, MA, 1981; pp 1–96.
15. Oki1, M.; Akintola, S. A.; Adediran, A. A.; Ikubanni, P. P.; Ogunsemi, B. T. Reinforcement bar Corrosion - Causes and Management. *J. Phys.: Conf. Ser.* **2019**, *1378*, 022095.
16. Schweitzer, P. A. *Corrosion Engineering Handbook - 3 Volume Set*; CRC Press: London, England, 2019, 20−40.
17. Perez, N. *Electrochemistry and Corrosion Science*; Kluwer Academic: Tucson, AZ, 2004, 155−165.
18. Palanisamy, G. Corrosion Inhibitors. In *Corrosion Inhibitors*; Intech Open, 2019.
19. Evans, U. R. Dry Oxidation and Wet Corrosion. *Nature* **1951**, *168*, 853–855.
20. Evans, U. R.; Taylor, C. A. J. Mechanism of Atmospheric Rusting. *Corros. Sci.* **1972**, *12*, 227–246.

Chapter 3

Factors Affecting Corrosion Inhibitors

Forouz Rastegari[1] and Ali Rastegari[*,2]

[1]Department of Chemistry, University of Isfahan, Isfahan 81746-73441, Iran
[2]Department of Pharmaceutics, Faculty of Pharmacy, Iran University of Medical Sciences, Tehran 1475886973, Iran

[*]Email: rastegari.a@iums.ac.ir.

Corrosion is a chemical reaction that converts metal units into a more chemically stable form. Corrosion inhibitors form an impermeable hydrophobic layer that prevents the damaging materials present in the media from having access to the metal surface, thus delaying the process of corrosion. The effectiveness of corrosion inhibitors was found to depend on the metals under investigation and the environmental conditions. In this chapter, we discuss the different factors that affect the performance of corrosion inhibitors, such as temperature, concentration fluid velocity, metal cations, electrolytes, and pH.

Introduction

Corrosion is a chemical process in which metals are converted into more chemically stable agents like oxide, hydroxide, or sulfide species. Generally, metals can be destroyed under certain conditions such as the presence of moisture/water (H_2O), bases, acids, salts, liquid chemicals, aggressive metal polishes, and gases (*1*).

Corrosion inhibitors can be categorized into inorganic inhibitors and organic inhibitors. Inorganic inhibitors work based on the oxidation of metal to an impermeable layer, and organic inhibitors work based on adsorption on the surface of metal by heteroatoms and/or double bonds, forming a hydrophobic layer that prevents accessibility of aggressive agents into media from the metal surface (*2*).

Corrosion prevention can alleviate different types of adverse outcomes, including economic damage and detrimental effects on water sources and the environment. Accordingly, corrosion scientists and material engineers try to decrease the adverse effects of corrosion in such things as machinery components and construction to reduce economic damage and improve safety (*1*).

Corrosion inhibitors delay the process of corrosion. The inhibition may be achieved by the addition of specific chemical agents that adsorb on the surface of metal and create a protective film that enhances the lifetime of the metal (*3, 4*). The chemical agents added to the corrosive media to inhibit the corrosion reaction on the metal surface are known as corrosion inhibitors (*5, 6*).

In general, a substance acting as a corrosion inhibitor should have the potential to make an efficient protective film on the surface of the metal. This protective film should not be totally impermeable, but it must show enough rigidity to inhibit the accessibility of corrosive ions from the metal surface. The performance of corrosion inhibitors depends on different factors, including the specific metal involved and the environmental conditions under which the metal is used (*7*).

This chapter discusses the effect of different parameters on corrosion inhibitors, such as temperature, concentration fluid velocity, metal cations, metals, electrolytes, pH, and humidity.

Effect of Temperature

Generally, the corrosion procedure accelerates with rising temperature, particularly in environments in which evolution of hydrogen is associated with corrosion (*8*). Previous studies have reported that temperature is a critical factor in the corrosion process. Furthermore, the association between temperature and corrosion will be more complicated during corrosion if oxygen plays a part in a cathodic reaction. This phenomenon is related to the lesser solubility of oxygen at high temperatures (*9*). Generally, the efficiency of most inhibitors decreases with rising temperature, because of the lower amount of exposure of metal surface adsorbed the inhibitors (*7*).

In 1967, Riggs and Every proposed that the corrosion rate for iron and steel could be defined as the sum of two rates (*10*), $K_1(1-\theta)+K_2\theta$, where θ is the portion of surface covered by an inhibitor, K_1 is the constant rate for an unprotected system, and K_2 is the constant of rate for an inhibited system. At high temperatures, the efficiency of the inhibitor will decrease, because of the desorption of adsorbed inhibitors and also the depletion of oxygen atoms.

Arrhenius plots are used to investigate the effect of temperature on the chemical reaction rates (eq 1). Studies have shown that the relationship is often similar and linear in the protected systems, and also in unprotected systems (*11*). The slope of the line will be equal to $-E_a/R$ (Figure 1).

$$\ln K = \ln A - \frac{\mathrm{Ea}}{R}\left(\frac{1}{T}\right) \quad (1)$$

Equation 1 is the Arrhenius equation.

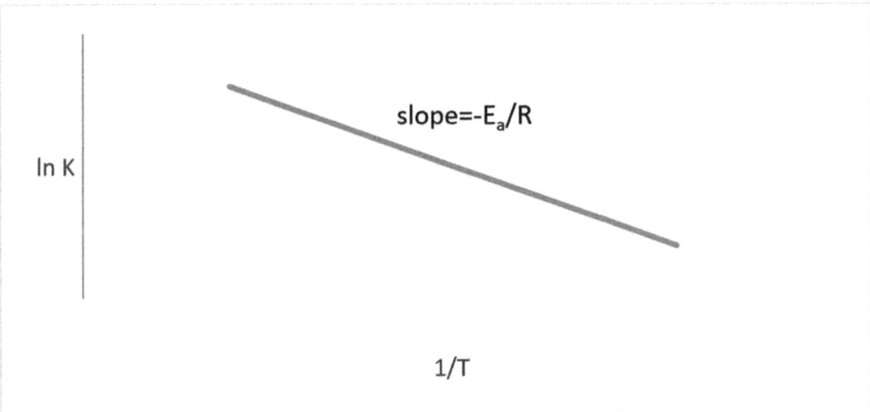

Figure 1. Arrhenius-type diagram model.

On the other hand, Putilova and co-workers (*11*) have shown that there are three kinds of relationships:

1. In the first category, inhibitors that delay corrosion at lower temperatures become less effective at higher temperatures. In fact, E_a is enhanced in the presence of inhibitors. For example, in 2005, Bouklah et al. examined pyridazine as an inhibitor for steel in 0.5 M H_2SO_4 solution at temperature range of 298–353 K. The inhibitory effect is reduced with rising temperature and E_a increases from 62.27 to 70.67 kJ/mol^{-1} (*12*). This enhancement of E_a in the presence of an inhibitor is related to electrostatic adsorption on the steel surface (*13*).
2. The second category includes the bulk of inhibitors. In this condition, the inhibiting effect is retained at higher temperatures. Also, the energy of activation stays the same as in the unprotected system. For example, the E_a is fixed (15.8 kcal/mole) for steel in the presence of HCl and under conditions of both inhibitor absence and presence of hexamine as inhibitor (*11*).
3. In the third category, inhibitors are defined with temperature coefficient value. Their temperature coefficient is lower than the corrosive procedure in the absence of inhibitors. These values are practical where postponing of corrosion is desired at high temperature. The activation energy will drop in the presence of this type of inhibitor. These inhibitors surprisingly reduce the reaction rate and energy of activation (*7*). In 1938, Machu demonstrated that in the presence of some types of inhibitors, for example, dibenzyl sulphide, dibenzyl sulphoxide, aniline, and gelatine for steel at 288, 320, and 348 K, are dropped the rate of corrosion and temperature coefficient (*14*).

Also, in 2014, Kumari and co-workers reported that 2-(3,4,5-trimethoxybenzylidene) hydrazinecarbothioamide (TMBHC) as a derivative of aromatic hydrazide acts as corrosion inhibitor on mild steel in 1 M HCl at temperatures between 30 and 60 C. The E_a decreased from 49.22 to 26.75 kJ mol^{-1} with greater concentration of inhibitor and increasing temperature owing to the gradual adsorption of inhibitor molecules on the soft surface of steel (*15*).

Effect of Inhibitor Concentrations

Temperature and concentration influence the protective effect of corrosion inhibitors. This effect is important to explain the mechanisms and kinetics of action and can aid in the selection of the types of inhibitors to use for specific practical situations (*16*).

In 1966, Putilova et al. showed that the corrosion rate of steel in HCl in the presence of butyric anhydride or gelatine as inhibitors, and the corrosion rate of steel in H_2SO_4 in the presence of gelatine or butylimine polymer as inhibitors, was enhanced under increasing concentrations of inhibitors (*11*). On the other hand, in 1997, Stoyanova and co-workers (*16*) showed that the inhibiting efficiency of linear and cyclic thiocarbamides was enhanced under increased concentration of inhibitors and increased temperature of the corrosion environment for steel. In fact, thiocarbamides in solution media reduce the amount of apparent energy of activation for reaction of hydrogen evolution. This procedure was proposed that the molecules are adsorbed via the sulfur atom that is the adsorption center, making donor–acceptor bond between ion pairs of the sulfur atom and the positive active centers of the metal surface. Also, in 2005, Ashassi-Sorkhabi et al. reported that amino acids, including alanine, leucine, valine, proline, methionine, and tryptophan, work as efficient inhibitors for the corrosion of aluminum in 1 M HCl and 1 M H_2SO_4 solution (*17*). The results obtained from this work are presented in Table 1, which shows that the effectiveness of corrosion inhibition increases with increasing inhibitor concentration.

Table 1. Corrosion of Aluminum in the Presence of Different Concentrations of Amino Acids (*17*)

The Name of Inhibitor	*Concentration of Corrosion Inhibitors*	*The Rate of Corrosion (mm/yr)*
Alanine	0.1	0.90
	0.01	1.77
	0.001	1.81
	0.0001	1.96
Valine	0.1	0.85
	0.01	1.61
	0.001	1.90
	0.0001	1.95
Leucine	0.1	0.77
	0.01	1.39
	0.001	1.48
	0.0001	1.92
Methionine	0.1	0.39
	0.01	0.41
	0.001	0.64
	0.0001	1.00
Proline	0.1	0.75
	0.01	0.99
	0.001	1.89
	0.0001	1.98
Tryptophan	0.1	0.22
	0.01	0.46
	0.001	0.54
	0.0001	1.45

Effect of Fluid Velocity

Fluid velocity can be explained as determination of the flow regime and the mass flow rate. As the fluid flows across the pipe, the rotational velocity of the sample is converted into rate of flow for the fluid using the following experimental equations (*18*):

$$q = 0.92nD_a^3(\frac{D_t}{D_a}) \quad (2)$$

$$A_p = \pi D_a W \quad (3)$$

where

q: total flow rate from the edges of impeller ($m^3 \cdot s^{-1}$)

n: number of rotations per second

D_a: the diameter of impeller

D_t: the solution vessel diameter

A_p: the cylinder area swept out by the tips of the impeller blades (m^2)

W: the impeller width

The fluid velocity can be calculated by dividing eq 2 by eq 3.

Usually, the effectiveness of inhibitors decreases under increasing fluid velocity. On the other hand, sufficient velocity is critical to disperse the inhibitor well and to propel suspended fragments along, which keeps them from precipitating onto the metal surface, which can create unwanted products that cause pitting corrosion or erosion (*19*). Thus, one of the important standards to consider when choosing from among inhibitors is whether or not the inhibitors have good action under actual flow environments (*20*). In 2009, Badiea and Mohana (*19*) demonstrated that the corrosion density for carbon steel in industrial water increased with rising flow velocity in the absence of inhibitors, but increases slightly when an inhibitor is used. They reported that the best inhibition efficiency is at 1.56 m/s. Increasing the velocity from a specified amount causes thinning of the laminar film layer along the metal surface, which decreases inhibition efficiency (*21*). Thus, oxygen can reach the metal surface, leading to more corrosion.

Effect of Metal Cations

A number of studies have found that metal cations can have a synergistic effect on the quality of corrosion inhibitors, particularly organic inhibitor compounds (*22*). Organic inhibitors are expensive and difficult to synthesize, but inorganic inhibitors have toxic effects on both humans and the environment, and so must be used sparingly (*23, 24*). In the same way, monovalent cations have no influence on organic inhibitors (*25*), so most investigations examine divalent cations, particularly zinc ions.

Organic inhibitors include organic polymers such as polyaniline (*26*), polydiphenylamine (*27*), polyaminoquinone (*28, 29*), and substituted polyaniline (*30, 31*), which have been found to be efficient corrosion inhibitors for iron in acidic environments. Additionally, a number of the cationic metals, such as Zn^{2+}, Mn^{2+}, and Ce^{2+} have a synergistic effect on inhibition of corrosion reaction in acidic environment (*32*). For example, in 2008, Sathiyanarayanan and co-workers published a paper in which they described the inhibitory effectiveness of polyaniline substance in the presence of Zn^{2+} and Mn^{2+} ions for iron in 0.5 M H_2SO_4. They reported that the efficiency of inhibition of polyaniline is 40° for 10 ppm and 72° for 100 ppm, and the inhibition effectiveness is enhanced in the presence of Zn^{2+} ions to 83° for 10 ppm and 84° for 100 ppm. Using Mn^{2+} ions enhances the inhibition effectiveness to 80° for 10 ppm and 96° for 100 ppm. They proposed that this result may be explained by the fact that interaction between cations and polymers is based on protonation of imine nitrogen atoms in polyaniline, which causes chemical and conformational changes to the polymer

(33). Thus, this interaction leads to more effective absorption of electron-rich benzenoid groups on the iron surface, thereby decreasing the corrosion rate significantly.

Effect of Metal

The nature of metal affects the action of corrosion inhibitors. In fact, an inhibitor that is fairly effective for one metal may not be suitable for another metal. Despite this, some corrosion inhibitors are effective on more than one metal. Table 2 presents a brief summary of the nature of metals' influence on inhibition efficiency.

Table 2. A Brief Summary of the Nature of Metals' Influence on Inhibition Efficiency

Metal	*Medium*	*Inhibitor*	*Inhibition Efficiency*
Mild steel	1 M HCl	Schiff bases	92-88° (34)
Mild steel	1.0 M H_2SO_4	Schiff bases	96° (35)
Copper	0.5 M H_2SO_4	Schiff bases	83-77° (36)
Magnesium	0.01 M HCl	Schiff bases	93° (37)
Aluminum	3.5 wt ° NaCl	Triazole	72-80° (38)
Magnesium	Borate buffer	Triazole	90° (39)
Carbon steel	1 M HCl	Triazole	91° (40)
Copper	0.5 M HCl	Triazole	96.09° (41)
Mild steel	1 M HCl	Triazole	80-98° (42)

Effect of Electrolytes

Generally, the types of electrolytes used have extensive effects on the protective efficiency of corrosion inhibitors. An electrolyte is a material that creates an electrically conducting solution after dissolving in a polar solvent. Corrosion is caused by an ionic flow in the solution and an electronic flow on the metal surface. In the presence of corrosion inhibitors, the types of electrolytes can increase or decrease corrosion by decreasing the diffusion rate, decreasing the electrical resistance, and increasing or decreasing the anodic or cathodic reaction (43). Table 3 shows a brief summary of this effect.

Table 3. Brief Summary of the Effects of Electrolytes on the Effectiveness of Corrosion Inhibitors

Metal	*Medium*	*Inhibitor*	*Inhibition efficiency*
Steel	1 M HCl	N-(5,6-diphenyl-4,5-dihydro-[1,2,4]triazin-3-yl)-guanidine (NTG)	>99.6° (44)
Steel	0.5 M H_2SO_4	NTG	>99 ° (44)
Mild steel	1 M HCl	Derivatives of quinoline	95.2° (45)
Mild steel	2 M HCl	Derivatives of quinoline	86.1° (46)
Carbon steel	2 M H_3PO_4	Derivatives of quinoline	88.3° (47)

Table 3. (Continued). Brief Summary of the Effects of Electrolytes on the Effectiveness of Corrosion Inhibitors

Metal	*Medium*	*Inhibitor*	*Inhibition efficiency*
Carbon steel	2 M HCl	Derivatives of quinoline	94.6° (*48*)
Copper	3° NaCl	NTG	>99.8° (*49*)
Copper	0.5 M H_2SO_4	NTG	92.95° (*50*)
Copper	0.5 M HCl	Glutamic acid	59.70° (*51*)
Copper	1 M HNO_3	Glutamic acid	37.55° (*52*)
Copper	2 M HNO_3	Imidazole derivatives	Up to >90° (*53*)
Copper	0.5 M H_2SO_4	Imidazole derivatives	99.03° (*54*)
Copper	1 M HCl	Imidazole derivatives	70.5° (*55*)
Carbon steel	2 M HCl	Pyrimidin-2-amine	85° (*56*)
Carbon steel	0.5 M H_2SO_4	Pyrimidin-2-amine	82° (*57*)
Carbon steel	0.5 M HCl	Pyrimidin-2-amine	96.4° (*58*)
Carbon steel	1.0 M acetic acid	Plastic waste	84–92° (*59*)
Carbon steel	0.1 M HNO_3	Plastic waste	58–64° (*60*)

Effect of pH

The potential of hydrogen, or pH, is a scale that characterizes the acidity or basicity of aqueous media (*61*). It represents the hydrogen ion concentration in an aqueous media (eq 4):

$$pH = -\log [H^+] \quad (4)$$

pH is measured on a scale of 1–14; 1–7 is the acidic range, 7–14 is the basic range, and a pH of 7 is neutral. The influence of pH on corrosion inhibitor efficiency is variable. For example, sodium benzoate is inactive in media with pH lower than 5.5 because insoluble iron benzoate cannot form a protective film under acidic conditions (*7*).

In 2010, Curkovic et al. found that the inhibiting effectiveness of imidazoles is improved by enhancing the pH value of the media from 20° in 0.5 M HCl (pH ~ 3) to 92° in 0.5 M NaCl (pH = 5.6) (*62*). This enhancement can be attributed to tight adsorption of neutral imidazole compared to protonated imidazole at higher pH values. In addition, electrochemical impedance spectroscopy results indicated that the corrosion inhibitors caused a delay in corrosion reaction owing to enhancing the resistance of charge transfer and also to faradaic resistance related to the stabilization of corrosion products.

Effect of Humidity

Surface wetness and the effect of relative humidity (RH) on corrosion has been investigated in many studies. Studies have shown that the corrosion rate is strongly affected by relative humidity. The corrosion and water absorption of salt strongly depend on humidity and could be important factors determining corrosion in a solid salt environment (*63*). For example, the corrosion of carbon steel occurs above 20° relative humidity and increases to a maximum point at 80° of relative humidity (7.04 mm/yr).

Furthermore, fluctuation in relative humidity could affect atmospheric corrosion. A recent study investigated the effect of changing relative humidity on atmospheric pitting corrosion of 304L stainless steel. Results demonstrated that relative humidity fluctuations (between 33 and 85° RH or between 33 and 12° RH) caused nucleation of many small pits. Exposure at constant 33° caused the growth of a single large penetrating pit (*64*).

References

1. Raja, P. B.; Ismail, M.; Ghoreishiamiri, S. Its; Mirza, J.; Ismail, M. C.; Kakooei, S.; Rahim, A. A. Reviews on corrosion inhibitors: a short view. *Chemical Engineering Communications* **2016**, *203* (9), 1145–1156.
2. Umoren, S. A.; Solomon, M. M. Synergistic corrosion inhibition effect of metal cations and mixtures of organic compounds: a review. *Journal of environmental chemical engineering* **2017**, *5* (1), 246–273.
3. Chen, Y.; Jiang, Y.; Chen, H.; Zhang, Z.; Zhang, J.; Cao, C. Corrosion inhibition of mild steel in acidic medium by linseed oil-based imidazoline. *Journal of the American Oil Chemists' Society* **2013**, *90* (9), 1387–1395.
4. Odewunmi, N. A.; Umoren, S. A.; Gasem, Z. M.; Ganiyu, S. A.; Muhammad, Q. L-citrulline: An active corrosion inhibitor component of watermelon rind extract for mild steel in HCl medium. *Journal of the Taiwan Institute of Chemical Engineers* **2015**, *51*, 177–185.
5. Papavinasam, S. Corrosion inhibitors. *Uhlig's Corrosion Handbook*, 2nd ed.; Ravie, R., Ed; Wiley: New York, 2000.
6. Riggs Jr, O. *Corrosion Inhibitors*; Nathan, C. C., Ed.; National Association of Corrosion Engineers, Houston, Tex, USA 1973.
7. Sanyal, B. Organic compounds as corrosion inhibitors in different environments—a review. *Progress in Organic Coatings* **1981**, *9* (2), 165–236.
8. Arab, S. T.; Al-Mhyawi, S. R. Effect of temperature on the corrosion inhibition on mild steel in 2.0 MH2so4 by some organic compounds containing S and N atoms in absence and presence of halides. *Oriental Journal of Chemistry* **2008**, *24* (2), 365.
9. Arab, S. T.; Al-Mhyawi, SR. Effect of temperature on the corrosion inhibition on mild steel in 2.0 MH2so4 by some organic compounds containing S and N atoms in absence and presence of halides. *Oriental Journal of Chemistry* **2008**, *24* (2), 365.
10. RIGGS, O. L., JR; Every, R. Study of organic inhibitors for hydrochloric acid attack on iron. *Corrosion* **1962**, *18* (7), 262t–26t.
11. Putilova, N.; Balezin, S.; Barannik, V., *Metallic Corrosion Inhibitors* (translated from Russian). Pergamon Press, London: 1966.
12. Bouklah, M.; Attayibat, A.; Kertit, S.; Ramdani, A.; Hammouti, B. A pyrazine derivative as corrosion inhibitor for steel in sulphuric acid solution. *Applied Surface Science* **2005**, *242* (3-4), 399–406.

13. Popova, A.; Sokolova, E.; Raicheva, S.; Christov, M. AC and DC study of the temperature effect on mild steel corrosion in acid media in the presence of benzimidazole derivatives. *Corrosion science* **2003**, *45* (1), 33–58.
14. Machu, W.; Korrosion, u. *Metallschutz* **1938**, *14* (324), 40.
15. Kumari, P. P.; Shetty, P.; Rao, S. A. Electrochemical measurements for the corrosion inhibition of mild steel in 1 M hydrochloric acid by using an aromatic hydrazide derivative. *Arabian Journal of Chemistry* **2017**, *10* (5), 653–663.
16. Stoyanova, A.; Sokolova, E.; Raicheva, S. The inhibition of mild steel corrosion in 1 M HCl in the presence of linear and cyclic thiocarbamides—Effect of concentration and temperature of the corrosion medium on their protective action. *Corrosion Science* **1997**, *39* (9), 1595–1604.
17. Ashassi-Sorkhabi, H.; Ghasemi, Z.; Seifzadeh, D. The inhibition effect of some amino acids towards the corrosion of aluminum in 1 M HCl+ 1 M H2SO4 solution. *Applied surface science* **2005**, *249* (1-4), 408–418.
18. Waren, L. J., C. S.; Hariott, P. *Unit Operations of Chemical Engineering*, fourth ed. McGraw Hill, New York 1985, 217, 219 and 690.
19. Badiea, A.; Mohana, K. Effect of temperature and fluid velocity on corrosion mechanism of low carbon steel in presence of 2-hydrazino-4, 7-dimethylbenzothiazole in industrial water medium. *Corrosion Science* **2009**, *51* (9), 2231–2241.
20. Chen, Y.; Hong, T.; Gopal, M.; Jepson, W. EIS studies of a corrosion inhibitor behavior under multiphase flow conditions. *Corrosion Science* **2000**, *42* (6), 979–990.
21. Sinnott, R. Safety and loss prevention. *Chemical Engineering* **1983**.
22. Arthur, D. E.; Jonathan, A.; Ameh, P. O.; Anya, C. A review on the assessment of polymeric materials used as corrosion inhibitor of metals and alloys. *International Journal of Industrial Chemistry* **2013**Dec, *4* (1), 1–9.
23. Twite, R.; Bierwagen, G. P. Review of alternatives to chromate for corrosion protection of aluminum aerospace alloys. *Progress in organic coatings* **1998**, *33* (2), 91–100.
24. U. S. Congress, Office of Technology Assessment, *Environmental Policy Tools: A User's Guide*. US Government Printing Office: 1995; Vol. 634.
25. Riggs, O. L., Jr The second anodic current maximum for type 430 stainless steel in 0.1 N H_2SO_4. *Corrosion* **1975**, *31* (11), 413–415.
26. Jeyaprabha, C.; Sathiyanarayanan, S.; Venkatachari, G. Polyaniline as corrosion inhibitor for iron in acid solutions. *Journal of applied polymer science* **2006**, *101* (4), 2144–2153.
27. Jeyaprabha, C.; Sathiyanarayanan, S.; Phani, K.; Venkatachari, G. Investigation of the inhibitive effect of poly (diphenylamine) on corrosion of iron in 0.5 M H_2SO_4 solutions. *Journal of Electroanalytical Chemistry* **2005**, *585* (2), 250–255.
28. Phani, K.; Pitchumani, S.; Muralidharan, S.; Ravichandran, S.; Iyer, S. Electrosynthesis of polyamino-benzoquinone (PAQ) polymers. *Journal of Electroanalytical Chemistry* **1993**, *353* (1-2), 315–322.
29. Jeyaprabha, C.; Sathiyanarayanan, S.; Phani, K.; Venkatachari, G. Influence of poly (aminoquinone) on corrosion inhibition of iron in acid media. *Applied surface science* **2005**, *252* (4), 966–975.
30. Maniyel, P. The inhibitive effect of para-amino benzoic acid and its polymer on corrosion of iron in 1 mol/L HCl solution. *Journal of material sciences and technology* **2006**, *22* (3), 301–305.

31. Manivel, P.; Venkatachari, G. The Inhibitive Effect of Poly (p-Anisidine) on Corrosion of Iron in 1M HCl Solutions. *Corrosion Science and Technology* **2005**, *4* (2), 51–55.
32. Sathiyanarayanan, S.; Jeyaprabha, C.; Muralidharan, S.; Venkatachari, G. Inhibition of iron corrosion in 0.5 M sulphuric acid by metal cations. *Applied surface science* **2006**, *252* (23), 8107–8112.
33. Sathiyanarayanan, S.; Jeyaprabha, C.; Venkatachari, G. Influence of metal cations on the inhibitive effect of polyaniline for iron in 0.5 M H_2SO_4. *Materials Chemistry and Physics* **2008**, *107* (2-3), 350–355.
34. Issaadi, S.; Douadi, T.; Zouaoui, A.; Chafaa, S.; Khan, M. A.; Bouet, G. Novel thiophene symmetrical Schiff base compounds as corrosion inhibitor for mild steel in acidic media. *Corrosion Science* **2011**, *53* (4), 1484–1488.
35. Al-Amiery, A. A.; Kadhum, A. A. H.; Kadihum, A.; Mohamad, A. B.; How, C. K.; Junaedi, S. Inhibition of mild steel corrosion in sulfuric acid solution by new Schiff base. *Materials* **2014**, *7* (2), 787–804.
36. RAVARI, F.; Dadgarinezhad, A.; Shekhshoaei, I. Investigation on two salen type schiff base compounds as corrosion inhibition of copper in 0.5 M H_2SO_4. *Gazi University Journal of Science* **2009**, *22* (3), 175–182.
37. Seifzadeh, D.; Basharnavaz, H.; Bezaatpour, A. A Schiff base compound as effective corrosion inhibitor for magnesium in acidic media. *Materials Chemistry and Physics* **2013**, *138* (2-3), 794–802.
38. Sherif, E.-S. M. Effects of 3-Amino-1, 2, 4-triazole-5-thiol on the inhibition of pure aluminum corrosion in aerated stagnant 3.5 wt.° NaCl solution as a corrosion inhibitor. *Int. J. Electrochem. Sci* **2012**, *7* (6), 4847–4859.
39. Kuznetsov, Y. I.; Semiletov, A.; Chirkunov, A. Inhibition of magnesium corrosion by triazoles. *International Journal of Corrosion and Scale Inhibition* **2016**, *5* (1), 31–49.
40. Rouifi, Z.; Benhiba, F.; El Faydy, M.; Laabaissi, T.; About, H.; Oudda, H.; Warad, I.; Guenbour, A.; Lakhrissi, B.; Zarrouk, A. Performance and computational studies of new soluble triazole as corrosion inhibitor for carbon steel in HCl. *Chemical Data Collections* **2019**, *22*, 100242.
41. Quraishi, M. Electrochemical and theoretical investigation of triazole derivatives on corrosion inhibition behavior of copper in hydrochloric acid medium. *Corrosion Science* **2013**, *70*, 161–169.
42. Desai, P.; Indorwala, N. Triazoles used as a Corrosion inhibitor for mild steel in Hydrochloric Acid. *Int. J. Curr. Microbiol. App. Sci* **2015**, *4* (2), 928–938.
43. Raja, P. B.; Sethuraman, M. G. Natural products as corrosion inhibitor for metals in corrosive media—a review. *Materials letters* **2008**, *62* (1), 113–116.
44. Khaled, K. New synthesized guanidine derivative as a green corrosion inhibitor for mild steel in acidic solutions. *Int. J. Electrochem. Sci* **2008**, *3* (1), 462–475.
45. Rbaa, M.; Benhiba, F.; Obot, I.; Oudda, H.; Warad, I.; Lakhrissi, B.; Zarrouk, A. Two new 8-hydroxyquinoline derivatives as an efficient corrosion inhibitors for mild steel in hydrochloric acid: synthesis, electrochemical, surface morphological, UV–visible and theoretical studies. *Journal of Molecular Liquids* **2019**, *276*, 120–133.

46. Saliyan, V. R.; Adhikari, A. V. Quinolin-5-ylmethylene-3-{[8-(trifluoromethyl) quinolin-4-yl] thio} propanohydrazide as an effective inhibitor of mild steel corrosion in HCl solution. *Corrosion Science* **2008**, *50* (1), 55–61.
47. El Faydy, M.; Lakhrissi, B.; Jama, C.; Zarrouk, A.; Olasunkanmi, L. O.; Ebenso, E. E.; Bentiss, F. Electrochemical, surface and computational studies on the inhibition performance of some newly synthesized 8-hydroxyquinoline derivatives containing benzimidazole moiety against the corrosion of carbon steel in phosphoric acid environment. *Journal of Materials Research and Technology* **2020**, *9* (1), 727–748.
48. Abou-Dobara, M.; Omar, N.; Diab, M.; El-Sonbati, A.; Morgan, S. M.; Salem, O.; Eldesoky, A. Polymer complexes. LXXV. Characterization of quinoline polymer complexes as potential bio-active and anti-corrosion agents. *Materials Science and Engineering: C* **2019**, *103*, 109727.
49. Khaled, K. Guanidine derivative as a new corrosion inhibitor for copper in 3° NaCl solution. *Materials Chemistry and Physics* **2008**, *112* (1), 104–111.
50. Khaled, K. Adsorption and inhibitive properties of a new synthesized guanidine derivative on corrosion of copper in 0.5 M H_2SO_4. *Applied Surface Science* **2008**, *255* (5), 1811–1818.
51. Zhang, D.-Q.; Cai, Q.-R.; He, X.-M.; Gao, L.-X.; Zhou, G.-D. Inhibition effect of some amino acids on copper corrosion in HCl solution. *Materials Chemistry and Physics* **2008**, *112* (2), 353–358.
52. Barouni, K.; Kassale, A.; Albourine, A.; Jbara, O.; Hammouti, B.; Bazzi, L. Amino acids as corrosion inhibitors for copper in nitric acid medium: Experimental and theoretical study. *J. Mater. Environ. Sci* **2014**, *5* (2), 456–463.
53. Fouda, A.; Mohamed, A.; Mostafa, H. Inhibition of corrosion of copper in nitric acid solution by some arylmethylene cyanothioacetamide derivatives. *Journal de Chimie Physique et de Physico-Chimie Biologique* **1998**, *95* (1), 45–55.
54. Yan, T.; Zhang, S.; Feng, L.; Qiang, Y.; Lu, L.; Fu, D.; Wen, Y.; Chen, J.; Li, W.; Tan, B. Investigation of imidazole derivatives as corrosion inhibitors of copper in sulfuric acid: combination of experimental and theoretical researches. *Journal of the Taiwan Institute of Chemical Engineers* **2020**, *106*, 118–129.
55. Larabi, L.; Benali, O.; Mekelleche, S.; Harek, Y. 2-Mercapto-1-methylimidazole as corrosion inhibitor for copper in hydrochloric acid. *Applied Surface Science* **2006**, *253* (3), 1371–1378.
56. Awad, H.; Gawad, S. A. Mechanism of inhibition of iron corrosion in hydrochloric acid by pyrimidine and series of its derivatives. *Anti-Corrosion Methods and Materials* **2005**.
57. Li, X.; Xie, X. Adsorption and inhibition effect of two aminopyrimidine derivatives on steel surface in H2SO4 solution. *Journal of the Taiwan Institute of Chemical Engineers* **2014**, *45* (6), 3033–3045.
58. Mert, B. D.; Mert, M. E.; Kardas, G.; Yazici, B. The experimental and quantum chemical investigation for two isomeric compounds as aminopyrazine and 2-amino-pyrimidine against mild steel corrosion. *Anti-Corrosion Methods and Materials* **2016**.
59. Devi, N.; Karthiga, N.; Keerthana, R.; Umasankareswari, T.; Krishnaveni, A.; Singh, G.; Rajendran, S. Extracts of leaves as corrosion inhibitors–An overview and corrosion inhibition by an aqueous extract of henna leaves (Lawsonia inermis). *Int. J. Corros. Scale Inhib* **2020**, *9* (4), 1169–1193.

60. Yüce, A. O. Corrosion inhibition behavior of Robinia pseudoacacia leaves extract as a eco-friendly inhibitor on mild steel in acidic media. *Metals and Materials International* **2020**, *26* (4), 456–466.
61. Jensen, W. B. The symbol for pH. *Journal of Chemical Education* **2004**, *81* (1), 21.
62. Curkovic, H. O.; Stupnisek-Lisac, E.; Takenouti, H. The influence of pH value on the efficiency of imidazole based corrosion inhibitors of copper. *Corrosion Science* **2010**, *52* (2), 398–405.
63. Toba, K.; Suzuki, T.; Kawano, K.; Sakai, J. Effect of relative humidity on ammonium chloride corrosion in refineries. *Corrosion* **2011**, *67* (5), 055005-1–055005-7.
64. Guo, L.; Street, S. R.; Mohammed-Ali, H. B.; Ghahari, M.; Mi, N.; Glanvill, S.; Du Plessis, A.; Reinhard, C.; Rayment, T.; Davenport, A. J. The effect of relative humidity change on atmospheric pitting corrosion of stainless steel 304L. *Corrosion Science* **2019**, *150*, 110–120.

Chapter 4

Experimental Methods of Corrosion Inhibition Assessment

Marziya Rizvi,[1,*] Husnu Gerengi,[2] and Preeti Gupta[1]

[1]Department of Chemistry, Babasaheb Bhimrao Ambedkar University, Lucknow 226003, Uttar Pradesh, India

[2]Corrosion Research Laboratory, Department of Mechanical Engineering, Duzce University, Duzce 81620, Turkey

[*]Email: marziyarizvi@gmail.com.

Corrosion testing is a vital step that must be completed before materials and products can be used in real-world applications. Electrochemical reactions accompanied by changes in both the metal and the environment in contact with the metal are known as corrosion. Although corrosion has a common mechanism on a microscopic level, various microstructures, designs, and compositions will undergo different types of corrosion. Experimental methods for the estimation of kinetic, thermodynamic, and mechanistic aspects of metal corrosion are described in this chapter. Special emphasis is paid to electrochemical experimentation to determine the relative stability of materials, annual loss of material, inhibition efficiency, and the effects of environmental conditions, preferably in situ, in vitro, and in vivo.

The Basic Corrosion Concept

The results of corrosion can be very expensive, not to mention lethal. It can cause buildings and bridges to fall, oil pipelines to fail, chemical tanks to leak, and bathroom pipes to fail and cause flooding. Corroded electrical contacts might lead to a fire, and corroded medical implants can cause blood poisoning. Heavy air pollution in some areas has corroded art works. Corrosion poses a threat to radioactive waste containers, jeopardizing the chances of safe disposal. Corrosion initiates when an electrochemical reaction occurs in a corrosion cell. Metals tend to oxidize easily and lose electrons. As oxygen is reduced, a metal oxide forms. It is surprising that experimentation to determine precise measurements of the thermodynamic and kinetic aspects of corrosion is rare, although such determinations might assist both novices and professionals with standardized procedures. Processes like linear potential scan voltammetry are already standardized (*1*, *2*). Consequently, this chapter provides an overview covering all currently accepted guidelines that have been established by the global corrosion research community. Corrosion of metals in harsh environments is a common occurrence in cleaning and pretreatment in industrial processes like acid pickling, cleaning, descaling, and acidizing. Corrosion inhibitors are commonly used to control the corrosion rate of metals in harsh aqueous environments. Selecting an inhibitor depends on the nature of metal substrate and the environment in which it will be used. A qualitative classification of inhibitors is presented in Figure 1.

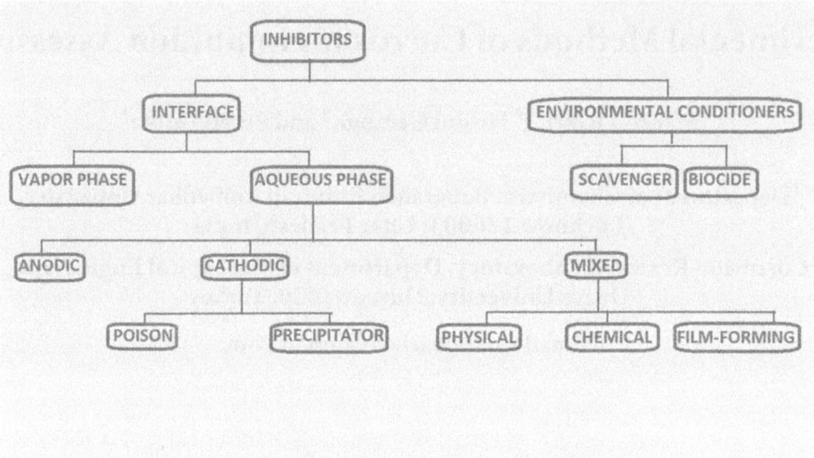

Figure 1. Qualitative classification of inhibitors.

The Corrosion Mechanism

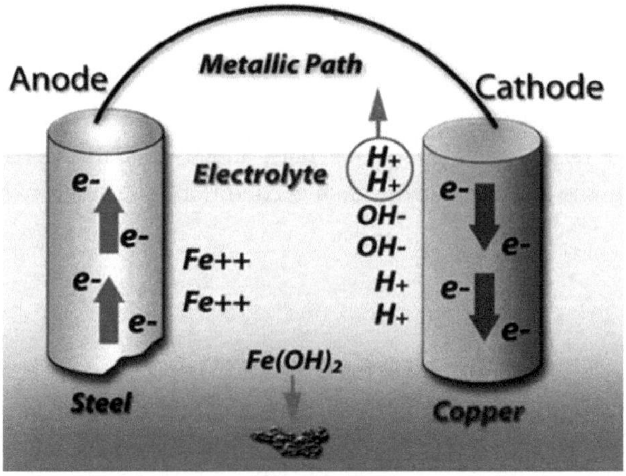

Figure 2. Typical corrosion cell.

An electrochemical cell is comprised of five parts: (1) anodic zones, (2) cathodic zones, (3) electrical contact between these zones, (4) an electrolyte, and (5) a cathodic reactant. Inside the electrochemical cell (Figure 2), electrons move from anodic to cathodic sites. The charged particles, ions, move across the conducting solution to balance the electron flow. Anions (from cathodic reactions) move toward the anode and cations (from the anode itself) drift toward the cathode. Resultantly, the anode corrodes and the cathode does not. There is also a voltage/potential difference between anode and cathode. Numerous discrete microcells develop on the metal surfaces from the constitutional phase difference; from stress variations; from coatings and imperfection levels like dislocations, grain boundaries, and kink sites; or from ionic conductivity alterations or compositional changes in the conducting solution. The corrosion process is a chemically spontaneous oxidation of the metal on reaction with the cathodic reactant. Every similar cell reaction results from a pair of simultaneous anodic and cathodic reactions going on at identical rates on the surface of metal.

Anodic Reaction

At the anode the metal is corroded. The anodic reaction is the metal's oxidation to its ionic form when an electric charge difference exists at the solid–liquid interface. Generally, anodic reaction is an oxidation reaction of a metal to its metal ions that pass into the conductive solution:

$$M \rightarrow M^{n+} + ne^- \quad (1)$$

where n is the metallic valence, e^- is the electron, M is metal, and M^{n+} its metal ion.

Cathodic Reactions

The cathodic reaction involves the environment and can be governed by the reaction

$$R^+ + e^- \rightarrow R^0 \quad (2)$$

where R^+ is the positive ion present in the electrolyte, e− is the metallic electron, and R^0 is the reduced species. Based on the environment, many cathodic reactions and electron-consuming reactions are possible. The main reactions are as follows. (1) In anaerobic acidic aqueous environments

$$2H^+ + 2e^- \rightarrow H_2 \quad (3)$$

(2) In anaerobic alkaline aqueous environments

$$2H_2O + 2e^- \rightarrow H_2 + 2OH^- \quad (4)$$

(3) In aerobic acidic aqueous environments

$$2O_2 + 4e^- + 4H^+ \rightarrow 4H_2O \quad (5)$$

(4) In aerobic alkaline aqueous environments

$$2O_2 + 4e^- + 2H_2O \rightarrow 4OH^- \quad (6)$$

Some other reactions commonly present in the chemical process are the following: (5) metal ion reductions

$$M^{3+} + e^- \rightarrow M^{2+} \quad (7)$$

(6) metal ion deposition

$$M^+ + e^- \rightarrow M \quad (8)$$

The results of the cathodic and anodic reactions react again to form solid depositions on the metal surface. The Fe^{2+} interacts with OH^- ions as

$$Fe^{2+} + 2OH^- \rightarrow Fe(OH)_2 \tag{9}$$

$Fe(OH)_2$ is reoxidized to $Fe(OH)_3$, an unstable product, which thus transforms to hydrated ferric oxide commonly known as red rust:

$$4Fe(OH)_2 + O_2 + 2H_2O \rightarrow 4Fe(OH)_3 : 2Fe(OH)_3 \rightarrow Fe_2O_3.3H_2O \tag{10}$$

Nevertheless, the phenomenon of corrosion continues devoid of any external flowing current; that is, on both sides, the rates should be equal. The electrode potential obtained at such a corroding interface is a comprehensive potential estimated by both reactions; it is commonly known as the corrosion potential E_{corr}. Such a category of comprehensive potential should be differentiated from the rest of mixed potentials as obtained in solutions containing many redox systems in an electrolyte, or from a redox potential. Most corrosion research goes beyond perceiving it toward corrosion protection, at least corrosion retardation. Considering the corrosion equations, several routes can be taken: inhibiting or retarding of a part of reaction, reducing the surface area available, or protection from reactants like H^+, H_2O, or O_2 are possible routes. Functional and practical perspectives would be to determine both thermodynamic and kinetic data first. ΔG values would be difficult to calculate and might be occasionally almost inaccessible. However, we can still compute the value of E_{corr}. Differentiating the values of E_{corr} as a function of experimental variables like composition of alloy, addition of inhibitor, variation in temperature, and so on, might furnish useful indicators for understanding and establishing corrosion inhibition.

Available Methods

The available experimental methods can be roughly divided into ex situ and in situ procedures. The former might furnish the results in the form of oxidation states of species in corrosion products, but generally, the latter are preferred. Specifically, the reactions permitted amidst the corrosion reaction intermediates or least stable products and the environment, like airborne dioxygen (3), permissible when transferring a sample being checked from its electrolyte/environment into, for example, the ultra high vacuum system required for multiple photoelectron spectroscopies like X-ray photoelectron spectroscopy, are the major causatives of profound experimental deviations, misinterpretations, and errors (3, 4). American Society for Testing and Materials (ASTM) standards for corrosion and wear render the befitting methods for conducting wear, abrasion, and other corrosion tests on specific alloys and metal surfaces (https://www.astm.org/Standards/corrosion-and-wear-standards.html).

Gravimetric Analysis

Immersion corrosion evaluation is used to estimate the resistance of a metal toward the aggressive and aqueous electrolyte. Corrosion testing coupons prepared in standard ways are incorporated to compute the gravimetric effects of corrosion. The gravimetric measurements are performed per ASTM designation G1-90 (Figure 3). After the initial weights and dimensions are measured, the freshly prepared steel coupons are fully immersed in 250 mL beakers containing at least 200 mL of test solution maintained at 30, 40, 50, and 60 °C, respectively, for a specific duration.

Figure 3. Immersion test for gravimetric analysis.

The temperature of the test solution is maintained in a water bath with controlled temperature. To keep the crevice corrosion at bay, the coupons are usually suspended inside the electrolyte with a nylon thread. After the immersion, the coupons are extracted from the solution, cleaned thoroughly using deionized water, and dried until a constant weight is achieved on a scale (Figure 4). The computation of the corrosion rate (CR; mg cm^{-2} h^{-1}) follows eq (10):

$$CR\ (\text{mg cm}^{--}\ \text{h}^{-}) = \frac{\Delta W}{A t} \tag{11}$$

where ΔW (mg) is weight loss, A (cm^2) is the area of specimen, and t (h) is exposure time. The surface coverage θ was calculated using the values obtained for CR as follows:

$$\theta = \frac{CR_0 - CR_i}{CR_0} \tag{12}$$

The inhibition efficiency η_{grav} (%) is calculated by incorporating the average CR as follows:

$$\eta_{grav}\ (\%) = \frac{CR_0 - CR_i}{CR_0} \times 100 \tag{13}$$

where CR_0 and CR_i represent the corrosion rates in free and inhibited acid solutions.

The ASTM G31 procedure normalizes corrosion rates to time units, normally in millimeters per year or mils per year (mpy). The terms of the test might vary, but results can generally be calculated within 24 h or 10 or 20 day exposure time periods. The pairing of electrolytes and the material of the coupon is often suggested on the basis of its application. Temperature of the electrolyte and oxygen availability changeable is also experiment specific. For example, TM0169-95 (281-228-6200), a National Association of Corrosion Engineers test method, offers standards for computing optimum test times for immersion service testing. After the experimental procedures, gravimetric tests render the corrosion weight. Surface analysis might also be conducted by stereo microscopy or scanning electron microscopy (SEM) to investigate the apparent posttest morphology better. Samples for electrochemical examination are generally a circle of about 38 mm (1.5 in.) diameter which is 3 mm (0.125 in.) thick and has an 8 mm (5/16 in.) or 11 mm (7/16 in.) diameter hole for suspension. The edges of samples are often sanded to smooth them out. The results of gravimetric analysis are the very first test to judge the amount of corrosion occurring in the test solution and the efficiency of coatings and inhibitors.

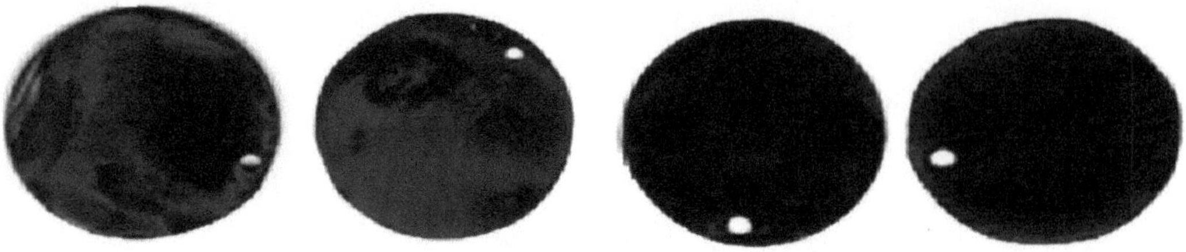

Figure 4. The coupons extracted from the setup for gravimetric analysis.

Electrochemical Measurements

In any 1 L corrosion cell, a metal specimen of working surface area of 1 cm^2 lodged in a sample holder is known as a working electrode (WE), a large area platinum wire is employed as a counter electrode (CE), and for reference electrode the Ag/AgCl electrode (saturated KCl) is the preferred one (RE). A Luggin-Haber capillary is also placed in the cell setup with its tip very close to the WE surface. This is done to minimize the ohmic potential or IR drop. The setup used in electrochemical corrosion experiments is shown in Figure 5.

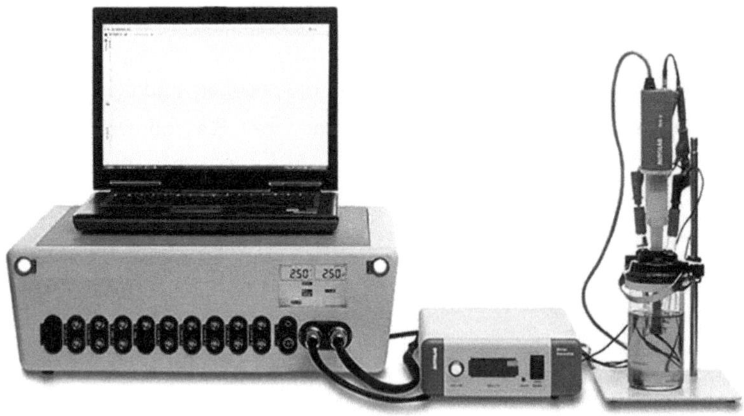

Figure 5. Autolab 128 N instrument used for electrochemical corrosion testing.

Before the electrochemical procedure is started the WE is immersed in the electrolyte and the rest potential is assessed until it reaches stability, and hence a steady-state potential is obtained. Usually, an hour-long immersion is apt to stabilize the potential and maintain a steady-state open circuit potential. The experiments are generally conducted at room temperature (30 °C $\pm$ 1 °C) without stirring in an aerobic environment. For potentiodynamic polarization measurements the WE is scanned in a potential range of −250 to +250 mV according to the steady-state open circuit potential at a scan rate of 0.00166 V/s. The anodic and cathodic polarization curves are studied and the values of E_{corr} (corrosion potential), i_{corr} (corrosion current density), β_a (anodic Tafel slope), and β_c (cathodic Tafel slope) are achieved during potentiodynamic polarization. These parameters determine not only the amount of corrosion occurring, but also the type of corrosion occurring and the factors associated with inhibitors and coatings that are affecting these parameters. The measured i_{corr} values assist in estimating the inhibition efficiencies η_{PDP} (%) using the following equation:

$$\eta_{PDP}(\%) = \frac{i_{corr} - i^{(i)}{}_{corr}}{i_{corr}} \times 100 . \quad (14)$$

In eq (13), $i^{(i)}{}_{corr}$ and i_{corr} depict the corrosion current densities in the electrolyte with and without inhibitor, respectively.

The impedance measurements for electrochrmical impedance spectroscopy (EIS), realized at corrosion potential, are conducted with AC signals of amplitude 10 mV within a specific frequency range, usually 10^{-2} to 10^4 Hz. The impedance results, R_{ct} (charge transfer resistance) achieved with the help of analyzing the Nyquist plots is employed in the calculation of % inhibition efficiency η_{EIS} ultimately:

$$\eta_{EIS}(\%) = \frac{R_{ct}^{(i)} - R_{ct}}{R_{ct}^{(i)}} \times 100 . \quad (15)$$

Here, $R_{ct}^{(i)}$ and R_{ct} are values of ct resistance in the presence and absence of inhibitor, respectively. The parameters obtained, such as ct resistance, double layer capacitance, and inhibition efficiency decide the amount of corrosion occurring on surface.

The conventional method of impedance measurement is frequency response analysis, which utilizes a perturbation in the form of a sequential sinusoidal signal of defined amplitude and frequency. EIS is a lengthy process, and so is applicable only to stationary systems. The electrochemical process of corrosion is not a stationary one; rather it is a dynamic process. Dynamic impedance spectroscopy (DEIS) depends on the simultaneous excitation of system with all the perturbation frequencies. It allows obtaining impedance change in the preset time dimension (multiple spectra in function of time). In this multifrequency perturbation signal generated continuously, several frequencies are selected and their sum is a perturbation signal. A multisinusoidal current excitation signal is used and the changes of potential are registered simultaneously as a function of time. The graphical representation of DEIS is shown in Figure 6.

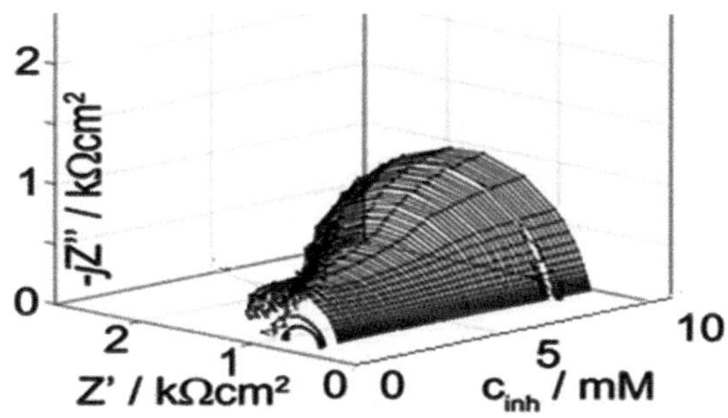

Figure 6. Evolution of DEIS spectra on addition of inhibitor after 1 h in a test solution.

Adsorption Isotherm

The adsorption isotherm studies are carried out to provide more information about the adsorption properties of inhibitor during the metallic corrosion in test solution. The surface coverage data θ obtained from gravimetric results or electrochemical measurements at different temperatures are visualized graphically to allow the apt-fitting isotherm models. Various adsorption isotherms are presented in Table 1, where θ is the section of the metallic area covered with the inhibitor resultant of adsorption, C is concentration of the used inhibitor, n is the number of water molecules replaced for every inhibitor molecule, f is the interaction parameter, and K_{ads} is the equilibrium constant of the adsorption–desorption process.

Table 1. Adsorption Isotherm Models That Can Be Used to Describe the Adsorption on the Metal Surface

Name	*Isotherm*		*Verification Plot*
Langmuir	$C/\theta = 1/K_{ads} + C$	(16)	C/θ vs $1/C$
Frumkin	$[\theta/(1-\theta]e^{-2f\theta}$	(17)	θ vs log C
Temkin	$\mathrm{Exp}(f\theta) = K_{ads}C$	(18)	θ vs log C
Freundluich	$\theta = K_{ads}C$	(19)	θ vs C
Bockris-Swinkels	$\theta/(1-\theta)^n[\theta + n(1-\theta)^{n-1}/n^n = Ce^{-Kads/55.5}$	(20)	$\theta/(1-\theta)$ vs log C
Virial Parson	$\theta e^{2f\theta} = K_{ads}\theta$	(21)	θ vs log(θ/C)

Determination of Kinetic and Thermodynamic Parameters

Free Energy of Adsorption

The adsorptive equilibrium constant K_{ads} (derived from the values of intercept of the adsorption isotherms) is related to the standard free energy of adsorption ΔG_{ads} according to eq (21):

$$\Delta G_{ads} = -RT \ln(1\text{x}10^{6} K_{ads}) \tag{22}$$

where 1×10^6 is the concentration of water molecules in parts per million, R is the universal gas constant (8.314 J K^{-1} mol^{-1}), and T stands for the absolute temperature (K) (*6*). A value of ΔG_{ads} obtained near −20 kJ mol^{-1} or less suggests physical adsorption, which occurs due to the electrostatic attraction between the charged molecules of inhibitor and charged metal surface, whereas ΔG_{ads} values near −40 kJ mol^{-1} or more suggests chemisorptions that revolve around sharing of charge or transfer of electrons from inhibitor molecule to metal surface (*7*).

Activation Energy

The corrosion reaction can be considered as an Arrhenius-type process, where activation energy can be achieved by

$$CR = A\exp(-\frac{E_a}{RT}) \tag{23}$$

where E_a is the apparent activation corrosion energy, T is the absolute temperature, A is the Arrhenius pre-exponential factor, and R is the universal gas constant. Such an equation is employed in the calculation of E_a and A values. The values of E_a and A in solutions are achieved by the linear regression of plot of log CR vs $1/T$ data. The activation energy of the metal in test solution without any inhibition is always the lowest. It increases as inhibition factors are increased.

Enthalpy and Entropy of Adsorption

The enthalpy of adsorption (ΔH) and entropy of adsorption (ΔS) was achieved by an alternative Arrhenius equation known as transition state plot, which is governed by the following equation:

$$CR = \frac{RT}{Nh} exp\left(\frac{\Delta S}{R}\right) exp\left(-\frac{\Delta H}{RT}\right) \quad (24)$$

where h represents Planck's constant, N represents Avogadro's number, ΔS represents the entropy of activation, and ΔH represents the enthalpy. From plot of log(CR/T) vs $1/T$, a slope of $\Delta H/R$ and an intercept of ln($R/Nh + \Delta S/R$) is obtained from which the data of ΔS and ΔH are calculated.

Determination of Synergism Parameter When Inhibitors and Additives Are Used Together

The synergism parameter S_1 was estimated by the following relationship proposed by Aramiki and Hackerman (5, 7) and used subsequently by several other researchers:

$$S_1 = \frac{1 - \theta_{1+2}}{1 - \theta'_{1+2}} \quad (25)$$

where $\theta_{1+2} = (\theta_1 + \theta_2) - (\theta_1 \times \theta_2)$, θ'_{1+2} represents the estimated surface coverage for inhibitor in addition to the surfactants, θ_1 is the surface coverage of pure inhibitor, and θ_2 is the surface coverage of pure surfactant. The value of $S_1 > 1$ indicates a synergistic effect. S_1 approaches 1 when there is no interaction between inhibitor and the surfactant, and $S_1 < 1$ suggests antagonistic behavior, which may be a characteristic of competitive adsorption.

Surface Morphological Examination

Surface characterization methods like Scanning Electron Microscopy (SEM) with Energy Dispersive X-Ray Analysis (EDX) SEM/EDX, Atomic force microscopy (AFM), and X-ray photoelectron spectroscopy were conducted to view and estimate qualitatively and quantitatively the corrosion on the metal sample in test electrolytes in terms of surface heterogeneity. SEM/EDX/AFM are run on metallic samples after gravimetric assays are done with and without optimum concentrations of inhibitors.

Binding Studies

UV-Visible Spectroscopic Analysis

To realize and confirm any kind of interaction between inhibitor and metal that leads to the formation of inhibitor/Fe complex, UV-visible spectra are conducted on the test solution containing inhibitor before and after immersion for specific time and temperature.

Fourier-Transform Infrared Spectroscopy

Fourier-transform infrared spectra of inhibitor alone and inhibitor adsorbed on the metal surface are usually recorded to check whether any binding is occurring at surface of metal. The spectra are usually recorded within 4000–500 cm^{-1} frequency. Initially, the powder is mixed with KBr and made into a disc. The other specimen is the thin adsorption layer formed on metallic surface after immersion for a specific time in the electrolyte with inhibitor.

Quantum Study

Quantum estimations are conducted using either of the following methods.

Semiempirical Method: Extended Hückel Calculation

The extended Hückel method and MM2 energy minimization geometrically optimize the structural parameters. All the computations can be done with any graphic user interface like Chem 3D. The quantum chemical values achieved were E_{LUMO} (energy of lowest unoccupied molecular orbital), E_{HOMO} (energy of highest occupied molecular orbital), ΔE (orbital energy gap), and dipole moment (μ). The E_{HOMO}, E_{LUMO}, and μ are recorded. The energy gap (ΔE), global hardness (η), global electronegativity (χ) of the molecule, and section of molecules transferred (ΔN) were estimated using the appropriate relations (*8, 9*); these parameters decide the feasibility of corrosion and corrosion protection:

$$\Delta E = E_{LUMO} - E_{HOMO} \tag{26}$$

$$\eta = \frac{1}{2}\left(E_{LUMO} - E_{HOMO}\right) \tag{27}$$

$$\chi = \frac{1}{2}\left(E_{HOMO} + E_{LUMO}\right) \tag{28}$$

$$\Delta N = \frac{\chi_{Fe} + \chi_{inh}}{2(\eta_{Fe} + \eta_{inh})} \tag{29}$$

Density Functional Theory Method

The density functional theory method is employed to derive the electron density distributions and obtain notable quantum chemical values of inhibitors. For example, the B3LYP/6-31G (d,p) model achieves the gas-phase optimized geometry of the inhibitor and Gaussian 09 software suite runs the calculations. Electron density distributions of the Fukui functions are viewed using interfaces like Multiwfn (*9, 10*). Electrophilic (f^{+}) and nucleophilic (f^{-}) Fukui indices of the molecules were obtained using the Mulliken atomic charge differences based on the relations (*11*)

$$f^{+} = q_{N+1} - q_{N} \tag{30}$$

$$f^{-} = q_{N} - q_{N-1}. \tag{31}$$

In eqs (29) and (30), q corresponds to the Mulliken charge on the atom for N number of electrons. $N + 1$ is related to an anion, with an electron added to the LUMO of the molecule. $N - 1$ relates to the cation with an electron removed from the HOMO of the molecule.

Monte Carlo Simulation

The interaction between inhibitor and, for example, the Fe (110) plane surface was analyzed using Monte Carlo simulations. In this simulation, the adsorption detector code running in the Material Studio 8.0 software from Biovia-Accelrys Inc. USA was adopted. For the simulation of all inhibitor molecules and entities, condensed-phase optimized molecular potentials for atomistic simulation studies was applied. The simulation of the corrosion inhibitor molecule on Fe (110) surface was done to locate the low-energy adsorption sites of inhibitors on metal surface.

Conclusions

The prime advantage offered by instrumental corrosion measurement techniques is that a test must not be interrupted or terminated to provide corrosion information. When an instrument is connected to a corrosion measurement probe, the corrosion data are read out on a meter or are recorded continuously. Corrosion measuring probes are available that can be conveniently inserted or withdrawn from pressurized process streams. Like coupons, corrosion sensing probes are not of exactly the same material or configuration as the corroding system and, as such, care must be taken in determining probe locations. Other instrumental techniques for measuring corrosion include ultrasonic and radioactive wall thickness measurement and atomic hydrogen detection with hydrogen diffusion probes. Ultrasonic thickness measurement is used to check the wall thickness of process vessels and piping. Radioactive measurements provide similar information. The use of both techniques is limited by the lack of sensitivity to small amounts of metal loss and inconvenience in gaining access to some surfaces of interest. A hydrogen probe detects and measures atomic hydrogen produced by various corrosion reactions. Hydrogen probes are used in refineries rather widely.

References

1. ASTM G5–94 (2011) *Standard reference test method for making potentiostatic and potentiodynamic anodic polarization measurements*, ASTM International. West Conshohocken, USA.
2. Holze, R. (2009) *Experimental electrochemistry: a laboratory textbook*. VCH-Wiley, Weinheim, Germany.
3. ASTM G1-90, "*Standard practice for preparing, cleaning, and evaluating corrosion test specimens*," ASTM, Philadelphia, PA, USA, 2000.
4. Solomon, M. M.; Gerengi, H.; Umoren, S. A. Carboxymethyl cellulose/ silver nanoparticles composite: synthesis, characterization and application as a benign corrosion inhibitor for St37 steel in 15% H2SO4 medium. *ACS Appl. Mater. Interfaces* **2017**, *9*, 6376–6389.
5. Avci, G. Inhibitor effect of N, N- methylenediacrylamide on corrosion behavior of mild steel in 0.5M HCl. *Mater. Chem. Phys.* **2008**, *112*, 234–238.
6. Bai, X. L.; Feng, L. J.; Wang, H. Y.; Lu, Y. B.; Lei, X. W.; Bai, F. L. Comparison of the synergistic effect of counter ions on the inhibition of mild steel corrosion in acid solution: electrochemical, gravimetric and thermodynamic studies. *RSC Adv.* **2015**, *5*, 4716–4726.
7. Aramiki, K.; Hackerman, N. Inhibition mechanism of medium size polymethyleneimine. *J. Electrochem. Soc.* **1969**, *116*, 568–574.

8. Yang, W.; Mortier, W. J. The use of global and local molecular parameters for the analysis of the gas-phase basicity of amines. *J. Am. Chem. Soc.* **1986**, *108*, 5708–5711.
9. Lu, T.; Chen, F. Multiwfn: A Multifunctional Wavefunction Analyzer. *J. Comput. Chem.* **2012**, *33*, 580–592.
10. Lu, T.; Chen, F. Quantitative analysis of molecular surface based on improved marching tetrahedra algorithm. *J. Mol. Graph. Model.* **2012**, *38*, 314–323.
11. Singh, A.; Lin, Y.; Obot, I. B.; Ebenso, E. E.; Ansari, K. R.; Quraishi, M. A. Corrosion mitigation of J55 steel in 3.5% NaCl solution by a macrocyclic inhibitor. *J. Appl. Surf. Sci.* **2015**, *356*, 341–347.

Chapter 5

Electrochemical Evaluation of Sustainable Corrosion Inhibitors via Dynamic Electrochemical Impedance Spectroscopy

Pawel Slepski,[1] Husnu Gerengi,[2,*] Gökhan Gece,[3] Ertuğrul Kaya,[2] Marziya Rizvi,[4] and Michal Szociński[1]

[1]Department of Electrochemistry, Corrosion and Materials Engineering, Faculty of Chemistry, Gdańsk University of Technology, 11/12 G. Narutowicza Str., Gdańsk 80-233, Poland

[2]Corrosion Research Laboratory, Department of Mechanical Engineering, Faculty of Engineering, Düzce University, 81620 Düzce, Turkey

[3]Department of Chemistry, Faculty of Engineering and Natural Sciences, Bursa Technical University, 16310 Bursa, Turkey

[4]Department of Chemistry, Babasaheb Bhimrao Ambedkar University, Lucknow - 226025, India

[*]Email: husnugerengi@gmail.com.

Finding suitable measurement methods for the effective management of electrochemical problems is of paramount importance, particularly for improving efficiency in corrosion protection. The need for accurate measurement techniques specific to nonstationary conditions has long been recognized, and promising approaches have emerged. This chapter introduces dynamic electrochemical impedance spectroscopy as a novel advancement in electrochemistry that can be used efficiently in galvanostatic and potentiostatic modes. The review focuses first on an explanation of the method and second on presenting a comprehensive corpus covering available studies that have applied dynamic electrochemical impedance spectroscopy for the purpose of preventing corrosion phenomena. This chapter defines the merits of this novel approach compared with the conventional electrochemical impedance spectroscopy method.

An Overview

The field of corrosion research inevitably requires a strong perception of the thermodynamic and kinetic properties of the metal–solution interface. Accordingly, the characterization of systems with such interfaces is a major concern in electrochemical sciences, and this has provoked a trend of focusing on frequency-related phenomena (i.e., small-signal alternative current studies). The overall electrical conductivity of a system is known to be decreased because of polarizations, and in particular, mechanical, electrical, crystallographic, and compositional fluctuations in physical properties are expected to occur at an interface. Hence, it is not surprising that when a system undergoes an applied potential variance, its interface will polarize in an unprecedented way (*1*). In this regard, impedance spectroscopy comes into prominence as a relatively robust characterization method that is used to investigate the charge dynamics in solid and liquid interfaces (*2*).

The standard approach to measuring impedance is to apply a single-frequency voltage or current to the interface and then to record the response signal phase shift and amplitude at that frequency. A frequency response analysis applied to a stationary system is an example of this approach to impedance measurement. However, it is often the case that examined electrochemical objects are time- and frequency-dependent, and this requires a different type of impedance measurement based on a time–frequency domain analysis to extract information for possible transient conditions.

In the 1990s, the first attempts were made to create an electrochemical impedance technique allowing investigation of nonstationary systems (*3–6*). Fourier transform (FT) is the most frequently used transformation in signal analysis. However, FT analysis is not suitable for analyzing signal changes over time, and thus it should be used only for stationary systems. Because most of the electrochemical objects of interest possess nonstationary characteristics, this deficiency is overcome by a method called windowing the signal. The short-time Fourier transform (STFT) returns information in a two-dimensional function of time and frequency (*7*). This method was described and implemented for the first time by Darowicki et al. (*8*, *9*) in an impedance analysis of a model nonstationary system. The essence of dynamic electrochemical impedance spectroscopy (DEIS) is the application of a multisinusoidal waveform of small amplitudes to an electrochemical object that is under nonstationary conditions and the analysis of the response and perturbation signals using the STFT method.

This is the first up-to-date and authoritative comprehensive review of the electrochemical behavior of several metals, alloys, and some organic compounds during DEIS experiments. We offer a mixture and comparison of findings with the objective of setting up a theoretical framework for the application of DEIS as a technique for the examination of challenging electrochemical as well as corrosion processes.

Understanding DEIS Methodology

As alluded to previously, electrochemical impedance spectroscopy (EIS) has been considered one of the basic and most reliable techniques for analyzing any electrochemical process. It is apt in this context to touch on some details of its methodology. Traditionally, the most popular method of impedance measurement is frequency response analysis, which utilizes a perturbation in the form of a sequential sinusoidal signal of defined amplitude and frequency. However, it is a time-consuming process; therefore, EIS can be used solely for static systems. In the case of dynamic processes, any impedance point on the impedance spectrum can represent another state of the investigated system. This is an overt constraint of this exceptional analysis method. The EIS technique is being used by almost every researcher who is studying corrosion. Not only is this technique cumbersome, but also, as mentioned before, the electrochemical process of corrosion is not stationary. It is a dynamic process. Increasing and decreasing the frequency range (longer or shorter impedance measurement process) alone will give no interpretable impedance spectrum that justifies the nonstationary nature of a corrosion reaction. Another proof of the nonstationary nature of the electrochemical process of corrosion is the nonoverlapping impedance spectra obtained. Many researchers have substantiated the nonstationary nature of electrochemical processes (*10–14*). To simplify this, imagine monitoring is performed on a corrosion cell to estimate the efficiency of a corrosion inhibitor. The classical EIS suggests that first, a steady state potential is acquired, perturbation is then generated, and the frequency response is recorded. During this time of attaining a steady potential, the layer of oxidation products often forms on the working electrode, and the impedance recorded via the EIS process is a sum of the impedance offered by the oxide layer and inhibitor. This mixed and time-dependent resistance is recorded as the inhibition efficiency of the inhibitor under analysis. The presented case shows that the system does not meet the stability condition and is not analyzable with EIS.

To overcome or reduce these restrictions in analyzing the complex and dynamic process of corrosion and to shorten the measurement time, researchers have introduced modifications to EIS in the form of a multifrequency perturbation signal. This technique depends on all the perturbation frequencies bringing the system to an excited state in a simultaneous manner. It allows obtaining impedance change in the preset time dimension (multiple spectra in function of time). Several frequencies are selected in this continuously generated multifrequency perturbation signal (Figure 1), and their sum is taken as the perturbation signal. A multisinusoidal current excitation signal is used, and the changes in potential are registered simultaneously versus time. The STFT allows changes in the impedance values to be measured and recorded over time.

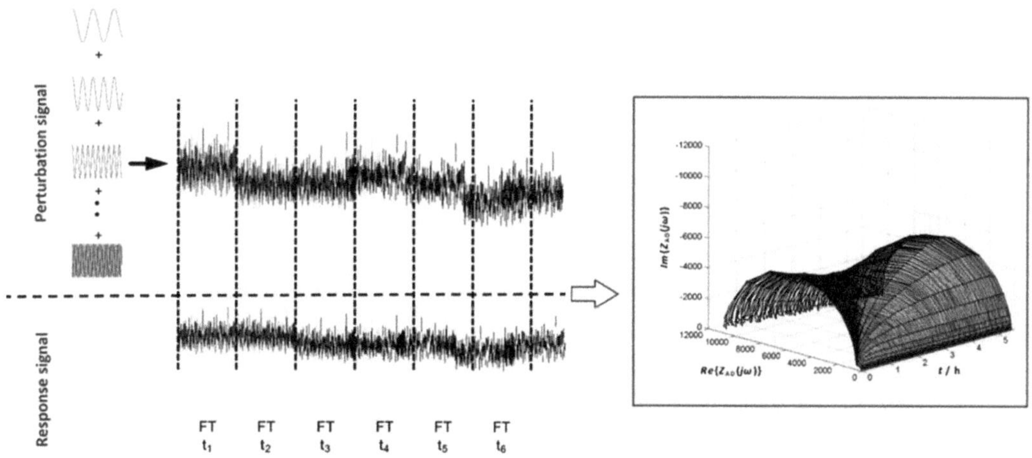

Figure 1. Multisinusoidal (perturbation) and response signals in DEIS (t_{1-6}: the time of subsequent spectra).

It is critical to select the appropriate sampling frequency component to obtain all of the bandwidth information. Moreover, in accordance with the Nyquist criterion, in order to accurately convey the prospective response signal, the sampling needs to be at a minimum double the highest perturbation frequency (*15*). Alternately, if the sampling frequency is too high, it might exceed the limits of the hardware. For experiments conducted in changing scenarios, impedance data allocation is assigned to the system's instantaneous state using a window function, which employs STFT (*16, 17*).

$$STFT\ (T,\omega) = \int s(\tau)\gamma(\tau - t)\exp(-j\omega\tau)\,d\tau \qquad (1)$$

Here, $STFT\ (T,\ \omega)$ represents the STFT, $s(\tau)$ the studied signal, and $\gamma(\tau)$ the window function. In practice, not continuous but discrete signals are recorded, and therefore, the STDFT can be utilized (*18*).

$$STDFT\{s[k,n]\} = \sum_{i-0}^{L-1} s[i]\gamma[i-k]\exp\left(\frac{-j2\pi ni}{L}\right) \quad (2)$$

Here, k represents the relative frequency resolution per unit of time, where $s(i)$ signifies the digital form of the analyzed signal. Accordingly,

- The entire investigated electrochemical process is perturbed by a continuous multifrequency signal.
- During the entire measurement process, an acquisition of the perturbance and response takes place.
- After or during the experiment, FT is operated upon a number of signal fragments of determined length and time.
- As an outcome of this process, complex values in the form of a time–frequency mesh are generated, which makes it possible to obtain local impedance spectra for determined time instants.

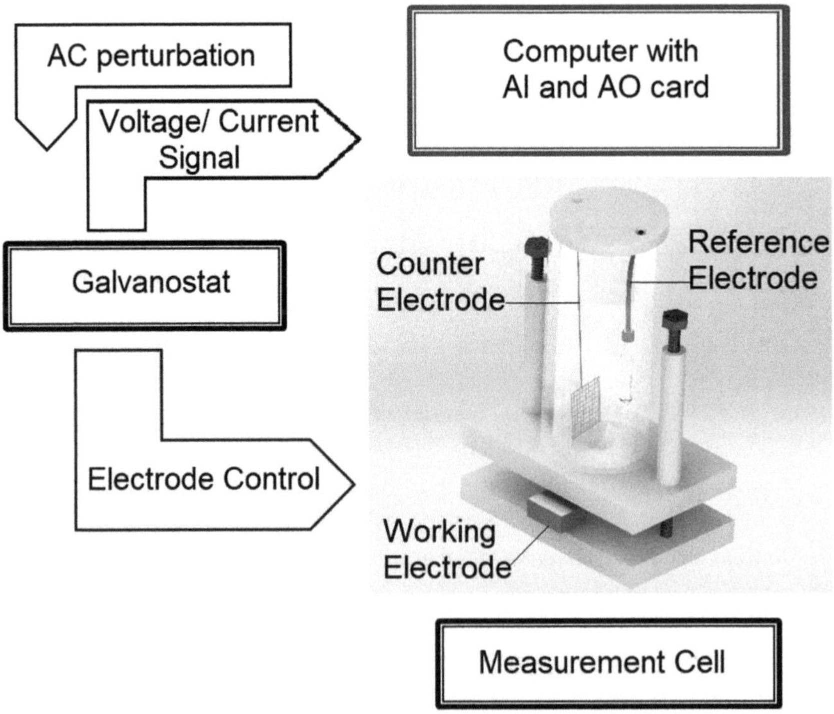

Figure 2. DEIS measurement setup.

Impedance is computed for a specific analysis window length. The range of analyzed frequency points is related to the analysis window length, thus making it a parameter of great consequence in DEIS. For the repetitive slicing sequence, the STDFT application renders the impedance spectra series within the finite time domain (*19, 20*). A narrow window is necessary to get a pseudostationary measurement, and its length confirms the lowest perturbation signal frequency. For such an outcome, a major shortcoming of the DEIS approach is that lower frequencies restrict the measurements. Because the forces at work are lower, a longer window is required for analysis even if lower frequencies are used. Although the DEIS performs summing and averaging functions because the FT has an averaging function, the spectra are averaged over the window length. Therefore, despite this limitation, useful instantaneous impedance spectra can be obtained, even if slight nonstationarity occurs inside the analysis window.

Figure 2 presents a depiction of the introduction of a digital analogue card with AI and AO channels to a regular EIS setup in a typical dynamic measurement. The corrosion cell is comprised of a three-electrode system: a working electrode, a counter electrode, and a reference electrode. The card generates current and measures the current and voltage signals. The galvanostatic conditions are maintained.

The corrosion cell is perturbed with the current signal set with different frequencies and with amplitudes that are similar or different. During the experiment, the current response signals and voltage perturbation are continuously recorded. The analysis window function is employed to slice a fragment from the recorded response signals. This extracted part is then exposed to conventional FT in order to obtain an instantaneous spectrum, the average of which is recorded over the time range corresponding to the length of the analysis window.

Review of Forms of Electrochemical Corrosion Measurement and the Evolution of DEIS: Merits and Advantages

Much of the earlier interest stems from research on the frequency response analysis of electrochemical systems (*21–23*). For instance, Burstein and Liu investigated the transitory repassivation of aluminum in solutions that do not induce breakdown of passivity or localized corrosion. They used time-resolved frequency response analysis with a guillotined aluminum electrode in a pH 6.9 buffered solution (0.3 M NaH_2PO_4, 0.3 M Na_2HPO_4) and a pH 8.4 unbuffered solution (0.1 M Na_2CrO_4) (*21*). The crucial result of these experiments was that the measurement of the electrode potential and its ensuing transformation over time was successful for both open-circuit currents and those applied galvanostatically. The seminal studies by Darowicki et al. (*8*, *9*, *24*, *25*) were the start. They compared the convergence of dynamic and static STFT spectrograms obtained by both theoretical mathematical analysis and introduced an electrical system model composed of a parallel-connected capacitor and diode. Almost all experimental knowledge of the nonstationary process properties of the investigated model comes from comparisons between these two approaches. The work was extended to consider measurements of the complex impedance versus the potential (103) Cd(II) reduction reaction for different possible scan rates using the hanging mercury electrode drop technique (*26*) and to consider the process of stress corrosion cracking initiation (cracking of the passive layer) of 304L stainless steel in a solution of 0.5 M NaCl (*27*). As reported for the first time (*27*), the benefit of utilizing DEIS was that, in contrast to other techniques, the authors were able to elicit the dynamics during the process of passive layer and stress corrosion cracking, allowing for distinct observation of the passive layer cracking process.

An additional demonstration of the salient feature of the DEIS technique was done by Darowicki et al. (*28*), who applied it in the assessment of the 304L stainless steel passive layer under static tensile stresses. Apart from FT, they used the analyzing window function to demonstrate the dynamics of the investigated process via instantaneous impedance spectra set. It was recognized that the parameters that best reflected the passive layer cracking dynamics were the charge transfer resistance (R_{ct}) and the capacitance of the internal portion of the passive layer. The unambiguous assignment of the passive layer cracking time took only about 18 s, and in their following study, the authors were able to show by impedance diagrams and related electrochemical parameters that the passive layer structure had suffered a change prior to the cracking process (*29*).

The first extensive contribution to the impedance measurements of pitting corrosion was that of Darowicki et al. (*30*) They conducted a combined studied on impedance and chronoamperometric techniques (i.e., chronoimpedantiometry) to obtain information about pitting corrosion formed in stainless steel 304 submerged in 0.5 M NaCl and discovered that the use of DEIS made it feasible to record the impedance changes over time during the instigation and propagation of the pits. In addition, they also determined the passive layer variations caused by changes in electrode potential and showed that the layer was fabricated, with the internal part being compact and thin and the external part porous and thick (*31*). Again, the influence of electrochemical and mechanical factors on the tightness of this passive film was explored via DEIS in the work of Orlikowski et al. (*32*) Partly because of the discrepancies involved in the mechanism of the pit initiation stage of stainless steel corrosion and despite the proposed models to clarify this abstruseness, Krakowiak et al. were able to resolve the issue of the stainless steel 304L pitting corrosion mechanism by determining the impedance changes over time in the prepitting state (*33*). According to the changes in the equivalent circuit of electrical parameters, the film-breaking model was ostensibly reported to succeed. Experimental evidence exists, however, for the three-stage pickling mechanism of oxidized 304 stainless steel in HF–H_2SO_4 electrolyte (*34*). Based on DEIS measurements, this three-step pickling mechanism was corroborated at first by the undercutting of external and internal oxide scales because the chromium-depleted layer had been actively dissolved. In the latter case, the increase of chromium content in the exposed metal resulted in transition of the corrosion potential from active to passive, and third, passivation of the metal exposed to the electrolyte was monitored because of the growth of a compact inner layer resulting from the increased chromium content in the metal. As a result, intergranular corrosion and honeycombed recesses were formed. Meanwhile, a few precursor works (*35–37*) appeared that evaluated dynamic impedance changes in the dissolution process of AISI 304 stainless steel under intergranular corrosion in a solution of 0.5 M SO_4^{2-} + 0.01 M KSCN at various pH levels. The overall rate of dissolution of the stainless steel was controlled by the process of intermediate $Fe(OH)_{n(ads)}$ adsorption (*35*). Indeed, direct alterations of the relative effective surface area were accounted for by the instantaneous impedance changes in the dissolution process taking place during the intergranular corrosion at different reactivation potentials (*36*), and regardless of the period of sensitization resulting from proceeding cathodic polarization, a decline in the dynamics of the charge transfer reaction changes with reactivation potential was observed for the sensitized samples (*37*).

Among the materials used in the plate and pipe manufacturing industry, the 300 series austenitic stainless steels have an immense impact. Alternatively, because of their high content of molybdenum and chromium compared with the 300 series austenitic stainless steels, super-austenitic stainless steels exhibit perfect localized corrosion resistance (*38*). It is, however, often true that they suffer from pitting corrosion. Recognizing this fact, Cheng et al. (*39*) attempted to assess pitting corrosion mechanisms of duplex stainless steels. Their careful DEIS experiments revealed the properties of the passive film formed on the 2205 duplex stainless steel single phase in 0.5 M $NaHCO_3$ + 0.5 M NaCl solution. They showed that the passive layers on the surface of the austenitic and ferritic phases differed and that it was easier for pitting corrosion to take place in the austenitic phase than in the ferritic phase. Because of advances in scanning probe microscopy, together with atomic force microscopy (AFM), DEIS has also been successfully applied to discern the mechanical and electrical properties of two-phase alloys such as 2205 duplex stainless steel and spheroidal graphite cast iron (*40*). As an example of rapidly changing systems, it was reported that the tip-surface contact

succeeded in illustrating the association between impedance and the force applied to the probe at each phase of the investigated materials. This has spurred research into the understanding of cavitation and the corrosion involved (*41*). Accurate knowledge of the changes in electrochemical parameters undergoing cavitation plays a prime role in understanding the synergistic mechanism and lowering of degradation resulting from corrosive elements. Thus, the DEIS method was tested in the galvanostatic mode by tracking the changes in electrochemical parameters during exposure of mild steel samples to cavitation erosion in 3% NaCl solution (*42*). The constant phase element parameters provided a completely plausible chronic state of mechanical erosion on the surface, thus increasing the surface roughness. Material failure from cavitation erosion–corrosion was further remedied using impedance monitoring via DEIS (*43*). The measurements were carried out on a selected group of alloys (i.e., mild steel, AA 7075, brass, and bronze) again in 3% NaCl, and the lowest cavitation erosion–corrosion resistance was found in the AA7075 and mild steel. However, the key conclusion was that, although the bronze and brass specimens were characterized as having superior erosion resistance, neither could avoid the dealloying process that impacted the level of synergism.

Another aspect of research (*44–49*) concerning the oxidation of some organic compounds on platinum focused on relieving hesitation for unresolved aspects of the adsorption processes and their dependency on system variables through the use of DEIS. Having established a practical experimental setup that could be used to study the formic acid electro-oxidation on polycrystalline platinum wire in sulfuric acid, Seland et al. overcame the challenge of obtaining the EIS spectra for every 2 mV around the DC voltammogram (*44*). Because none of their impedance spectra showed evidence of diffusion, and based on the relaxation times that were observed at potentials prior to the voltammetric peak, they deduced that $CO_{(ads)}$ and $OH^-_{(ads)}$ or oxide could have been the two kinetically significant species in the formic acid adsorption mechanism on the surface of the platinum. Complementary to this study, Lasia et al. (*45*) explored the inherent connection between cyclic voltammetry and DEIS methods. They obtained reproducible impedances for ethanol and acetaldehyde oxidation on a polycrystalline platinum electrode in 0.5 M H_2SO_4 during continuous slow cycling. The sharp peaks observed on the cyclic voltammograms of ethanol by other researchers were linked to the presence of negative resistances (*46*). A minor adaptation of the same approach was introduced by Darab et al. (*47*) for the characterization of a membrane–electrode assembly in a proton exchange membrane fuel cell both before and after cycling in a single-cell configuration. DEIS was shown to be a remarkable method of gathering impedance spectra for all cell voltages and to be capable of revealing previously undetectable minute electrode modifications via cyclic voltammograms in addition to being used for CO stripping and transmission electron microscopy. The authors also reported that the cycling protocol preceded a change in the fuel cell reaction mechanism. These changes were suggested for the formation of particularly active sites of growing crystal facets or at the interface between agglomerated particles. A similar and interesting effect was also reported by Holm et al. (*48*) for the methanol oxidation reaction on platinum at 21–140 °C using the combined methods of cyclic voltammetry, chronoamperometry, and DEIS. They observed that the inductive loop in the DEIS spectra was related to the change in CO coverage that led to a net positive effect on the current. The methanol adsorption and surface reaction between the adsorbed intermediates controlled the overall current. More recently, the same group demonstrated mechanistic modeling in their study on cyclic voltammetry and DEIS data for methanol oxidation on platinum in 0.5M H_2SO_4 at 100 °C (*49*). They surmised on the basis of their results that the water adsorption could be reversed and that the reaction on the surface was a chemical reaction between adsorbed CO and adsorbed OH^-. Their study also suggested water activation as the source of high overpotentials for the indirect mechanism of methanol oxidation. The findings of Darowicki et al. using the DEIS technique further concluded that the influence of parameters such as temperature, load, and methanol flow rate were decisive for the methanol oxidation reaction at the Pt-Ru anode (*50*).

Determining in detail the cracking of the passive layer in the time and strain domain has generated considerable interest because no electrochemical technique or mechanical test has been able to accomplish this. However, it was only in 2006 that the electrochemical monitoring of aluminum passive film durability and protective characteristics under tensile stress through DEIS were reported by Darowicki et al. (*51*) In this case, a time lag was noted between the decrease in passive layer resistance and the rise in DC current, suggesting that the corrosion processes were ongoing. Electrochemical parameter values (i.e., resistance and capacitance) confirmed the dynamic nature of the passive layer cracking process, and variations in the parameters clearly indicated a reduction in the tightness and durability of the passive layer. The Darowicki study, as a prime source of motivation, deserves particular mention in discussions of the electrochemical description of the Portevin-Le Chatelier (PLC) effect, the serration process, and dynamics versus electrode potential (*52*). The authors conducted detailed DEIS measurements for A9 5251 aluminum alloy in a buffered solution (H_3BO_3 + Na_3BO_3; pH 7.4) and added 0.002 M NaCl during the tensile tests. Their investigations showed that the PLC effect on the electrochemical surface was linked to the mechanical conditions and to the degree of anodic polarization. The decisive factor in determining the occurrence of the PLC effect on the electrochemical surface was the dynamics of the change from a passive to an active state and vice versa. Later, Wysocka et al. (*53*) also demonstrated the potential of DEIS for tracking the kinetics and mechanism of the AA1050 anodic dissolution in an aqueous alkaline solution in the 14–8.5 pH range. On the basis of the results, it was observed that in the 14–10 pH range, the AA1050 was in the active state and exposed to anodic dissolution, thus forming $Al(OH)_4^-$. Impedance analysis confirmed the formation of a stable passive layer only at the lowest pH (8.5). Gerengi et al. (*54*) successfully employed DEIS techniques to study the corrosion behavior of AA6060 and AA6082 aluminum alloys in pure water and artificial acid rain. They showed that in aluminum alloys, the intermetallic particle types are crucial to the pit morphology, and the pH of the studied environment was a causal factor in screening the electrochemical corrosion process according to the DEIS data. Strengthened by the outcomes, the authors then studied the corrosion kinetics of 6060 and 6082 alloys using artificial acid rain (*55*). The authors conducted DEIS for that purpose over a 48-h period and found that the AA6060 showed more resistance to corrosion than did the AA6082 in the acid rain medium.

To obtain a better understanding of the charge distribution in the passive layer of nickel in 1 M NaOH solution (pH 13.8), Darowicki et al. (*56*) carried out an investigation on the determination of charge space capacity versus potential. They compared the Mott–Schottky relationships achieved at specific frequencies from the imaginary part of impedance (reactance) with the outcomes determined using the DEIS technique. Lack of explicit conditions of frequency marked for charge space capacity determination was again seen as a drawback of the one-frequency measurement techniques. Nevertheless, based on the experiments, the DEIS performance was regarded as exceptional. Taking advantage of the DEIS method, Orlikowski et al. analyzed the characteristics of magnesium surface layers and their effect on the corrosion process (*57*). Their findings demonstrated that the properties of the surface layers of magnesium changed considerably at a pH higher than 12.50.

In a relevant article, Burczyk et al. (*58*) proposed a new technique combining local EIS with mapping in a dynamic mode. This proposed technique allowed the spectra measurement of two model couples (i.e., zinc/copper and cadmium/copper) in 1 M Na_2SO_4 solution in a better, time-saving manner and provided broad impedance property at the same time. Research efforts have also been devoted to the novel method of dynamic impedance spectra extraction known as dynamic multifrequency analysis (DMFA), which depends on quadrature digital filters in place of the conventional STFT. Although STFT is viewed as a type of digital lock-in amplifier, DMFA extensively exploits the Fourier spectrum to separate only the frequency ranges relative to the impedance time changes (*59*). Koster et al. estimated the dynamic impedance spectra of the hydrogen evolution reaction on polycrystalline platinum via DMFA, encompassing a wide frequency range (from 1 MHz down to 8 Hz) at various dilutions (2–20 mM) of perchloric acid. They investigated the possibility of detecting the various basic steps of the reaction mechanism along with the kinetic parameters and their quantification via an equivalent circuit based on the Volmer-Tafel mechanism (*60*). It is worth mentioning that this approach facilitated the extraction of reliable kinetic parameters, despite the high ohmic resistance of the main electrolyte and the lack of a supporting electrolyte. In this context, Erinmwingbovo et al. examined nickel hexacyanoferrate thin film in terms of its reaction kinetics by applying DMFA (*61*). Their results proved that the DMFA method together with an optimized electrochemical cell setup was capable of analyzing the kinetics of an unstable electrochemical system. This process would be tedious using conventional EIS because of the drift effects. A similar but more elaborate strategy was also introduced by Erinmwingbovo et al. (*62*) They performed DMFA in an aqueous environment to investigate the kinetics of the reversible insertion of lithium ions into $LiMn_2O_4$ thin film via multilayer pulsed laser deposition. The resulting impedance spectra demonstrated that a two-step intercalation mechanism was involved in the reversible insertion of Li^+ in 200 nm $LiMn_2O_4$ thin films. The first step of this mechanism is the (de)solvation, and the reversible insertion is the second step.

A report by Darowicki et al. demonstrated the behavior of the anodic excursion peak, which represents the rise in the anodic current during the cathodic polarization sweep for lead in 0.5M H_2SO_4 solution (*63*). Using DEIS, they determined that cracking had not caused the anodic excursion peaks and that they could be associated with the oxidation of the anodic layer because of the higher nonstoichiometric coefficient of the oxides. The most notable development of alternative cell impedance measurement was achieved by the Slepski group (*64*). It was prompted by the observation that the measurement could be realized only in galvanostatic mode because of the current control applied to the cyclic charging–discharging process. Apparently, this technique was optimized for an online examination of impedance variations of cyclically charged or discharged cells by applying a proper perturbation signal together with a time–frequency analysis. Thus, the authors employed this method for actual diagnostics of NiMH batteries (*65*). Particularly, the results obtained when the cell was discharged under various conditions of the electrode indicated that the impedance of both the positive and the negative electrodes could not be overlooked. In another relevant theoretical analysis of impedance under potentiostatic conditions, the Slepski group aimed to describe the impedance of a cation electron-coupled transfer reaction (*66*). They found that in the potentiodynamic mode, the polarization depended on the resistance of the potential scan rate and adsorption capacity.

Table 1. Evolution of DEIS Measurement

Year	*Evolution of the Application of DEIS*	*Reference*
1991	The electrode surface potential was measured, and its ensuing development over time in an open circuit and under galvanostatic condition was studied using frequency response analysis.	(*21*)
2003	Dynamic and static STFT spectrograms, obtained by both theoretical and mathematical analyses, and a model electrical system were compared in order to extract all experimental knowledge of the nonstationary process properties of the investigated model.	(*25*)
2003	Dynamic changes were elicited during the cracking and stress corrosion of the passive layer. The initiation point of the cracking in the passive layer was clearly observed using DEIS.	(*27*)
2004	Apart from Fourier transformation, the analyzing window function was incorporated to achieve an instantaneous impedance spectra set depicting the dynamics of the process. The authors were able to show the morphology of the passive layer prior to cracking via impedance diagrams and related electrochemical parameters.	(*29*)
2004	The combined techniques of impedance and chronoamperometry were applied to obtain information about corrosion. DEIS enabled the recognition of the variations of impedance over time during pitting initiation and development. Variations in the passive layer created by changes in electrode potential were recorded.	(*31*)
2005	The electrochemical and mechanical factors influencing the tightness of the passive film were evaluated using DEIS.	(*32*)
2006	Using DEIS in the electrochemical evaluation of the protection and durability of passive film under tensile stress, a time lag was noted between the reduced passive layer resistance and the increased DC current that demonstrated the corrosion activity.	(*51*)
2006	The lack of explicit specifications for the frequency range for charge space capacity determination was considered a disadvantage of single-frequency measurement techniques, and DEIS was considered an exceptional approach. Darowicki et al. compared the results of the Mott–Schottky relationships at specific imaginary impedance frequencies with the results of the DEIS method.	(*56*)
2008	With the advancement in scanning probe microscopy techniques, DEIS has also been successfully applied together with AFM. As an example of rapidly changing systems, it was reported that the tip-surface contact succeeded in illustrating the association between impedance and the force applied to the probe at each phase of the investigated materials.	(*41*)
2009	The current control in the process of cyclic charging and discharging allows the DEIS measurements to be performed solely in the galvanostatic mode. A complete perturbation signal in combination with a time–frequency analysis renders this a unique approach for on-line assessment of differences in impedance of cyclically charged and discharged cells.	(*64*)
2011	In order to better understand the synergism and inhibition, DEIS was tested in the galvanostatic mode by observing the changes of electrochemical parameters.	(*42*)

Table 1. (Continued). Evolution of DEIS Measurement

Year	*Evolution of the Application of DEIS*	*Reference*
2012	Lasia et al. explored the inherent connection between cyclic voltammetry and DEIS methods.	(*45*)
2012	Using the DEIS technique, Darowicki et al. reported further findings that the influence of parameters such as temperature, load, and flow rate were decisive factors in redox reactions.	(*50*)
2015	Huang et al. introduced a new application of the DEIS technique eliminating the state-of-charge change during DEIS testing. In that approach, first, the impedance was measured continuously at the same frequency over one charging–discharging cycle. The measurement was then repeated at a number of other designated frequencies. They successfully extracted the impedance spectrum for different frequencies during charging and discharging at each specific state of charge.	(*68*)
2015	In addition, the Huang group revealed a significant improvement to their DEIS procedure during pulse charge and discharge. They found the low frequency limit to be an important parameter that could ensure a quasistationary state during DEIS measurement, and a positive correlation between the lower frequency limit and the direct current pulse current was estimated.	(*69*)
2017	Research efforts have also been devoted to using DMFA to extract the dynamic impedance spectra. This technique uses quadrature digital filters instead of the conventional STFT. Although the STFT is seen as a digital version of a lock-in amplifier, the DMFA benefits from using the entire Fourier spectrum in the selection of specific frequency ranges only, depending on impedance time variation.	(*59*)
2018	Burczyk et al. described a new technique combining local EIS and dynamic mode mapping that enabled spectra to be measured in a more time-efficient manner and at the same time provided wide-ranging impedance characteristics.	(*58*)
2020	Janicka et al. used multisinusoidal stimulation to study the variation in impedance parameters by adding a variable component to the direct current signal during the change of the cell load.	(*71*)

Throughout many collective works, there has been a strong underlying emphasis on the competence of DEIS to examine the current dependency of the R_{ct} and to distinguish the R_{ct} between charge and discharge in lithium-ion batteries. For instance, such differences were explored by Huang et al. using DEIS for a $LiMn_2O_4$/Li half-cell (0.8 mAh) in the galvanostatic mode while charging or discharging via a series of direct currents (*67*). The authors reported that with the decrease of the R_{ct}, the direct current increased, and they explained this by using the Butler-Volmer equation. The R_{ct} during charging was observed to be generally lower than it was during discharging. Shortly thereafter, Huang et al. introduced a new application of the DEIS technique that eliminated the state-of-charge change during the DEIS testing of lithium-ion batteries (*68*). In that approach, the impedance was measured continuously at a single frequency in a charging–discharging cycle, and the calculation was then repeated at a number of other selected frequencies. They successfully extracted the impedance spectrum of different frequencies at each specific state-of-charge during charging and discharging. Furthermore, the Huang group disclosed a significant improvement to their DEIS procedure during pulse charging and discharging that showcased the utilization of these tests for a pouch-type, three-electrode, lithium-ion battery at different states-of-charge and temperatures (*69*). They found that lower frequency demarcation was an important factor for ensuring a quasistationary state during the DEIS measurement and predicted the positive correlation between the lower frequency limit and the direct current pulse current. As inferred from the previously mentioned contributions, DEIS has been applied in the field of lithium-ion batteries and in proton exchange membrane fuel cells, which are near to being fully exploited as their advantages of high efficiency and positive environmental impact have emerged (*70*). Taking into account the disadvantage of the EIS method in which frequency-by-frequency measurements are required when the fuel cell is under a stable load, an alternative approach to sequential excitation that can be performed during dynamic changes was instituted by Janicka et al. (*71*) They used multisinusoidal stimulation to investigate the variance in the impedance parameter by introducing an additional changeable value to the signal of the direct current during the cell load change. In proof-of-concept experiments performed during the operation of the fuel cell, remarkable assistance was provided by DEIS in evaluating the impact of air humidity on the fuel cell electrical load when under dynamic operation. Table 1 presents the evolution and advancements in the application of DEIS in evaluating electrochemical processes.

Beyond these, a cluster of references consider the use of DEIS in a variety of fields—including electropolymerization, organic coatings, organic film degradation, surface electrochemical reactions, and quasireversible redox systems (*72–76*). Insofar as such DEIS studies paved the way for further investigations on the processes of surface oxidation, boron-doped diamond (BDD) electrodes have been the subject of close scrutiny in electrochemistry research (*77*). To gain further insight into the formation of oxygen terminations on BDD under deep anodic polarization conditions, Ryl et al. studied the kinetics of oxidation in situ by means of DEIS in 1 M H_2SO_4 solution (*78*). Their findings demonstrated that the DEIS technique was robust enough to examine in situ changes in electric parameters resulting from BDD surface oxidation. Moreover, they performed an in-depth analysis of the BDD termination bond oxidation phenomenon and determined that a complex process takes place depending on three factors: the anodic oxidation level, boron incorporation, and crystallographic orientation (*79*). The way in which the heterogeneous electric property segmentation influences the electrochemical properties of polycrystalline BDD electrodes was also evaluated using DEIS in another study (*80*). As reported, the DEIS was used in the monitoring of the process of oxidation, which demonstrated that uniformity of the sample decreased when the polarization voltage was equal to or above 1.6 V against the Ag|AgCl. The nonhomogeneous feature of the electrochemical activity agreed with the results of multifrequency nanoscale impedance microscopy studies.

To conclude this discussion, one of the latest contributions of the Scohy group deserves to be mentioned. Although it is not concerned with the BDD electrodes so far, it sets the basis for the DEIS approach for unearthing the initial levels of surface oxidation along with the mechanism of the oxygen evolution reaction on the Ir(111) single-crystal electrode and electrochemically activated Ir(111) (*81*). The EIS and DEIS methods were compared, and the results clearly substantiated the precedence of DEIS in terms of saving time. Moreover, they highlighted the great potential of DEIS to provide a better view of the occurrences at the Ir(111) electrode–electrolyte interface.

DEIS as the Most Advanced Method for Studying the Corrosion Processes

Since the realization that DEIS can be successfully implemented in several electrochemical processes, its use in corrosion inhibitor studies has grown exponentially. Gerengi et al. presented one of the earliest applications of DEIS in studying the inhibition effect of a well-known corrosion inhibitor (i.e., benzotriazole [BTA]) for corrosion of brass-118 (*82*) and brass-MM55 alloy (*83*) in artificial seawater. Considering the new mode of EIS measurement, which permits impedance spectra to be studied in a combined time–frequency domain, impedance changes were adopted as a time function for the BTA inhibitory effect. The DEIS was found to outperform the conventional EIS approach in ascertaining the influence of BTA concentration because of the formation of its thinner homogeneous layer. Similar results were also observed for the inhibition of corrosion in an alloy of copper–manganese–aluminum by BTA using DEIS and conventional techniques (i.e., linear polarization resistance and Tafel extrapolation measurements) in artificial seawater (*84*). It was again noted that BTA was an excellent inhibitor in accord with all the methods used for inhibiting corrosion in a copper–manganese–aluminum alloy. In explaining this effect, the production of the polymeric species [Cu (I) BTA]n was put forward according to the DEIS results. Gerengi et al. (*85*) extended the application of the DEIS method for BTA as an inhibitor for St37 steel corrosion in an environment of 0.1 M HCl. In agreement with previous works studying BTA, further evidence was provided of the positive convenience of DEIS for determining the electrochemical changes on steel surfaces that occur over time in the presence of BTA. Based on energy dispersive X-ray spectroscopy data, the effectiveness of this compound for St37 steel corrosion was supported when its molecules were adsorbed on the surface and the reaction sites were blocked.

Another corrosion inhibitor, in use since the 1970s but progressively improved throughout the 1990s and 2000s, is 8-hydroxyquinoline (8-HQ) (*86–88*). Given that 8-HQ has attracted considerable research, its corrosion inhibition properties, which are contingent upon its molecular structure, were studied for the first time by Gerengi et al. for the corrosion of copper in 0.1 M HCl using DEIS as well as other electrochemical and theoretical methods (*89*). This study proposed that 8-HQ possessed the capacity to inhibit the dissolution of copper in its protonated quinolinium form and exhibited a permanency characteristic that supports the emergence of robust Coulombic electrostatic attraction forces between 8-HQ and the copper surface. The best results for 8-HQ corrosion inhibition of AA2024-T4 in a solution of 3.5% NaCl were also achieved using DEIS (*90*).

Aside from its ubiquitous use in industrial applications such as cosmetology, pharmacy, agriculture, biotechnology, and waste management, chitosan and its derivatives are now exhibiting their indispensable value in corrosion science as well (*91, 92*). The successful applications of DEIS in corrosion inhibition studies of chitosan date back to the original contributions from Gerengi et al. (*93–95*) The Gerengi group developed a strategy in which small amounts of potassium iodide generated an in situ synergistic effect with unmodified chitosan to enhance the inhibition of St37 steel corrosion in 15% H_2SO_4 solution (*93*). The DEIS spectra of the mixture of chitosan–iodide and chitosan clearly supported the adsorption of chitosan molecules into the film and enhanced its stability over time when compared with EIS data. Following this report, the Gerengi group ably applied the DEIS method this time to shed light on boosting the inhibitive performance of unmodified chitosan on St37 steel in a solution of 15% H_2SO_4 by dispersing silver nanoparticles into its matrix (*94*). By careful employment of various electrochemical measurements, the authors discovered that by incorporating silver nanoparticles in the composite, the inhibition efficiency of the chitosan was doubled. Shortly after, Solomon et al. evaluated the inhibition performance of the composite in question in the corrosion of the same steel in a solution of 15% HCl. Interestingly, the addition of natural honey as the reducing and stabilizing agent in this study significantly facilitated the synthesis of the composite, and the prepared composite was found to be effective in retarding the corrosion, particularly at higher temperatures. For example, 1000 ppm silver nanoparticles–chitosan composite manifested 87.88% protection at 25 °C, whereas 97.09% protection was recorded at 60 °C (*95*).

The follow-up studies included some organic acids such as citric acid, maleic acid, malic acid, tartaric acid, succinic acid, and tricarballylic acid as well as a monocarboxylic amino acid (serine) that are capable of forming complexes with multivalent metal ions, serving the purpose of accelerating the removal of corrosive products (*96, 97*). Wysocka et al. confirmed the use of the DEIS method as a valuable and efficient tool for monitoring the inhibition effect of citric acid for AA5754 alloy corrosion in a mixture of 0.05 M $NaHCO_3$ + 0.1 M NaOH (bicarbonate buffer) with a pH of 11 (*96*). It is fair to say that the DEIS method led the way in demonstrating that the citric acid interacted spontaneously with the aluminum surface because of its relative supremacy for constructing the Langmuir isotherm compared with the conventional tools. Nevertheless, the DEIS method was able to determine the exact concentration of citric acid necessary for the formation of a three-dimensional layer on the AA5754 alloy surface. A similar conclusion was drawn again by Wysocka et al., in which they properly described the contributions of the carboxyl and carbonyl groups present in the structures of the aforementioned carboxylic acids to the higher corrosion resistance of AA5754 alloy in a bicarbonate buffer (*97*). Ryl et al. used DEIS to measure the electrochemical behavior of AA5754 in alkaline electrolyte (i.e., bicarbonate buffer solution) (*98*). Logically, the hydroxyl functional group position within some dihydroxybenzene isomers (catechol, quinol, and resorcinol) influenced the altered adsorption mechanism that resulted. Indeed, not only was the chemical adsorption of resorcinol on the aluminum surface verified, but the dependence of its inhibition efficiency on the mechanism for keto-enol tautomerism occurring in the aqueous alkaline medium was also revealed. Furthermore, the ample corrosion protection of catechol and quinol even at lower corrosion concentrations was explained in terms of the ligand exchange model of adsorption. Insights into the high inhibition efficiency of bee pollen for AA5754 corrosion under the same experimental conditions were also achieved by the same group (*99*).

Boron minerals, particularly borax and some other alkali metal borates present in mixtures, exhibit corrosion inhibition activity by forming buffer solutions (*100, 101*). This provided a compelling reason to examine the inhibition property of some boron compounds, and for this purpose, the Gerengi group employed copper chrome boron, a wood preservative that combines H_3BO_3, $CuSO_4{\cdot}5H_2O$, and $K_2Cr_2O_7$, to elucidate its inhibition effect on St37 steel alloy in 3.5% NaCl solution using DEIS (*102*). The R_{ct} as well as the constant phase element data led to the conclusion that increasing concentrations of copper chrome boron were able to inhibit the corrosion of the alloy because it formed a protective surface layer. In their systematic work on ionic liquids with an anionic borate part, Gerengi et al. uncovered the inhibiting powers of 1-ethyl-3-methylimidazolium tetrafluoroborate (*103*) and 1-butyl-1-methylpiperidinium tetrafluoroborate (*104*), respectively, toward the corrosion of St37 steel in 0.1 M HCl solution. The introduction of small amounts of potassium iodide into the corrosive environment was found to notably invigorate the nature of the interaction between the 1-ethyl-3-methylimidazolium tetrafluoroborate molecule and the steel surface (*103*). A positive inhibition effect was also observed with the 1-butyl-1-methylpiperidinium tetrafluoroborate per se, indicating the involvement of the cationic organic and the anionic parts of this molecule, which distorted the anodic iron dissolution reactions through adsorption (*104*).

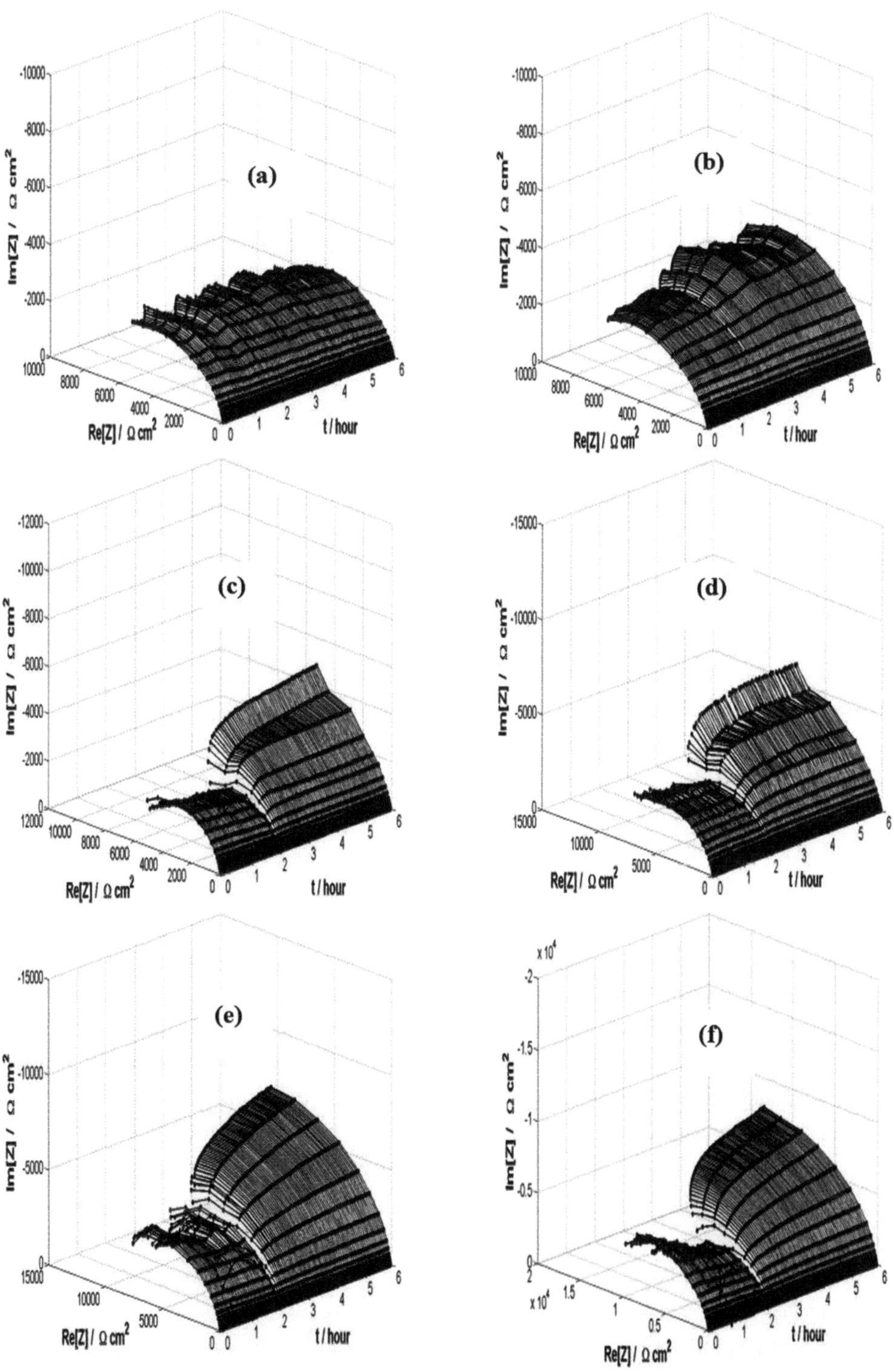

Figure 3. DEIS spectra of AA7075-T6 aluminum alloy in 3.5% NaCl solution for 6 h at 25 °C: (a) with no additions, and with different amounts of ethylene glycol: (b) 10%, (c) 20%, (d) 30%, (e) 40%, and (f) 50%. Reprinted with permission from reference (111). Copyright 2018 Elsevier.

Although carbohydrate carboxymethyl cellulose and other such polymer exudate gums have been used extensively to inhibit corrosion and protect the targeted metal substrates in their service environments (*105*), the application of DEIS in determining the properties of these polymers in terms of inhibition and stability is relatively underdeveloped. To date, only limited examples have been reported (*106, 107*). One of them was specifically designed to improve the inhibition capability of carboxymethyl cellulose (*106*), and the other was created to enhance the anticorrosive property of gum Arabic (*107*) for use on St37 steel in a solution of 15% sulfuric acid. Solomon et al. accomplished a remarkable enhancement of the polymer in boosting its stability and inhibitory ability at elevated temperatures by the infusion of silver nanoparticles into the carboxymethyl cellulose membrane (*106*). Such a reclamation was also made possible as a result of a gum arabic and silver nanoparticle composite that was reported to effectively inhibit the corrosion of steel in solutions of 15% sulfuric acid and 15% hydrochloric acid, respectively (*107*).

Because of its exceptional features and versatile applications in the commercial sector (*108*), ethylene glycol has been in vogue with corrosion scientists since the 1950s (*109, 110*). In the search for a long-term corrosion inhibitor, using the AA7075-T6 alloy, the Gerengi group deployed DEIS to note its electrochemical property variations in a solution of 3.5% NaCl in the presence and absence of ethylene glycol over time (*111*). The nexus between the highest studied concentration (50%) of ethylene glycol and the protection of the alloy surface in a solution of 3.5% NaCl was determined to be 70% (*111*) (Figure 3).

Armed with the knowledge that the use of organic compounds with pyridinium moiety had achieved positive results in the inhibition of corrosion, Gerengi et al. extended their work on N-phenyl-1-(4-((11-(pyridin-1-ium-1yl) undecanoyl)oxy)phenyl) methanimine oxide bromide, which was the first amphiphilic nitrone reported as a new corrosion inhibitor for St37 steel in the literature (*112*). The DEIS results showed that the R_{ct} of the steel sample had increased as a result of the nitrone present in the 1M HCl medium. The inhibition mechanism of this compound was also explained through STDFT calculations.

A number of researchers over the years have focused on plants in their attempts to access biologically active compounds through extraction. An extensive and diverse assortment of plant extracts has been found with the potential for beneficial applications, one of which is corrosion protection (*113*). Because all parts of a plant can be used to obtain extracts, the investigation of the inhibition efficiency of such extracts using DEIS has been another topic of interest. To that end, research on the protective effect of mimosa extract derived from black wattle tree (*Acacia mearnsii*) bark on aluminum alloy using a solution of artificial acid rain was performed by Gerengi et al. (*114*) They showed not only the noticeable effect of the mimosa extract in retarding corrosion in the AA6060 alloy but also the superiority of DEIS in producing better and more accurate results. Gerengi et al. (*115*) also used DEIS in their study that demonstrated that the leaf extract of the oriental persimmon (*Diospyros kaki*) applied to the surface of St37 steel was effective in inhibiting corrosion in a solution of 0.1 M HCl. The DEIS findings are shown in Figure 4.

Furthermore, Slepski et al. used DEIS to explore the influence of beet molasses, a byproduct of the sugar refining process, on the etching of carbon steel in a hydrochloric acid environment (*116*). They confirmed the inhibiting properties of the molasses according to the decrease in corrosion current and the reduced R_{ct}. However, another investigation using toxic honey ("mad" honey) produced from rhododendron flower nectar also revealed a significant improvement in corrosion protection of Al-2007 alloy in a medium of 3.5% NaCl (*117*).

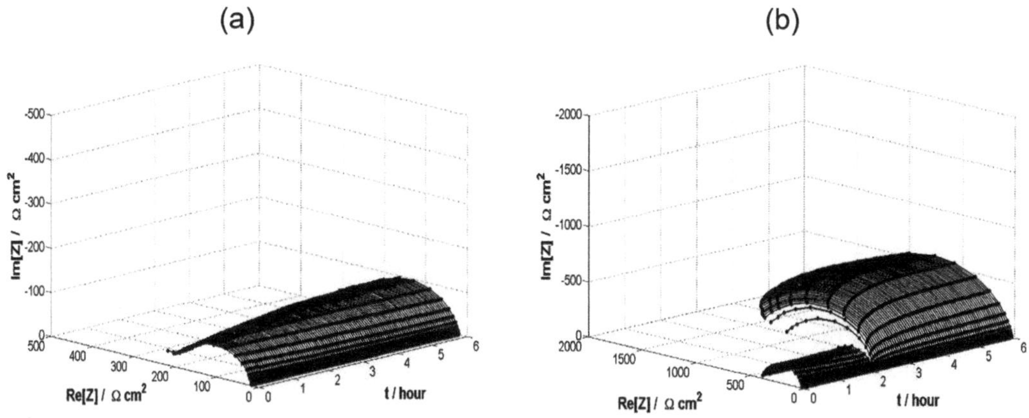

Figure 4. St37 steel DEIS spectra: (a) in 0.1 M HCl and (b) in 0.1 M HCl+225 ppm Diospyros kaki leaf extract. Reprinted with permission from reference (115). Copyright 2016 Elsevier.

Given the limited literature on the use of DEIS in corrosion inhibitor studies, it has become increasingly essential for influential researchers in this area to demonstrate the true novelty of this approach. The best outcome might be interpreted as recognition that rapid studies of nonstationary systems and the electrode kinetics changes are possible (*118–121*). Also noted is that higher accuracy for adsorption isotherm determination can be achieved, the most efficient inhibitor concentrations can be determined, and the Langmuir isotherm model applicability by capacitance changes can be readily procured. Table 2 lists some significant studies that applied DEIS as an effective tool to measure corrosion and inhibitor efficiency in various corrosive environments.

Table 2. Some Notable Studies That Used DEIS to Assess Corrosion

Inhibitor	*Corrosive Medium*	*Substrate*	*IE (%)*	*Reference*
BTA	0.1 M HCl	St37 steel	78.5	(*85*)
8-Hydroxyquinoline	0.1 M HCl	Copper	62	(*89*)
8-Hydroxylquinoline	3.5% NaCl	AA2024-T4 alloy	83	(*90*)
Chitosan + I^-	15% H_2SO_4	St37 steel	99.72	(*93*)
Citric acid	Bicarbonate buffer (0.05 M $NaHCO_3$ + 0.1 M NaOH)	Aluminum	99.8	(*96*)
Resorcinol	Bicarbonate buffer (pH = 11)	Aluminum alloy 5754	98.7	(*98*)
Bee pollen	Bicarbonate buffer solution	Aluminum	99.7	(*99*)
Mad honey	3.5% NaCl	2007-Aluminum alloy	54	(*117*)
Ethanolic extract of propolis	3.5% NaCl	SAE 1010 carbon steel	76.4	(*120*)

Conclusions

It is not an overstatement that good success has transformed DEIS into an indispensable experimental method. Enormous strides have been made in the field of impedance measurements in little over a decade, yet there remain a number of challenges to be overcome. The DEIS technique shows great promise and remarkable potential for development; however, in order to establish its true efficacy, thorough and meticulously regulated experiments are still needed. Although DEIS has been recognized as a formidable technique for comprehending a number of important electrochemical processes, an important development would be to secure its position in the study of kinetics during nonstationary processes. The central focus of this review was to highlight the advantages of employing DEIS, and these are briefly summarized:

- A single impedance spectrum can be attained in a minimum amount of time.
- An average instantaneous impedance spectrum for a nonstationary system can be obtained using Fourier transformation executed on a fragment of a signal.
- Because the multisinusoid signal is superimposable in nature, DEIS can be combined with other techniques (e.g., Tafel polarization, cyclic voltammetry, AFM scanning).
- Impedance spectra acquired via the DEIS technique can be a function of varying parameters including time, potential, current, and temperature.
- The continuous nature of the multisinusoid perturbation ensures fewer electrochemical disturbances upon changing the frequency.

Finally, perhaps the true potential of this powerful tool can be realized only when more researchers begin to apply DEIS in their electrochemical work.

References

1. Macdonald, J. R.; Johnson, W. B. In *Fundamentals of Impedance Spectroscopy*, 2nd ed, Barsoukov, E., Macdonald, J. R., Eds.; John Wiley & Sons Inc., 2005, pp 1–26.
2. Kashkovskiy, R.; Strelnikova, K.; Fedotova, A. Application of Electrochemical Impedance Spectroscopy to Study Hydrogen Sulphide Corrosion of Steel and Its Inhibition: A Review. *Corros. Eng. Sci. Technol.* **2019**, *54* (6), 493–515. DOI: 10.1080/1478422X.2019.1619979.
3. Savova-Stoynov, B.; Stoynov, Z. B. Four-Dimensional Estimation of the Instantaneous Impedance. *Electrochim. Acta* **1992**, *37* (12), 2353–2355. DOI: 10.1016/0013-4686(92)85131-4.
4. Stoynov, Z. B. Rotating Fourier Transform—New Mathematical Basis for Non-Stationary Impedance Analysis. *Electrochim. Acta* **1992**, *37* (12), 2357–2359. DOI: 10.1016/0013-4686(92)85132-5.
5. Stoynov, Z. B. Nonstationary Impedance Spectroscopy. *Electrochim. Acta* **1993**, *38* (14), 1919–1922. DOI: 10.1016/0013-4686(93)80315-Q.
6. Semenikhin, O. A.; Ovsyannikova, E. V.; Alpatova, N. M.; Rotenberg, Z. A. Dynamic Impedance Measurements on a Thin-Film poly-3-methylthiophene Electrode: Memory Effects and Space Charge Formation. *Electroanal. Chem.* **1996**, *408* (1–2), 67–75. DOI: 10.1016/0022-0728(95)04505-8.
7. Gabor, D. Theory of Communication. *J. Inst. Elec Eng.* **1946**, *93*, 429–457.
8. Darowicki, K. Theoretical Description of the Measuring Method of Instantaneous Impedance Spectra. *J. Electroanal. Chem.* **2000**, *486* (2), 101–105. DOI: 10.1016/S0022-0728(00)00110-8.
9. Darowicki, K.; Orlikowski, J.; Lentka, G. Instantaneous Impedance Spectra of a Non-Stationary Model Electrical System. *J. Electroanal. Chem.* **2000**, *486* (2), 106–110. DOI: 10.1016/S0022-0728(00)00111-X.
10. Urquidi-Macdonald, M.; Real, S.; Macdonald, D. D. Pplication of Kramers-Kronig Transforms in the Analysis of Electrochemical Impedance Data: II. Transformations in the Complex Plane. *J. Electrochem. Soc.* **1986**, *133* (10), 2018–2024. DOI: 10.1149/1.2108332.
11. Urquidi-Macdonald, M.; Real, S.; Macdonald, D. D. Applications of Kramers—Kronig Transforms in the Analysis of Electrochemical Impedance Data—III. Stability and Linearity. *Electrochim. Acta* **1990**, *35* (10), 1559–1566. DOI: 10.1016/0013-4686(90)80010-L.

12. Macdonald, D. D.; Urquidi-Macdonald, M. Application of Kramers-Kronig Transforms in the Analysis of Electrochemical Systems: I . Polarization Resistance. *J. Electrochem. Soc.* **1985**, *132* (10), 2316–2319. DOI: 10.1149/1.2113570.
13. Gabrielli, C.; Keddam, M.; Takenouti, H. In *Electrochemical Impedance Analysis and Interpretation*, Scully, J. R., Silverman, D. C., Kendig, M. W., Eds.; STP 1188, ASTM, 1993, p 140.
14. Gabrielli, C. *Use and Applications of Electrochemical Impedance Spectroscopy*, Technical report; Schlumberger Technologies, 1990.
15. Házì, J.; Elton, D. M.; Czerwinski, W. A.; Schiewe, J.; Vicente-Beckett, V. A.; Bond, A. M. Microcomputer-Based Instrumentation for Multi-Frequency Fourier Transform Alternating Current (Admittance and Impedance) Voltammetry. *J. Electroanal. Chem.* **1997**, *437* (1–2), 1–15. DOI: 10.1016/S0022-0728(96)05038-3.
16. Lyons, R. G. *Understanding Digital Signal Processing*; Longman; New York, 1997.
17. Cohen, L. *Time–Frequency Analysis*; Prentice Hall, 1995.
18. Cohen, L.; Lee, C. Time-Frequency Signal Analysis. In *Wiley Halsted Press, Lingman Cheshire*, Boashash, B., Ed., 1992.
19. Krakowiak, S.; Darowicki, K.; Ślepski, P. Impedance of Metastable Pitting Corrosion. *J. Electroanal. Chem.* **2005**, *575* (1), 33–38. DOI: 10.1016/j.jelechem.2004.09.001.
20. Qian, S.; Chen, D. *Joint Time–Frequency Analysis: Methods and Applications*; Prentice Hall PTR, 1996.
21. Burstein, G. T.; Cinderey, R. J. The Potential of Freshly Generated Metal Surfaces Determined from the Guillotined Electrode—A New Technique. *Corros. Sci.* **1991**, *32* (11), 1195–1211. DOI: 10.1016/0010-938X(91)90132-9.
22. Burstein, G. T.; Liu, C. Time-Resolved Electrochemical Impedance of the Guillotined Aluminium Electrode. *Electrochim. Acta* **1994**, *39* (7), 873–882. DOI: 10.1016/0013-4686(94)85101-8.
23. Burstein, G. T.; Liu, C. The Cathodic Reaction During Repassivation of Aluminium in Open Circuit. *Corros. Sci.* **1995**, *37* (7), 1151–1162. DOI: 10.1016/0010-938X(95)00023-D.
24. Darowicki, K.; Ślepski, P. Influence of the Analyzing Window on Electrode Impedance Measurement by the Continuous Frequency Scanning Method. *J. Electroanal. Chem.* **2002**, *533* (1–2), 25–31. DOI: 10.1016/S0022-0728(02)01085-9.
25. Darowicki, K.; Zielinski, A.; Slepski, P. Continuous-Frequency Method of Measurement of Electrode Impedance. *Instrum. Sci. Technol.* **2003**, *31* (1), 53–63. DOI: 10.1081/CI-120018407.
26. Darowicki, K.; Ślepski, P. Dynamic Electrochemical Impedance Spectroscopy of the First Order Electrode Reaction. *J. Electroanal. Chem.* **2003**, *547* (1), 1–8. DOI: 10.1016/S0022-0728(03)00154-2.
27. Darowicki, K.; Orlikowski, J.; Arutunow, A. Investigations of the Passive Layer Cracking by Means of Dynamic Electrochemical Impedance Spectroscopy. *Electrochim. Acta* **2003**, *48* (28), 4189–4196. DOI: 10.1016/S0013-4686(03)00604-2.
28. Darowicki, K.; Orlikowski, J.; Arutunow, A. Comparative Electrochemical Analysis of the Passive Layer Cracking Process on Aluminum Alloys Performed by Means of DC and AC Techniques. *J. Solid State Electrochem.* **2004**, *8* (6), 352–359. DOI: 10.1007/s10008-003-0470-0.
29. Darowicki, K.; Orlikowski, J.; Arutunow, A. Analysis of Electrochemical Parameters in Time Domain During the Passive Layer Cracking Occurring on the 304L Stainless Steel in Chlorides Solution Under Tensile Stresses. *Electrochim. Acta* **2004**, *49* (28), 5069–5078. DOI: 10.1016/j.electacta.2004.05.045.
30. Darowicki, K.; Krakowiak, S.; Slepski, P. The Time Dependence of Pit Creation Impedance Spectra. *Electrochem. Commun.* **2004**, *6* (8), 860–866. DOI: 10.1016/j.elecom.2004.06.010.
31. Darowicki, K.; Krakowiak, S.; Ślepski, P. Evaluation of Pitting Corrosion by Means of Dynamic Electrochemical Impedance Spectroscopy. *Electrochim. Acta* **2004**, *49*, 17–18, 2909–2918. DOI: 10.1016/j.electacta.2004.01.070.
32. Orlikowski, J.; Darowicki, K.; Arutunow, A.; Jurczak, W. The Effect of Strain Rate on the Passive Layer Cracking of 304L Stainless Steel in Chloride Solutions Based on the Differential Analysis of

Electrochemical Parameters Obtained by Means of DEIS. *J. Electroanal. Chem.* **2005**, *576* (2), 277–285. DOI: 10.1016/j.jelechem.2004.10.025.
33. Krakowiak, S.; Darowicki, K.; Slepski, P. Impedance Investigation of Passive 304 Stainless Steel in the Pit Pre-Initiation State. *Electrochim. Acta* **2005**, *50* (13), 2699–2704. DOI: 10.1016/j.electacta.2004.11.015.
34. Li, L. F.; Caenen, P.; Celis, J. P. Chemical Pickling of 304 Stainless Steel in Fluoride- and Sulfate-Containing Acidic Electrolytes. *J. Electrochem. Soc.* **2005**, *152* (9), B352. DOI: 10.1149/1.1990127.
35. Arutunow, A.; Darowicki, K. DEIS Assessment of AISI 304 Stainless Steel Dissolution Process in Conditions of Intergranular Corrosion. *Electrochim. Acta* **2008**, *53* (13), 4387–4395. DOI: 10.1016/j.electacta.2008.01.063.
36. Arutunow, A.; Darowicki, K. DEIS Evaluation of the Relative Effective Surface Area of AISI 304 Stainless Steel Dissolution Process in Conditions of Intergranular Corrosion. *Electrochim. Acta* **2009**, *54* (3), 1034–1041. DOI: 10.1016/j.electacta.2008.08.045.
37. Arutunow, A.; Darowicki, K. Impact of Sensitization on Dissolution Process of AISI 304 Stainless Steel During Intergranular Corrosion Evaluated Using DEIS Technique. *J. Solid State Electrochem.* **2009**, *13* (11), 1687–1694. DOI: 10.1007/s10008-008-0642-z.
38. Cardoso, J. L.; Cavalcante, A. L. S. N.; Vieira, R. C. A.; de Lima-Neto, P.; da Silva, M. J. G. Corrosion Behavior of Austenitic Stainless Steels in CO_2-Saturated Synthetic Oil Field Formation Water. *J. Mater. Res.* **2016**, *31*, 1755–1763.
39. Cheng, X.; Li, C.; Dong, C.; Li, X. Constituent Phases of the Passive Film Formed on 2205 Stainless Steel by Dynamic Electrochemical Impedance Spectroscopy. *Int. J. Miner. Metall. Mater.* **2011**, *18* (1), 42–47. DOI: 10.1007/s12613-011-0397-x.
40. Arutunow, A.; Tobiszewski, M. T.; Darowicki, K. Electrical and Mechanical Characterization of Two-Phase Alloys by Means of Scanning Probe Microscopy in Dynamic Impedance Spectroscopy Mode. *J. Alloys Compd.* **2015**, *619*, 172–176. DOI: 10.1016/j.jallcom.2014.08.214.
41. Ryl, J.; Darowicki, K. Impedance Monitoring of Carbon Steel Cavitation Erosion Under the Influence of Corrosive Factors. *J. Electrochem. Soc.* **2008**, *155* (4), 44. DOI: 10.1149/1.2840619.
42. Ryl, J.; Darowicki, K.; Slepski, P. Evaluation of Cavitation Erosion–Corrosion Degradation of Mild Steel by Means of Dynamic Impedance Spectroscopy in Galvanostatic Mode. *Corros. Sci.* **2011**, *53* (5), 1873–1879. DOI: 10.1016/j.corsci.2011.02.004.
43. Ryl, J.; Wysocka, J.; Slepski, P.; Darowicki, K. Instantaneous Impedance Monitoring of Synergistic Effect Between Cavitation Erosion and Corrosion Processes. *Electrochim. Acta* **2016**, *203*, 388–395. DOI: 10.1016/j.electacta.2016.01.216.
44. Seland, F.; Tunold, R.; Harrington, D. A. Impedance Study of Formic Acid Oxidation on Platinum Electrodes. *Electrochim. Acta* **2008**, 53 (23), 6851–6864. DOI: 10.1016/j.electacta.2007.12.021.
45. Tremblay, M. L.; Guay, D.; Lasia, A. Dynamic Impedance Studies of Ethanol Oxidation at Polycrystalline Pt. *Bulg. Chem. Commun.* **2012**, *44*, 333–337.
46. Désilets, C.; Lasia, A. Dynamic Impedance Study of Ethanol and Acetaldehyde Oxidation at Platinum in Acid Solutions. *Electrochim. Acta* **2012**, *78*, 286–293. DOI: 10.1016/j.electacta.2012.05.102.
47. Darab, M.; Dahlstrøm, P. K.; Thomassen, M. S.; Seland, F.; Sunde, S. Dynamic Electrochemical Impedance Spectroscopy of Pt/C-Based Membrane-Electrode Assemblies Subjected to Cycling Protocols. *J. Power Sources* **2013**, *242*, 447–454. DOI: 10.1016/j.jpowsour.2013.05.105.
48. Holm, T.; Dahlstrøm, P. K.; Sunde, S.; Seland, F.; Harrington, D. A. Dynamic Electrochemical Impedance Study of Methanol Oxidation at Pt at Elevated Temperatures. *Electrochim. Acta* **2019**, *295*, 139–147. DOI: 10.1016/j.electacta.2018.10.071.
49. Holm, T.; Sunde, S.; Seland, F.; Harrington, D. A. Understanding Reaction Mechanisms Using Dynamic Electrochemical Impedance Spectroscopy: Methanol Oxidation on Pt. *Electrochim. Acta* **2019**, 323. DOI: 10.1016/j.electacta.2019.134764, 134764.

50. Darowicki, K.; Janicka, E.; Slepski, P. Study of Direct Methanol Fuel Cell Process Dynamics Using Dynamic Electrochemical Impedance Spectroscopy. *Int. J. Electrochem. Sci.* **2012**, 7, 12090–12097.
51. Darowicki, K.; Orlikowski, J.; Arutunow, A.; Jurczak, W. The Effect of Tensile Stresses on Aluminium Passive Layer Durability. *Electrochim. Acta* **2006**, *51* (27), 6091–6096. DOI: 10.1016/j.electacta.2005.12.054.
52. Darowicki, K.; Orlikowski, J. Impedance Analysis of Portevin-Le Chatelier Effect on Aluminium Alloy. *Electrochim. Acta* **2007**, *52* (12), 4043–4052. DOI: 10.1016/j.electacta.2006.11.022.
53. Wysocka, J.; Krakowiak, S.; Ryl, J.; Darowicki, K. Investigation of the Electrochemical Behaviour of AA1050 Aluminium Alloy in Aqueous Alkaline Solutions Using Dynamic Electrochemical Impedance Spectroscopy. *J. Electroanal. Chem.* **2016**, *778*, 126–136. DOI: 10.1016/j.jelechem.2016.08.028.
54. Gerengi, H.; Slepski, P.; Ozgan, E.; Kurtay, M. Investigation of Corrosion Behavior of 6060 and 6082 Aluminum Alloys Under Simulated Acid Rain Conditions. *Mater. Corros.* **2015**, *66* (3), 233–240. DOI: 10.1002/maco.201307287.
55. Gerengi, H.; Bereket, G.; Kurtay, M. A Morphological and Electrochemical Comparison of the Corrosion Process of Aluminum Alloys Under Simulated Acid Rain Conditions. *J. Taiwan Inst. Chem. Eng.* **2016**, *58*, 509–516. DOI: 10.1016/j.jtice.2015.05.023.
56. Darowicki, K.; Krakowiak, S.; Ślepski, P. Selection of Measurement Frequency in Mott–Schottky Analysis of Passive Layer on Nickel. *Electrochim. Acta* **2006**, *51* (11), 2204–2208. DOI: 10.1016/j.electacta.2005.04.079.
57. Orlikowski, J.; Darowicki, K. Investigations of Pitting Corrosion of Magnesium by Means of DEIS and Acoustic Emission. *Electrochim. Acta* **2011**, *56* (23), 7880–7884. DOI: 10.1016/j.electacta.2010.12.021.
58. Burczyk, L.; Darowicki, K. Local Electrochemical Impedance Spectroscopy in Dynamic Mode of Galvanic Coupling. *Electrochim. Acta* **2018**, *282*, 304–310. DOI: 10.1016/j.electacta.2018.05.192.
59. Koster, D.; Du, G.; Battistel, A.; La Mantia, F. L. Dynamic Impedance Spectroscopy Using Dynamic Multi-Frequency Analysis: A Theoretical and Experimental Investigation. *Electrochim. Acta* **2017**, *246*, 553–563. DOI: 10.1016/j.electacta.2017.06.060.
60. Koster, D.; Zeradjanin, A. R.; Battistel, A.; La Mantia, F. L. Extracting the Kinetic Parameters of the Hydrogen Evolution Reaction at Pt in Acidic Media by Means of Dynamic Multi-Frequency Analysis. *Electrochim. Acta* **2019**, *308*, 328–336. DOI: 10.1016/j.electacta.2019.04.013.
61. Erinmwingbovo, C.; Koster, D.; Brogioli, D.; La Mantia, F. L. Dynamic Impedance Spectroscopy of Nickel Hexacyanoferrate Thin Films. *ChemElectroChem* **2019**, *6* (21), 5387–5395. DOI: 10.1002/celc.201900805.
62. Erinmwingbovo, C.; Siller, V.; Nuñez, M.; Trócoli, R.; Brogioli, D.; Morata, A.; La Mantia, F. L. Dynamic Impedance Spectroscopy of LiMn2O4 Thin Films Made by Multi-Layer Pulsed Laser Deposition. *Electrochim. Acta* **2020**, *331*. DOI: 10.1016/j.electacta.2019.135385, 135385.
63. Darowicki, K.; Andrearczyk, K. Determination of Occurrence of Anodic Excursion Peaks by Dynamic Electrochemical Impedance Spectroscopy, Atomic Force Microscopy and Cyclic Voltammetry. *J. Power Sources* **2009**, *189* (2), 988–993. DOI: 10.1016/j.jpowsour.2009.01.039.
64. Slepski, P.; Darowicki, K.; Andrearczyk, K. On-Line Measurement of Cell Impedance During Charging and Discharging Process. *J. Electroanal. Chem.* **2009**, *633* (1), 121–126. DOI: 10.1016/j.jelechem.2009.05.002.
65. Slepski, P.; Darowicki, K.; Janicka, E.; Sierczynska, A. Application of Electrochemical Impedance Spectroscopy to Monitoring Discharging Process of Nickel/Metal Hydride Battery. *J. Power Sources* **2013**, *241*, 121–126. DOI: 10.1016/j.jpowsour.2013.04.039.
66. Darowicki, K.; Zieliński, A.; Ryl, J.; Slepski, P. Impedance of Cation-Coupled Electron Transfer Reaction: Theoretical Description of One Pathway Process. *Electrochim. Acta* **2013**, *87*, 930–939. DOI: 10.1016/j.electacta.2012.09.019.

67. Huang, J.; Zhang, J.; Li, Z.; Song, S.; Wu, N. Exploring Differences Between Charge and Discharge of LiMn2O4/Li Half-Cell with Dynamic Electrochemical Impedance Spectroscopy. *Electrochim. Acta* **2014**, *131*, 228–235. DOI: 10.1016/j.electacta.2014.02.030.
68. Huang, J.; Li, Z.; Zhang, J. Dynamic Electrochemical Impedance Spectroscopy Reconstructed from Continuous Impedance Measurement of Single Frequency During Charging/discharging. *J. Power Sources* **2015**, *273*, 1098–1102. DOI: 10.1016/j.jpowsour.2014.07.067.
69. Huang, J.; Ge, H.; Li, Z.; Zhang, J. Dynamic Electrochemical Impedance Spectroscopy of a Three-Electrode Lithium-Ion Battery During Pulse Charge and Discharge. *Electrochim. Acta* **2015**, *176*, 311–320. DOI: 10.1016/j.electacta.2015.07.017.
70. Darowicki, K.; Janicka, E.; Mielniczek, M.; Zielinski, A.; Gawel, L.; Mitzel, J.; Hunger, J. Implementation of DEIS for Reliable Fault Monitoring and Detection in PEMFC Single Cells and Stacks. *Electrochim. Acta* **2018**, *292*, 383–389. DOI: 10.1016/j.electacta.2018.09.105.
71. Janicka, E.; Mielniczek, M.; Gawel, L.; Darowicki, K.; Landowska, P. The Impact of Air Humidity on the Operation of Proton Exchange Membrane Fuel Cells Determined Using Dynamic Electrochemical Impedance Spectroscopy. *Electrochim. Acta* **2020**, *341*. DOI: 10.1016/j.electacta.2020.136036, 136036.
72. Darowicki, K.; Kawula, J. Impedance Characterization of the Process of Polyaniline First Redox Transformation After Aniline Electropolymerization. *Electrochim. Acta* **2004**, *49* (27), 4829–4839. DOI: 10.1016/j.electacta.2004.05.035.
73. Darowicki, K.; Ślepski, P.; Szociński, M. Application of the Dynamic EIS to Investigation of Transport Within Organic Coatings. *Prog. Org. Coat.* **2005**, *52* (4), 306–310. DOI: 10.1016/j.porgcoat.2004.06.007.
74. Itagaki, M.; Ono, A.; Watanabe, K.; Katayama, H.; Noda, K. Analysis on Organic Film Degradation by Dynamic Impedance Measurements. *Corros. Sci.* **2006**, *48* (11), 3802–3811. DOI: 10.1016/j.corsci.2006.01.011.
75. Sacci, R. L.; Seland, F.; Harrington, D. A. Dynamic Electrochemical Impedance Spectroscopy, for Electrocatalytic Reactions. *Electrochim. Acta* **2014**, *131*, 13–19. DOI: 10.1016/j.electacta.2014.02.120.
76. Pajkossy, T. Dynamic Electrochemical Impedance Spectroscopy of quasi-Reversible Redox Systems. Properties of the Faradaic Impedance, and Relations to Those of Voltammograms. *Electrochim. Acta* **2019**, *308*, 410–417. DOI: 10.1016/j.electacta.2019.03.197.
77. Muzyka, K.; Sun, J.; Fereja, T. H.; Lan, Y.; Zhang, W.; Xu, G. Boron-Doped Diamond: Current Progress and Challenges in View of Electroanalytical Applications. *Anal. Methods* **2019**, *11* (4), 397–414. DOI: 10.1039/C8AY02197J.
78. Ryl, J.; Bogdanowicz, R.; Slepski, P.; Sobaszek, M.; Darowicki, K. Dynamic Electrochemical Impedance Spectroscopy (DEIS) as a Tool for Analyzing Surface Oxidation Processes on Boron-Doped Diamond Electrodes. *J. Electrochem. Soc.* **2014**, *161* (6), H359–H364. DOI: 10.1149/2.016406jes.
79. Ryl, J.; Burczyk, L.; Bogdanowicz, R.; Sobaszek, M.; Darowicki, K. Study on Surface Termination of Boron-Doped Diamond Electrodes Under Anodic Polarization in H 2 SO 4 by Means of Dynamic Impedance Technique. *Carbon* **2016**, *96*, 1093–1105. DOI: 10.1016/j.carbon.2015.10.064.
80. Ryl, J.; Burczyk, L.; Zielinski, A.; Ficek, M.; Franczak, A.; Bogdanowicz, R.; Darowicki, K. Heterogeneous Oxidation of Highly Boron-Doped Diamond Electrodes and Its Influence on the Surface Distribution of Electrochemical Activity. *Electrochim. Acta* **2019**, *297*, 1018–1027. DOI: 10.1016/j.electacta.2018.12.050.
81. Scohy, M.; Montella, C.; Claudel, F.; Abbou, S.; Dubau, L.; Maillard, F.; Sibert, E.; Sunde, S. Investigating the Oxygen Evolution Reaction on Ir(111) Electrode in Acidic Medium Using Conventional and Dynamic Electrochemical Impedance Spectroscopy. *Electrochim. Acta* **2019**, *320*. DOI: 10.1016/j.electacta.2019.07.047, 134536.

82. Gerengi, H.; Darowicki, K.; Bereket, G.; Slepski, P. Evaluation of Corrosion Inhibition of brass-118 in Artificial Seawater by Benzotriazole Using Dynamic EIS. *Corros. Sci.* **2009**, *51* (11), 2573–2579. DOI: 10.1016/j.corsci.2009.06.040.
83. Gerengi, H.; Darowicki, K.; Slepski, P.; Bereket, G.; Ryl, J. Investigation Effect of Benzotriazole on the Corrosion of Brass-MM55 Alloy in Artificial Seawater by Dynamic EIS. *J. Solid State Electrochem.* **2010**, *14* (5), 897–902. DOI: 10.1007/s10008-009-0923-1.
84. Gerengi, H.; Slepski, P.; Bereket, G. Dynamic Electrochemical Impedance Spectroscopy and Polarization Studies to Evaluate the Inhibition Effect of Benzotriazole on Copper–Manganese–Aluminium Alloy in Artificial Seawater. *Mater. Corros.* **2013**, *64*, 1024–1031.
85. Gerengi, H. The Use of Dynamic Electrochemical Impedance Spectroscopy in Corrosion Inhibitor Studies. *Prot. Met. Phys. Chem. Surf.* **2018**, *54* (3), 536–540. DOI: 10.1134/S2070205118030267.
86. Niki, K.; Delnick, F. M.; Hackerman, N. Behavior of Salicylaldoxime and 8-Hydroxyquinoline at a Copper Electrode. *J. Electrochem. Soc.* **1975**, *122* (7), 855–861. DOI: 10.1149/1.2134361.
87. Cicileo, G. P.; Rosales, B. M.; Varela, F. E.; Vilche, J. R. Inhibitory Action of 8-Hydroxyquinoline on the Copper Corrosion Process. *Corros. Sci.* **1998**, *40* (11), 1915–1926. DOI: 10.1016/S0010-938X(98)00090-0.
88. Verma, C.; Ebenso, E. E.; Quraishi, M. A. Corrosion Inhibitors for Ferrous and Non-Ferrous Metals and Alloys in Ionic Sodium Chloride Solutions: A review. *J. Mol. Liq.* **2017**, *248*, 927–942. DOI: 10.1016/j.molliq.2017.10.094.
89. Gerengi, H.; Mielniczek, M.; Gece, G.; Solomon, M. M. Experimental and Quantum Chemical Evaluation of 8-Hydroxyquinoline as a Corrosion Inhibitor for Copper in 0.1 M HCl. *Ind. Eng. Chem. Res.* **2016**, 55 (36), 9614–9624. DOI: 10.1021/acs.iecr.6b02414.
90. Gerengi, H.; Solomon, M. M.; Kurtay, M.; Bereket, G.; Goksen, K.; Yıldız, M.; Kaya, E. Electrochemical and Morphological Assessments of Inhibition Level of 8-Hydroxylquinoline for AA2024-T4 Alloy in 3.5% NaCl Solution. *J. Adhes. Sci. Technol.* **2018**, 32 (2), 207–223. DOI: 10.1080/01694243.2017.1350524.
91. Crini, G. Historical Review on Chitin and Chitosan Biopolymers. *Environ. Chem. Lett.* **2019**, *17* (4), 1623–1643. DOI: 10.1007/s10311-019-00901-0.
92. Ashassi-Sorkhabi, H.; Kazempour, A. Chitosan, Its Derivatives and Composites with Superior Potentials for the Corrosion Protection of Steel Alloys: A Comprehensive Review. *Carbohydr. Polym.* **2020**, *237*, 116110. DOI: 10.1016/j.carbpol.2020.116110.
93. Solomon, M. M.; Gerengi, H.; Kaya, T.; Kaya, E.; Umoren, S. A. Synergistic Inhibition of St37 Steel Corrosion in 15% H2SO4 Solution by Chitosan and Iodide Ion Additives. *Cellulose* **2017**, *24* (2), 931–950. DOI: 10.1007/s10570-016-1128-2.
94. Solomon, M. M.; Gerengi, H.; Kaya, T.; Umoren, S. A. Enhanced Corrosion Inhibition Effect of Chitosan for St37 in 15% H2SO4 Environment by Silver Nanoparticles. *Int. J. Biol. Macromol.* **2017**, *104* (A), 638–649. DOI: 10.1016/j.ijbiomac.2017.06.072.
95. Solomon, M. M.; Gerengi, H.; Kaya, T.; Umoren, S. A. Performance Evaluation of a Chitosan/Silver Nanoparticles Composite on St37 Steel Corrosion in a 15% HCl Solution. *ACS Sustainable Chem. Eng.* **2017**, 5 (1), 809–820. DOI: 10.1021/acssuschemeng.6b02141.
96. Wysocka, J.; Krakowiak, S.; Ryl, J. Evaluation of Citric Acid Corrosion Inhibition Efficiency and Passivation Kinetics for Aluminium Alloys in Alkaline Media by Means of Dynamic Impedance Monitoring. *Electrochim. Acta* **2017**, *258*, 1463–1475. DOI: 10.1016/j.electacta.2017.12.017.
97. Wysocka, J.; Cieslik, M.; Krakowiak, S.; Ryl, J. Carboxylic Acids as Efficient Corrosion Inhibitors of Aluminium Alloys in Alkaline Media. *Electrochim. Acta* **2018**, *289*, 175–192. DOI: 10.1016/j.electacta.2018.08.070.

98. Ryl, J.; Brodowski, M.; Kowalski, M.; Lipinska, W.; Niedzialkowski, P.; Wysocka, J. Corrosion Inhibition Mechanism and Efficiency Differentiation of Dihydroxybenzene Isomers Towards Aluminum Alloy 5754 in Alkaline Media. *Materials (Basel)* **2019**, *12* (19), 3067. DOI: 10.3390/ma12193067.
99. Ryl, J.; Wysocka, J.; Cieslik, M.; Gerengi, H.; Ossowski, T.; Krakowiak, S.; Niedzialkowski, P. Understanding the Origin of High Corrosion Inhibition Efficiency of Bee Products Towards Aluminium Alloys in Alkaline Environments. *Electrochim. Acta* **2019**, *304*, 263–274. DOI: 10.1016/j.electacta.2019.03.012.
100. Gece, G.; Bilgiç, S. A Computational Study of Two Hexitol Borates as Corrosion Inhibitors for Steel. *Int. J. Corros. Scale Inhib.* **2017**, *6*, 476–484.
101. Kipçak, A. S.; Moroydor Derun, E.; Pİşkİn, S. SYnthesis and Characterization of Magnesium Borate Minerals of Admontite and Mcallisterite Obtained via Ultrasonic Mixing of Magnesium Oxide and Various Sources of Boron: A Novel Method. *Turk. J. Chem.* **2014**, *38*, 792–805. DOI: 10.3906/kim-1307-61.
102. Gerengi, H.; Tascioglu, C.; Akcay, C.; Kurtay, M. Impact of Copper Chrome Boron (CCB) Wood Preservative on the Corrosion of St37 Steel. *Ind. Eng. Chem. Res.* **2014**, *53* (49), 19192–19198. DOI: 10.1021/ie5033342.
103. Gerengi, H.; Solomon, M. M.; Umoren, S. A.; Ugras, H. I.; Yıldız, M.; Slepski, P. Improved Performance of 1-Ethyl-3-Methylimidazolium Tetrafluoroborate AT Steel/HCl Interface by Iodide Ions. *J. Bio Tribo Corros.* **2018**, *4*, 1–11.
104. Yıldız, M.; Gerengi, H.; Solomon, M. M.; Kaya, E.; Umoren, S. A. Influence of 1-Butyl-1-Methylpiperidinium Tetrafluoroborate on St37 Steel Dissolution Behavior in HCl Environment. *Chem. Eng. Commun.* **2018**, *205* (4), 538–548. DOI: 10.1080/00986445.2017.1407759.
105. Umoren, S. A.; Eduok, U. M. Application of Carbohydrate Polymers as Corrosion Inhibitors for Metal Substrates in Different Media: A Review. *Carbohydr. Polym.* **2016**, *140*, 314–341. DOI: 10.1016/j.carbpol.2015.12.038.
106. Solomon, M. M.; Gerengi, H.; Umoren, S. A. Carboxymethyl Cellulose/Silver Nanoparticles Composite: Synthesis, Characterization and Application as a Benign Corrosion Inhibitor for St37 Steel in 15% H 2 SO 4 Medium. *ACS Appl. Mater. Interfaces* **2017**, *9* (7), 6376–6389. DOI: 10.1021/acsami.6b14153.
107. Solomon, M. M.; Gerengi, H.; Umoren, S. A.; Essien, N. B.; Essien, U. B.; Kaya, E. Gum Arabic-Silver Nanoparticles Composite as a Green Anticorrosive Formulation for Steel Corrosion in Strong Acid Media. *Carbohydr. Polym.* **2018**, *181*, 43–55. DOI: 10.1016/j.carbpol.2017.10.051.
108. Yue, H.; Zhao, Y.; Ma, X.; Gong, J. Ethylene Glycol: Properties, Synthesis, and Applications. *Chem. Soc. Rev.* **2012**, *41* (11), 4218–4244. DOI: 10.1039/c2cs15359a.
109. Mercer, A. D.; Wormwell, F. Corrosion Inhibitors for Cast Iron and Other Metals in Ethylene Glycol Solutions and in Mains Water. *J. Appl. Chem.* **1959**, *9*, 577–588.
110. Mercer, A. D.; Brasher, D. M. Corrosion Inhibitors for Ethylene Glycol Solutions, with Particular Reference to Cadmium-Plated and Terne-Plated Steel. *J. Appl. Chem.* **1959**, *9*, 589–594.
111. Gerengi, H.; Solomon, M. M.; Kaya, E.; Bagci, F. E.; Abai, E. J. An Evaluation of the Anticorrosion Effect of Ethylene Glycol for AA7075-T6 Alloy in 3.5% NaCl Solution. *Measurement* **2018**, *116*, 264–272. DOI: 10.1016/j.measurement.2017.11.030.
112. Gerengi, H.; Solomon, M. M.; Öztürk, S.; Yıldırım, A.; Gece, G.; Kaya, E. Evaluation of the Corrosion Inhibiting Efficacy of a Newly Synthesized Nitrone Against St37 Steel Corrosion in Acidic Medium: Experimental and Theoretical Approaches. *Mater. Sci. Eng. C Mater. Biol. Appl.* **2018**, *93*, 539–553. DOI: 10.1016/j.msec.2018.08.031.
113. Miralrio, A.; Espinoza Vázquez, A. E. Plant Extracts as Green Corrosion Inhibitors for Different Metal Surfaces and Corrosive Media: A Review. *Processes* **2020**, *8* (8), 942. DOI: 10.3390/pr8080942.

114. Gerengi, H.; Jazdzewska, A.; Kurtay, M. A Comprehensive Evaluation of Mimosa Extract as a Corrosion Inhibitor on AA6060 Alloy in Acid Rain Solution: part I. Electrochemical AC Methods. *J. Adhes. Sci. Technol.* **2015**, *29* (1), 36–48. DOI: 10.1080/01694243.2014.973159.
115. Gerengi, H.; Uygur, I.; Solomon, M. M.; Yıldız, M.; Göksu, H. Evaluation of the Inhibitive Effect of Diospyros kaki (Persimmon) Leaves Extract on St37 Steel Corrosion in Acid Medium. *Sustain. Chem. Pharm.* **2016**, *4*, 57–66. DOI: 10.1016/j.scp.2016.10.003.
116. Slepski, P.; Gerengi, H.; Jazdzewska, A.; Orlikowski, J.; Darowicki, K. Simultaneous Impedance and Volumetric Studies and Additionally Potentiodynamic Polarization Measurements of Molasses as a Carbon Steel Corrosion Inhibitor in 1M Hydrochloric Acid Solution. *Constr. Build. Mater.* **2014**, *52*, 482–487. DOI: 10.1016/j.conbuildmat.2013.11.059.
117. Gerengi, H.; Goksu, H.; Slepski, P. The Inhibition Effect of Mad Honey on Corrosion of 2007-Type Aluminium Alloy in 3.5% NaCl Solution. *Mater. Res.* **2014**, *17* (1), 255–264. DOI: 10.1590/S1516-14392013005000174.
118. Zhou, J.; Li, X.; Du, C.; Li, Y.; Li, T.; Pan, Y. Anodic Electrochemical Behavior of X80 Pipeline Steel in NaHCO3 Solution. *Acta Metall. Sin.* **2010**, *46*, 251–256.
119. Ryl, J.; Gawel, L.; Cieslik, M.; Gerengi, H.; Lentka, G.; Slepski, P. Instantaneous Impedance Analysis of Non-Stationary Corrosion Process: A Case Study of Carbon Steel in 1M HCl. *Int. J. Electrochem. Sci.* **2017**, *12*, 6908–6919. DOI: 10.20964/2017.07.15.
120. Rizvi, M.; Gerengi, H.; Yıldız, M.; Kekeçoğlu, M.; Pehlivan, M. M. Investigation of "Propolis" as a Green Inhibitor of SAE 1010 Carbon Steel Corrosion in 3.5% NaCl Environment. *Ind. Eng. Chem. Res.* **2020**, *59* (19), 9328–9339. DOI: 10.1021/acs.iecr.0c01239.
121. Kurtay, M.; Gerengi, H.; Chidiebere, A.; Yıldız, M. Corrosion Inhibition of Reinforcement Steel in Mixture Water by Caffeine and L-Arginine. *J. Adhes. Sci. Technol.* **2021**, 1–34. DOI: 10.1080/01694243.2021.1916251.

Chapter 6

Computational Methods of Corrosion Inhibition Assessment

Hassane Lgaz,[1] Abdelkarim Chaouiki,[2] Rachida Lamouri,[3] Rachid Salghi,[4] and Han-Seung Lee[1,*]

[1]Department of Architectural Engineering, Hanyang University-ERICA, 1271 Sa 3-dong, Sangrok-gu, Ansan 15588, Korea

[2]Materials Electrochemistry Group, School of Materials Science and Engineering, Yeungnam University, Gyeongsan 38541, Republic of Korea

[3]Department of Physics, Yeungnam University, Gyeongsan 38541, Republic of Korea

[4]Laboratory of Applied Chemistry and Environment, ENSA, University Ibn Zohr, PO Box 1136, 80000 Agadir, Morocco

[*]Email: ercleehs@hanyang.ac.kr.

Recently, the development of computational tools has gained significant interest. This has made computer-based simulation and the investigation of chemical problems parallel and independent fields of research of experimental work. Driven by this, computational methods such as density functional theory, molecular dynamics, and Monte Carlo simulations have been extensively used in corrosion inhibition studies to gain insight into the corrosion inhibition mechanism. This chapter provides a brief theoretical overview of these methods and their usefulness in corrosion inhibition research.

Introduction

Computational chemistry is a subfield of chemistry devoted to understanding and investigating chemical problems on a computer using mathematical methods in combination with the principles and laws of physics. Recent advances in computer hardware performance enable the investigation of complex chemical issues and can answer questions that experiments cannot, thereby making molecular modeling like an "experimental" tool for computational chemists.

In corrosion inhibition research, computational chemistry methods are of particular interest to corrosion researchers. Several computational methods have been applied to investigate the molecular geometry and the reactivity of corrosion inhibitor molecules and their interaction with different metal surfaces. The main goal is to gain insight into how inhibitor molecules interact with the metal surface, a relevant factor in designing effective corrosion-inhibiting materials.

The most widely used computational chemistry methods in corrosion inhibition research are density functional theory (DFT) calculations, molecular dynamics (MD), Monte Carlo (MC)

simulations, and ab initio DFT calculations. DFT calculations are mainly used to investigate an inhibitor molecule's molecular electronic properties and to predict the active sites potentially responsible for its interactions with a metal surface (*1–4*). Atomistic simulations (i.e., MD and MC simulations) are used to simulate the motion of a molecule as it interacts with a metal substrate in the vacuum phase or in the presence of a simulated solvent (*5*). Ab initio DFT calculations are very promising computational approaches used to study the adsorption properties of inhibitor molecules over metal surfaces (*6, 7*). However, these calculations are relatively slow and require high-performance computing resources, especially for large-sized molecules, which are not accessible for all researchers, thereby limiting their application.

This chapter provides a brief overview of the computational methods that are most widely used in corrosion inhibition research, namely, DFT-based calculations and atomic simulations using MD and MC methods. The aim is to give the reader a chance to understand their application in corrosion inhibition studies that will be reviewed in the next chapters of this book. Computational methods currently used in a limited number of studies will not be discussed in this chapter.

DFT-Based Calculations: Basics and Definition of Terms

DFT is one of the most commonly used quantum methods in solid-state physics and quantum chemistry. It is considered a reliable method for determining quantum quantities that can help understand reaction mechanisms in systems containing a large number of electrons, such as corrosion inhibitors (*8–10*). DFT was established based on the theoretical formalism proposed by Pierre Hohenberg, Walter Kohn, and Lu Sham in the mid-1960s (*11–13*). This formalism is based on quantum mechanics and involves only a limited number of input data. It allows the resolution of the Schrödinger equation only dealing with electron density instead of the all-electron wave function for a given multi-body.

In corrosion inhibition research, the combination of DFT and the hard and soft acid–base theory offers a better description and interpretation of the structure–properties relationship (*14, 15*). This combination assists in evaluating existing corrosion inhibitors, which helps in designing new derivatives with enhanced performance. The main advantage of computational methods, such as DFT, is in the analysis of different parameters that might be time consuming and often impossible to evaluate experimentally.

Theoretical Framework

Solving the Schrödinger equation with N electrons requires approximate methods to most accurately reproduce the physical quantities containing the possible information. However, it is possible to reformulate the problem by using appropriate theorems and approximations. Nowadays, DFT would have aroused little curiosity except within the framework of the theorem established by Kohn-Sham, which made it useful through approximations of the functional of the ground-state to describe real multi-electron systems.

The development of the DFT dates back to 1964 and 1965 with the publication of the Hohenberg and Kohn theorems (*16, 17*). The two theorems are given as follows:

Theorem 1: The total ground-state energy of an interacting N-electron system in the presence of an external potential is a unique function of the electronic density ρ:

$$E = E(\rho) \tag{1}$$

The first theorem highlighted a unique correspondence between the external potential and electronic density.

Theorem 2: The real ground-state density is nothing else than the density that minimizes the energy of the system, and all other properties are functional of the ground-state density:

$$E(\rho_0) = \min\left(E(\rho)\right) \tag{2}$$

The second theorem reduced the problem with 3N-variables to a problem of a scalar function in three-dimensional space, allowing the simplification of the solution of the Schrödinger equation.

In fact, the main goal of DFT is to determine the ground-state properties of a given system that is composed of a fixed number of electrons interacting with the nuclei using only information about the electronic density ρ. Therefore, we must first define the electronic density of the system to determine its ground-state properties.

In 1965, Walter Kohn and Lu Sham showed that the Schrödinger equations might be reformulated to equivalent equations with an actual potential in which quasiparticles move. These equations are called the Kohn-Sham equations (*18*):

$$\left\{-\frac{1}{2}\nabla^2 + V_{KS}(r)\right\}\varphi_i(r) = \varepsilon_i\varphi_i(r) \tag{3}$$

$$V_{KS}(r) = V_{ext}(r) + V_{XC}(r) + V_H(r) \tag{4}$$

where $V_{ext}(r)$ is the external potential created by the nuclei; $V_{XC}(r)$ is the potential of the exchange–correlation energy, given as

$$V_{XC}(r) = \frac{\partial E_{XC}[\rho(r)]}{\partial \rho(r)} \tag{5}$$

and $V_H(r)$ is the classical Hartree potential, given as

$$V_H(r) = \int \rho(r)\frac{1}{|r-r'|}\cdot dr' \tag{6}$$

Thus, the total energy of the system is obtained by solving the Kohn-Sham equation as follows:

$$E(r) = \sum_{ioccup}\varepsilon_i - \int \frac{\rho(r)\cdot\rho(r')}{|r-r'|}\cdot dr\cdot dr' + E_{XC}(\rho)\cdot \int V_{XC}(r)\rho(r)\cdot dr \tag{7}$$

Note that the Kohn-Sham equation is similar to the Schrödinger equation. Still, it is only an equivalence, because, in particular, the eigenvalues of the Kohn-Sham equation are pure vibrational parameters and therefore do not strongly represent the eigenenergies of the system. The eigenvectors are not the functions of real electronic waves. The only ambiguity in the Kohn-Sham approach is the term of the exchange–correlation. The formal complexity of this term makes the resolution of the Kohn-Sham equations difficult; nevertheless, this functional can be subjected to local or near-local approximations of the density. The most basic types of exchange–correlation approximation are the local density approximation (LDA) and the generalized gradient approximation (GGA). The LDA, introduced by Slater in 1951 (*19*), allows for transforming the DFT, an exact N-body theory, into an approximated, but very useful and much-used, theory. In the LDA, it is assumed that the electronic density can be processed locally as a uniform gas. In other words, this approach consists of carrying out the two following hypotheses:

- The exchange–correlation effects are dominated by the density at the point $\vec{r}$.
- The density $\rho(\vec{r})$ is a slowly varying function with respect to $\vec{r}$.

Thus, the fundamental assumption contained in the formalism of the LDA consists of considering that the contribution of exchange–correlation energy to the total energy of the system can be added cumulatively from each portion of the nonuniform gas as if it were locally uniform. Despite the usefulness of the LDA in describing the electronic structure of systems, it fails to evaluate different parameters, such as bandgap energy, molecular atomization energies, and bond lengths.

To improve the processing of the exchange–correlation energy, GGA formalism was introduced in the 1980s. In GGA formalism, the exchange–correlation energy function depends not only on the electronic density $\rho(\vec{r})$ but also on its gradient $\nabla[\rho(\vec{r})]$. Thanks to this modification, the exchange–correlation energy function considers the nonuniformity of the electron gas. Its contribution to the system's total energy can be added cumulatively from each nonuniform portion of the gas as if it were locally nonuniform (*20*). The introduction of these GGA-type functionalities led to the widespread use of DFT within the chemical community in the 1990s. The use of the GGA-type functional significantly increases the accuracy of the calculations compared with the description provided by the LDA, particularly for the binding energy of the molecules. In 1998, the Becke-3-parameter-Lee-Yang-Parr (B3LYP) hybrid density functional method was introduced for better accuracy by including a portion of exact exchange from the Hartee-Fock theory to the existing form of the exchange–correlation energy (*21*). Since its identification, the B3LYP method continues to dominate as one of the most used computational chemistry methods.

The exchange–correlation energy in the B3LYP method can be written as follows (*1*):

$$E_{XC}^{B3LYP} = a_0 E_X^{HF} + a_1 E_X^{LSD} + a_2 E_X^{GGA} + a_3 E_C^{LYP} \quad (8)$$

where E_X^{HF} is the exchange energy based on the Hartee-Fock method; E_X^{LSD} is the local exchange energy; E_X^{GGA} is the calculated exchange energy within the GGA; and E_C^{LYP} is the energy of the correlation functional of Lee, Yang, and Parr. This method has always yielded a good description of the thermochemistry of atoms and molecules, chemical reaction stability, and adsorption ability of molecules on different types of surfaces.

DFT calculations have had a great impact on the progress of understanding many interdisciplinary problems by assessing the fundamental properties of the organic and inorganic compounds. These methods have proven their efficiency in the prediction of materials properties and behaviors under different conditions, as well as the discovery of new functional molecules and materials from the electronic structure calculations which have paved the way for the development of many sectors from medical technology to environment-related applications. In recent years, DFT-based simulations have been particularly valuable in condensed matter physics and chemistry. Figure 1 illustrates the remarkable increase in the number of publications on the topics "density functional" and "DFT" that occurred from 1997 to 2020. This indicates that the DFT concept has been useful in many cases for giving and reproducing a good description of systems ranging from atoms and molecules to complex materials.

Materials surface corrosion is considered a major issue that limits the application of some materials. To study corrosion inhibition, different chemical parameters must be involved, which remains difficult to achieve experimentally. However, the DFT approach has showed good results in describing inhibitors–surface interactions and corrosion inhibition efficiency.

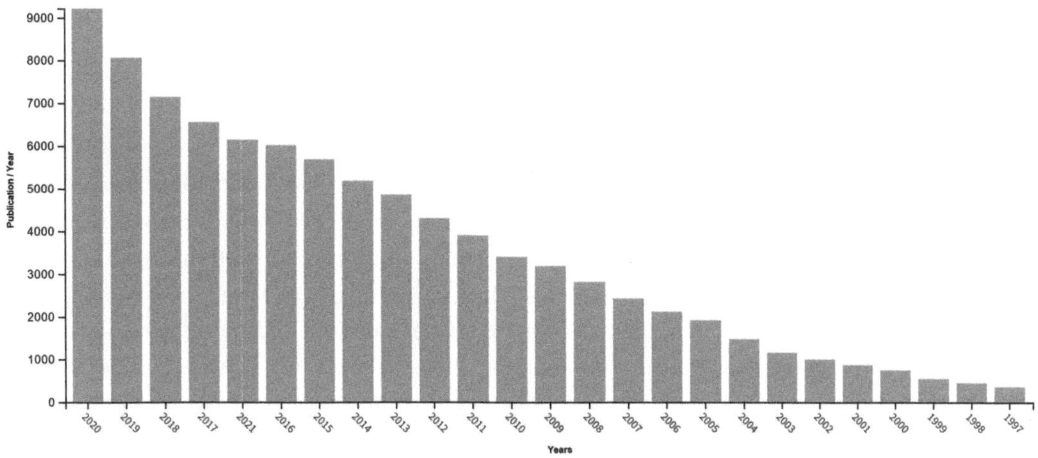

Figure 1. Number of publications per year based on the Web of Science database on the topics "density functional" AND "DFT."

DFT-Derived Parameters for Corrosion Inhibition Studies

DFT method-based computational chemistry is a rapidly growing field in corrosion science and is widely used to demonstrate the relationship between the corrosion inhibitors' electronic properties/molecular structures and their mechanism of corrosion inhibition. It can be said that the ability to evaluate the performance of corrosion inhibitors with respect to their geometric changes is critical for the study of the adsorption mechanism. The evaluation of molecular properties is closely related to the type of system that can be studied with DFT. Based on the molecular structure of inhibitors, a theoretical calculation-based DFT has proved to be very successful in estimating the reactivity of molecules for many systems and has yielded significant results. Interestingly, a number of publications have been performed in recent years within the framework of DFT, providing relevant insight into the molecular/electronic properties of organic compounds (see Figure 2).

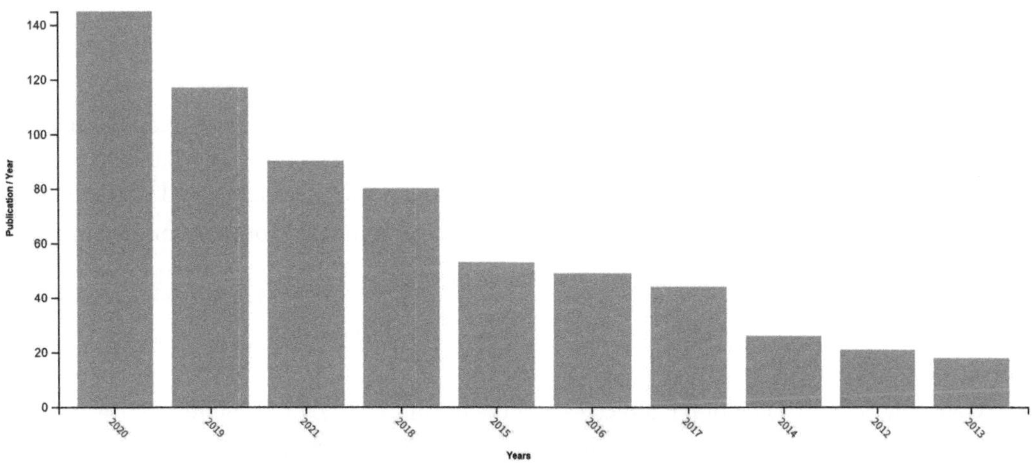

Figure 2. Number of publications per year based on the Web of Science database on the topics "Density functional theory" AND "corrosion inhibition."

The chemical descriptors presented below have provided important information regarding molecular reactivity, starting from structures to properties and reactivity of inhibitor molecules.

Researchers have used these descriptors to find correlation between the molecular structure of inhibitors and their experimental performance.

Predicting Energy and Density Distribution of Molecular Orbitals

One of the most successful models used to describe the density distribution over the molecular orbitals of inhibitor molecules and their energy is the frontier molecular orbitals theory. According to the frontier molecular orbitals theorem (22), the identification of the adsorption sites in corrosion inhibitor molecules is more useful by evaluating the highest-occupied (HO) and lowest-unoccupied (LU) molecular orbital (MO) distributions. Both the HOMO and the LUMO are generally seen as factors that are strongly related to the donor–acceptor abilities of electrons. What is significant about this definition is that the HOMO defines the ability of an organic molecule to donate electrons and to bond to a metal surface. However, the LUMO can act as an electron acceptor from a metallic surface. Therefore, as global reactivity parameters, the HOMO and the LUMO play important roles in the interaction between a metal surface and the inhibitor molecule. Consistent with the density distribution of the HOMO and the LUMO in terms of energy, it is well established that any inhibitor molecule with a high HOMO energy value might be effective because its electron-donating ability would be high. On the other hand, a lower value of LUMO energy indicates that the molecule is more likely to accept electrons. From this, the highest HOMO and lowest LUMO energy values can provide insight to elucidate the inhibition mechanism and determine which inhibitor will be more likely to be adsorbed onto a metal surface, protecting it from corrosion damage. Other parameters related to MO distributions are widely considered in predicting the inhibition performances of molecules against the corrosion of metal surfaces. According to the Koopmans' theorem (23), the approximate prediction of ionization energy and electron affinity values of organic compounds can be used. The negative values of HOMO and LUMO energies correspond to ionization potential (IP) and electron affinity (EA), respectively, as follow:

$$IP = -E_{HOMO} \tag{9}$$

$$EA = -E_{LUMO} \tag{10}$$

In addition, the energy gap (ΔE) between LUMO and HOMO energies is a crucial factor for understanding the reactivity and stability of a molecule within a corrosion inhibition process. Knowledge of the relative reactivity of a compound is essential for correlating and comparing the results and is used extensively in predicting the degree of adsorption of an organic compound onto a metallic surface.

$$\Delta E = E_{LUMO} - E_{HOMO} \tag{11}$$

Generally, a lower value of ΔE indicates that the organic compound is more reactive and thus more easily donates its electron, making it more effective for inhibiting corrosion of a metal. For example, the link between ΔE and corrosion inhibition performance was established in 2020 by Chaouiki et al. (24) Chaouiki et al. showed that the higher tendency of a corrosion inhibitor (noted Q1) (Figure 3) to be adsorbed onto an iron surface is consistent with its lower ΔE value. Hence, the ΔE value reflects good agreement with inhibition performance obtained from electrochemical measurements.

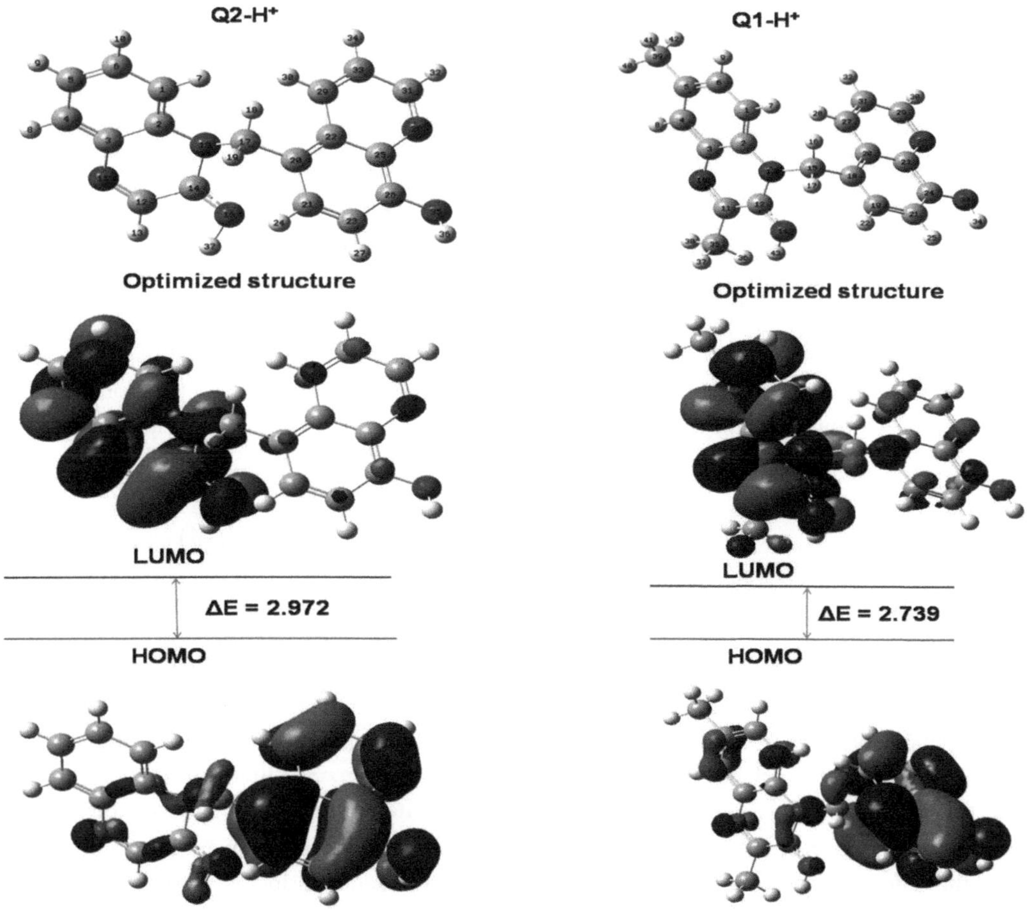

Figure 3. Optimized molecular structure, HOMO, and LUMO of Q1 and Q2 in their protonated forms using DFT/B3LYP/6-31+G(d,p) in an aqueous solution. Reproduced with permission from reference (24) under Creative Commons CC BY 4.0 license. Copyright 2021 MDPI.

Other Electronic Descriptors Related to Global and Local Reactivity

Using the DFT concept, other quantum chemical parameters—such as chemical potential (μ), absolute electronegativity (χ), global hardness (η), and global softness (σ)—are defined as the derivatives of the total electronic energy (E_{TOTAL}) with respect to number of electrons at a constant external potential. All mentioned parameters can be computed according to the following formulas:

$$\mu = -\chi = \left[\frac{\partial E}{\partial N}\right]_{v(r)} = -\left(\frac{IP + EA}{2}\right) \tag{12}$$

$$\eta = \frac{1}{2}\left[\frac{\partial^2 E}{\partial N^2}\right]_{v(r)} = \frac{IP - EA}{2} \tag{13}$$

$$\sigma = 1/2\eta \tag{14}$$

It is well known that electronegativity is reported as the negative value of chemical potential. The basic concept behind electronegativity descriptors is that each element has an intrinsic electronegativity. A stronger interaction between the surface and an inhibitor will occur if the difference of electronegativity between the organic compound and the metal surface is low (25). In addition, the approximate definition of μ and η of a chemical system has been given by Pearson (26). According to Pearson, soft molecules are molecules having high polarizability and low hardness values. As can be understood from the given information, soft molecules might be suitable corrosion inhibitors. In accordance with this, many studies have demonstrated that a molecule with a relatively low hardness value will be adequate to inhibit metal corrosion.

Electron donating power (ω^-) and electron-accepting power (ω^+), introduced by Gazquez (27), are other useful descriptors of corrosion inhibitors in terms of being donors–acceptors of electrons. The theoretical formulas for these parameters are expressed in terms of IP and EA as follows:

$$\omega^+ = \frac{(IP+3EA)^2}{16(IP-EA)} \quad (15)$$

$$\omega^- = \frac{(3IP+EA)^2}{16(IP-EA)} \quad (16)$$

Generally, high electro-donating power and low electro-accepting power values for an organic inhibitor are indicators of its efficiency against metal corrosion. Another important parameter reflecting the electronic flow in a system containing an inhibitor and a metallic surface is the fraction of electrons transferred. As can be understood from the information given by Pearson's theory (14), the determination of ΔN is based on chemical equalization principles regarding hardness and electronegativity concepts. Thus, ΔN is mathematically described according to the following equation:

$$\Delta N = \frac{\emptyset_{metal}-\chi_{inh}}{2(\eta_{metal}+\eta_{inh})} \quad (17)$$

where Ø is the work function calculated for the metal surface, and χ_{inh} and η_{inh} represent the electronegativity of the inhibitor and the global hardness of iron and the inhibitor, respectively.

From the simplest formalism of ΔN, it is well established that an inhibitor molecule can transfer its electron if $\Delta N > 0$, and vice versa if $\Delta N < 0$. According to this, inhibitor inhibition increases with the increasing electron-donation ability to the metal surface (28).

Indeed, other interesting descriptors related to the local charge populations and the condensed Fukui functions are helpful in providing important clues about the behavior of the reactive region of the inhibitor compound and thus its adsorption process. Detailed information about some of these descriptors and their applications in corrosion inhibition studies is described based on various theorems. For instance, according to the finite difference approximation and Mulliken's population analysis, the identification of reactive regions and the nucleophilic and electrophilic attacks in a chemical species can be analyzed by evaluating the Fukui indices. Accordingly, the Fukui function *f*(*r*) is defined as follows:

$$f_K^+ = q_K(N+1) - q_K(N) \qquad \text{(for nucleophilic attack)} \quad (18)$$

$$f_K^- = q_K(N) - q_K(N-1) \qquad \text{(for electrophilic attack)} \tag{19}$$

$$f_K^0 = \frac{q_K(N+1) - q_K(N-1)}{2} \qquad \text{(for radical attack)} \tag{20}$$

where $q_k(N)$, $q_k(N+1)$, and $q_k(N-1)$ denote the charge values of the neutral, anionic, and cationic forms of atom k, respectively.

Briefly, Table 1 shows an example of a Fukui index calculated for four organic compounds as corrosion inhibitors (*29*). Fukui indices provide information about the local reactivity region of a molecule resulting from the properties of the nucleus and the nature of the attack and, therefore, the most likely center for interaction with a metal surface.

Table 1. Fukui Functions and Dual Descriptor of Investigated Hydrazine Derivatives Calculated at the DFT/GGA/DNP Level. Source: Reproduced with permission from reference (29). Copyright 2021 Elsevier.

Atoms	*HDZ1*[a]			*HDZ2*[a]			*HDZ3*[a]			*HDZ4*[a]		
	fk+	*fk−*	*Δf(k)*	*fk+*	*fk−*	*Δf(k)*	*fk+*	*fk−*	*Δf(k)*	*fk+*	*fk−*	*Δf(k)*
C(2)	0.6	2.9	−2.3	0.8	1.4	−0.6	1.1	2.4	−1.3	1.2	3.0	−1.8
C(3)	0.3	1.3	−1.0	0.4	0.8	−0.4	0.5	1.0	−0.5	0.5	1.2	−0.7
C(4)	−0.2	4.1	−4.3	0.1	0.4	−0.3	0.9	2.6	−1.7	0.6	4.2	−3.6
C(5)	−0.3	1.5	−1.8	0.1	0.5	−0.4	0.2	1.0	−0.8	0.1	1.3	−1.2
C(6)	−0.9	3.7	−4.6	−0.9	−1.0	0.1	−0.5	1.5	−2.0	−0.5	3.2	−3.7
N(7)	−0.5	10	−10.5	2.0	11.1	−9.1	1.5	10.2	−8.7	0.2	8.1	−7.9
C(8)	−0.1	4.6	−4.7	−0.3	0.7	−1.0	0.4	2.5	−2.1	0.2	4.7	−4.5
C(10)	0.8	4.0	−3.2	0.9	2.1	−1.2	1.2	2.9	−1.7	1.6	3.9	−2.3
C(12)	1.3	7.8	−6.5	1.3	2.5	−1.2	2.2	5.3	−3.1	2.4	7.8	−5.4
C(20)	−1.1	2.8	−3.9	0.2	5.2	−5.0	0.4	4.3	−3.9	1.4	3.3	−1.9
C(21)	4.2	−0.7	4.9	5.4	1.2	4.2	5.9	1.2	4.7	8.1	0.2	7.9
C(22)	0.7	2.8	−2.1	1.3	4.0	−2.7	1.7	3.7	−2.0	2.9	3.2	−0.3
C(24)	1.4	5.7	−4.3	1.9	8.9	−7.0	1.8	7.5	−5.7	3.2	5.9	−2.7
C(26)	2.2	3.1	−0.9	4.0	3.8	0.2	4.3	3.3	1.0	5.7	2.9	2.8
C(28)	1.0	3.3	−2.3	2.6	6.2	−3.6	2.3	4.9	−2.6	2.8	2.9	−0.1
C(30)	0.1	0.8	−0.7	1.9	3.4	−1.5	2.1	2.0	0.1	2.5	1.0	1.5
O(31)	7.3	1.9	5.4	7.7	2.7	5.0	7.4	2.2	5.2	10.2	3.4	6.8
N(32)	2.6	−0.7	3.3	2.0	2.0	0.0	1.6	1.2	0.4	2.5	−0.1	2.6
N(33)	9.0	−2.4	11.4	6.4	−1.0	7.4	5.9	−1.4	7.3	5.1	−2.0	7.1
C(34)	6.0	1.1	4.9	8.2	2.9	5.3	6.5	2.3	4.2	7.8	1.2	6.6
C(36/37/36/36)	5.9	−0.7	6.6	3.8	1.7	2.1	1.7	0.7	1.0	1.5	0.0	1.5

Table 1. (Continued). Fukui Functions and Dual Descriptor of Investigated Hydrazine Derivatives Calculated at the DFT/GGA/DNP Level

Atoms	*HDZ1*[a]			*HDZ2*[a]			*HDZ3*[a]			*HDZ4*[a]		
	fk+	*fk−*	*Δf(k)*	*fk+*	*fk−*	*Δf(k)*	*fk+*	*fk−*	*Δf(k)*	*fk+*	*fk−*	*Δf(k)*
C(37/39/37/37)	8.7	2.2	6.5	3.4	1.7	1.7	3.3	1.1	2.2	1.1	0.4	0.7
C(38/41/39/38)	2.6	0.2	2.4	5.3	2.0	3.3	3.3	1.4	1.9	0.6	0.4	0.2
C(40/42/41/39)	3.0	0.7	2.3	4.6	2.3	2.3	5.7	1.9	3.8	1.1	0.4	0.7
C(42/43/43/40)	5.7	1.5	4.2	3.1	1.1	2.0	2.9	0.7	2.2	1.4	0.3	1.1
C(44/45/45)	3.5	0.9	2.6	3.8	0.3	3.5	2.9	0.0	2.9			
C(46/47/47)	3.3	0.7	2.6	2.2	0.2	2.0	2.0	−0.2	2.2			
C (48)	1.5	−0.3	1.8									

[a] The actual values have been multiplied by 100 for an easier comparison. H atoms are not shown.

Atomistic Simulations

MD Simulations

MD simulation is one of the most used computer simulation methods for analyzing molecular-scale phenomena. In this method, principles of classical mechanics are combined with concepts from computer science, mathematics, chemistry, and physics and are used to predict the physical movements of a system composed of particles, namely, atoms and molecules. In a given system where these particles can interact for a fixed time, MD simulation provides a view of the time evolution and atomic-scale dynamics of a model system, such as responses to system changes, time-dependent fluctuations, and transport coefficients. The MD trajectories of a system of interacting particles are determined by the numerical integration of Newton's equations of motion (*30*, *31*), and molecular mechanics force fields are often used to calculate interatomic interactions between particles.

Theoretical Background

MD simulation allows the tracing of the movement of particles by solving, step by step, Newton's second law and finding the molecular trajectory forces between particles from an initial configuration. The system's time evolution is then obtained by recalculating the position and velocity vector of each atom at every time-step. The total energy of the system is mathematically described as a function of all atomic coordinates. The forces Fi acting on particles are derived from a given potential energy $E(r_1, r_2, \ldots, r_N)$, where r_i is the vector position of the *j*th atom. The force acting on a particle changes when a particle interacting with it moves or changes its position. Given the complexity of the model system, the numerical integration of eq 21 is required to obtain atom trajectories that are raw data for predicting a system's properties.

$$F_i = m_i \times \frac{d^2 r_i}{dt^2} = -\nabla_i E(r_1, r_2, \ldots, r_N) \quad (21)$$

where r is the spatial gradient; E is the system's empirical potential; and r_i and m_i denote the spatial coordinates and the mass of the ith atom, respectively.

Energy Minimization, Ensemble, and Periodic Boundary Condition

Geometry optimization or total energy minimization is the first step in MD simulation, which aims to find a stable structure in which atoms reach the lowest energy configuration, that is, with adjusted coordinates and net forces acting on them equal zero (*32*, *33*). In this step, only total energies are considered in calculating the energy of the system and the forces acting on the atoms. Several minimization algorithms—such as steepest descents, conjugate gradients, and Quasi-Newton methods—are used in this procedure (*34*, *35*). A reliable and robust minimization algorithm would definitely generate excellent results. Compared with other algorithms, the conjugate gradients algorithm is generally accepted as an effective minimization method for a large number of simulation systems.

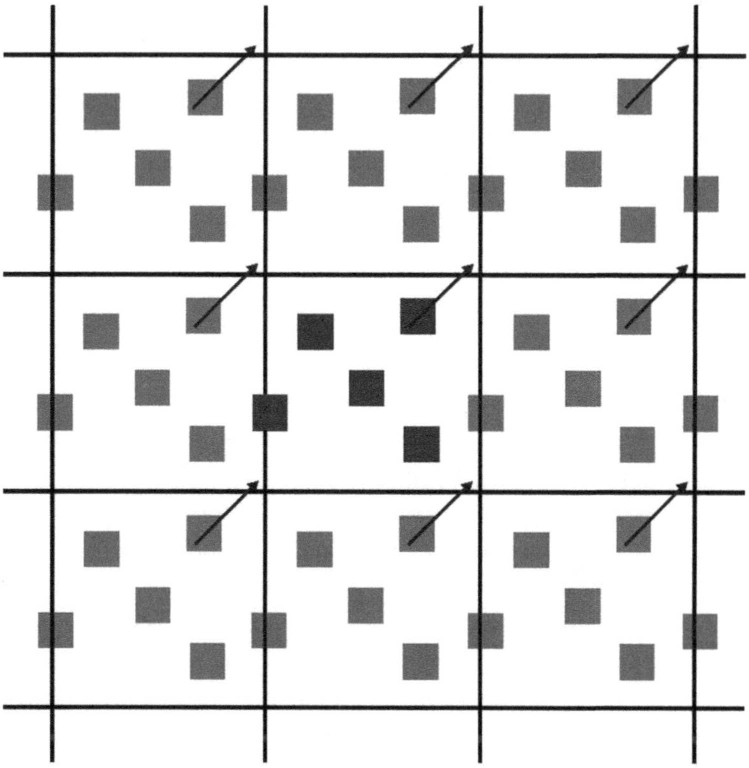

Figure 4. Periodic boundary conditions.

An essential aspect of MD simulations is the possibility for controlling some parameters such as entropy changes, volume, and pressure. The temperature can also be controlled in the canonical ensemble. These parameters are dealt with using an ensemble, which is defined as the set of all possible systems with identical thermodynamic or macroscopic attributes but different microscopic states. For a model system with N particles, temperature (T), pressure (P), volume (V), or number of particles (i.e., N) can be fixed while leaving one parameter variable. Some frequently used ensembles include the isobaric–isothermal ensemble, microcanonical ensemble, grand canonical ensemble, and canonical ensemble (NVT).

Periodic boundary conditions are essential when a small simulation model is used. For example, consider that all atoms of a model system are arranged in a small cube. Many of these atoms will move to the outer faces, thus affecting the calculated properties. Periodic boundary conditions consist of surrounding the simulation cube with replicas of itself. Then, when an atom leaves the main simulation box, it can interact with the nearest atom or virtual box, as shown in Figure 4. This approach has been used in all MD simulations of corrosion inhibition (*36–43*).

Force Fields

In MD simulation, the reliability of results depends on the force field used, which is an empirical function derived from experimental results and ab initio calculations. The force field is a set of parameters and formulas that relate potential energy to a configuration of a molecular system. It is of crucial importance to choose an appropriate force field for a specific system. To this end, several force fields have been developed, such as Assisted Model Building with Energy Refinement (AMBER) force fields and the latest version, AMBER14SB + parmbsc1 (*44, 45*), as well as the universal force field (*46, 47*) and the consistent-valence force field (*48–50*). In most corrosion inhibition studies, the Condensed-phase Optimized Molecular Potentials for Atomistic Simulation Studies force field is used (*51*). This force field was parameterized using ab initio calculations and experimental data for organic compounds, halogens, and metals. In classical MD simulations, all these force fields are unable to describe bond breaking/formation and electron transfer. To bridge this gap, attempts have been made to develop the reactive force field for some unique systems. Further, to reduce the computational cost, force fields are simplified so that three- and more-body interactions are not considered during parameterization (*52*).

MC Simulations

Generally, MC methods include any probabilistic technique that measures a numeric value by a random process. The MC method is based on the drawing of random numbers and was introduced to condensed matter physics in 1953 by Metropolis (*53*). Its principle is to determine representative configurations by randomly sampling the electronic configuration space (*54*). This method, like the MD method, allows for the investigation of physical quantities at the atomic level. The main difference between MD and molecular MC is that the MD method can deal with thermodynamic equilibrium and nonequilibrium phenomena, whereas molecular MC can only calculate the thermodynamic properties of a system at equilibrium through a series of random trial moves without giving any information about the time evolution of the simulated system. Several methods have been developed based on the MC concept—including the Metropolis MC, kinetic MC, and quantum MC—and are actively used in various physical and chemical applications. Metropolis MC is the most widely used method in corrosion inhibition research.

Parameters Derived from MD and MC Simulations of Corrosion Inhibition

Many promising insights have been gained from MD and MC simulations in corrosion inhibition studies over the last few years. Given the adsorptive nature of most corrosion inhibitors, these methods have been successfully used to investigate the adsorption properties of corrosion inhibitors and their interaction with different metals (*2, 5*). The main goals have been to gain insight into the corrosion inhibition mechanism and to find a link between experimental results and some key parameters derived from atomistic simulations.

The interaction strength between inhibitor molecules and metal surfaces can be estimated from MD simulations by determining the interaction energy (E_{inter}), which can be defined as the required energy for one mole of an inhibitor molecule to be adsorbed onto a metal surface (55). It is generally accepted that a more negative E_{inter} indicates a stronger interaction between an inhibitor molecule and a metal surface. This parameter can be calculated for simulated systems under vacuum and in the presence of a solvent using eqs 22 and 23 (56, 57). The negative value of the E_{inter} gives the binding energy as shown in eq 24 (58, 59).

$$E_{inter} = E_{Total} - \left(E_{Surface} + E_{inh}\right) \tag{22}$$

$$E_{inter} = E_{Total} - \left(E_{Surface+solvent} + E_{inh}\right) \tag{23}$$

$$E_{binding} = -E_{inter} \tag{24}$$

where E_{Total}, $E_{Surface}$, $E_{Surface+solution}$, and E_{inh} denote the total energy of the simulated system, surface without solution, surface with solution, and inhibitor molecule alone, respectively.

Static properties from MD simulations have also been used to interpret the adsorption of inhibitor molecules over metal surfaces. The most widely used thermodynamic quantity is the radial distribution function (RDF), noted as *g*(*r*). The RDF, defined as the probability distribution of an atom in a spherical volume with a radius of *r* in a random system of the same density, can be obtained by a structural analysis of MD trajectories. The interpretation of RDF results for corrosion inhibition studies is based on the estimation of bond length. The most prominent first peak located at a distance around 1–3.5 Å indicates a small bond length and a potential formation of a covalent bond between inhibitor molecule and metal surface, whereas a peak located at a distance of greater than 3.5 Å indicates a physical interaction (29). Theoretically, the RDF is defined by Hansen and McDonald's equation as follows (29).

$$g(r) = \frac{1}{\langle \rho_B \rangle_{local}} \times \frac{1}{N_A} \sum_{i \in A}^{N_A} \sum_{j \in B}^{N_B} \frac{\delta(r_{ij} - r)}{4\pi r^2} \tag{25}$$

where $\langle \rho_B \rangle_{local}$ represents the particle density of B average overall shells beside particle A.

Diffusion properties from MD simulations can also be applied to investigate the ability of an inhibitor molecule to protect a metal surface. This concept is based on the general principle that an effective corrosion inhibitor can hinder the diffusion of corrosive particles, thereby limiting their effect on the metal surface. Based on this concept, researchers have applied the mean square displacement (MSD) to calculate the diffusion coefficient of corrosive particles inside a simulated inhibition film and compare it with that in water (59). Suppose the diffusion coefficient of a corrosive particle in an inhibition film is higher or like its diffusion in water. In that case, it can easily penetrate the inhibitor film, and vice versa if it is low. eq 26 is the general formula of the diffusion coefficient, and its value can be obtained through a linear fit of the MSD plot and eq 27 (59):

$$D = \frac{1}{6} \lim_{t \to \infty} \frac{dMSD(t)}{dt}; \text{ where } MSD(t) = \left[\frac{1}{N} \sum_{i=1}^{N} \left| R_i(t) - R_i(0) \right|^2 \right] \tag{26}$$

$$D = \frac{m}{6} \tag{27}$$

$R_i(t)$ denotes the position vector of the *i*th particle at time *t*; *N* represents the total number of diffusion particles; the term $|R_i(t) - R_i(0)|^2$ is the ensemble average of the MSD; and *m* is the slope of the MSD plot.

Using the same simulation procedure as for getting the MSD, the free volume principle can be applied to investigate a simulated inhibition film. The importance of this property is related to the effect of pores and cavities inside an inhibitor film on the diffusion of corrosive particles. An inhibitor film with large cavities, that is, one with a high free volume percentage, can facilitate the diffusion of corrosive particles, whereas smaller cavities can limit their diffusion (*57*, *59*). The fractional free volume (FFV) of an inhibition film is calculated as follows:

$$FFV(\%) = \frac{V_{free}}{V_{free} + V_{occ}} \tag{28}$$

where V_{free} and V_{occ} are the free volume accessible to a probe particle and the total occupied volume of the inhibition film, respectively.

In the case of MC simulations of corrosion inhibition, the outputs mainly include a few energies, such as the total energy (which is the sum of energies of all adsorbate components, the deformation energy, and the adsorption energy). A representative example is shown in Table 2. Khaled and El-Maghraby calculated the MC simulation results reported in Table 2 for three furan derivatives (Inhibitors A–C) used as corrosion inhibitors for mild steel in 1.0 M HCl (*60*).

Table 2. Outputs and Descriptors Calculated by the MC Simulation for Adsorption of Furan Derivatives on Iron (0 0 1). Source: Reproduced with permission from reference (*60*). Inh, inhibitor. Copyright 2021 Elsevier.

Inhibitor	*Total energy (kJ mol⁻¹)*	*Adsorption energy (kJ mol⁻¹)*	*Rigid adsorption energy (kJ mol⁻¹)*	*Deformation energy (kJ mol⁻¹)*	*dE_{ads}/dNi (kJ mol⁻¹)*
Inh. C	−195.5	−24.3	−22.0	−2.3	−24.3
Inh. B	−280.3	−26.1	−24.6	−1.5	−26.1
Inh. A	−310.2	−28.2	−27.31	−0.99	−28.2

Atomistic Simulation Today

The growth of interest in atomistic simulations of corrosion inhibition has been fueled by the increased development of hardware and computer techniques, which has allowed the development of a wide range of commercial and noncommercial software. This has allowed researchers to perform a wide range of simulations using a longer simulation time and multiple simulations runs in a shorter period. The number of publications using atomistic simulations in the field of corrosion inhibition is now in the hundreds. For instance, Figure 5 shows the number of publications per year of corrosion inhibition studies using MD simulations. So, a comprehensive listing in this chapter would not be helpful for readers of this book. Instead, most of the works using atomistic simulations will be reviewed in the following chapters of this book when discussing the different types of corrosion inhibitors. However, we have chosen to describe two specific examples from our work where atomistic simulations, especially MD simulations, have been used to study the adsorption properties of corrosion inhibitors.

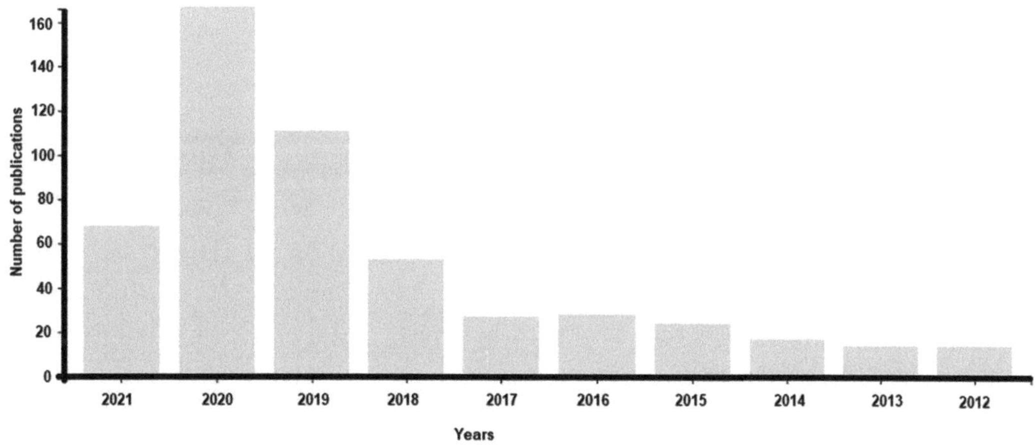

Figure 5. A graphical representation showing 554 records for the topics "molecular dynamics" AND "corrosion inhibition" based on the Web of Science database.

In recent work, Lgaz et al. applied MD simulations to investigate the adsorption properties of four organic compounds (HDZ^1–HDZ^4) that belong to hydrazone family used as corrosion inhibitors for mild steel in HCl (*29*). To this end, the authors constructed a simulation box composed of two layers: One layer included one inhibitor molecule, water molecules, and corrosive particles (i.e., 9 Cl^- and 9 H_3O^+), and the other was a pure iron surface [Fe(110) plan] as a representative of mild steel. The MD simulations have been performed using the Forcite code implemented in Materials Studio software for both neutral and protonated molecules. Some simulation details include the NVT, the Condensed-phase Optimized Molecular Potentials for Atomistic Simulation Studies force field, and a simulation time of 2000 ps. As shown in Figure 6, the authors found that, upon adsorption, neutral inhibitor molecules adopted a flat orientation over the iron surface, an orientation that can maximize their interaction with the iron atoms in the surface. In addition, MD outputs showed that the HDZ^1 had the strongest negative E_{inter} , followed by HDZ^2, HDZ^3, and then HDZ^4, which is consistent with experimental results. Results for protonated molecules were also in agreement with neutral ones but showed less E_{inter} . For instance, the E_{inter} of neutral HDZ^1 was –912.023 kJ/mol, whereas that of protonated form was decreased to –822.231 kJ/mol.

The authors also performed a structural analysis of MD trajectories and determined the total RDF of all adsorbed molecules, as represented in Figure 7. The authors found that all first $g(r)$ peaks for both neutral and protonated forms were located at a distance of less than 3.5 Å. The authors concluded that both forms of adsorbed molecules had a strong affinity toward iron atoms, thereby having a great chance of covalent bond formation between the molecules' active sites and any vacant d orbitals of iron.

In another work, Lgaz et al.(*59*) reported on the experimental and theoretical investigation of Fenugreek gum (FG) as a corrosion inhibitor for mild steel in an HCl medium. In the MD simulation part, the authors constructed a simulation box with a size of $25.68 \times 25.68 \times 25.68$ Å^3, including 30 FG monomers, Cl^- and H_3O^+, for studying these diffusion particles inside the constructed inhibitor film. MD simulations have been performed using the Forcite module of Materials Studio using a 500 ps for equilibrium followed by a 1000 ps for production under the NVT ensemble and a time-step of 1 fs. The constructed inhibition films and MSD plots obtained after MD simulations are shown in Figure 8.

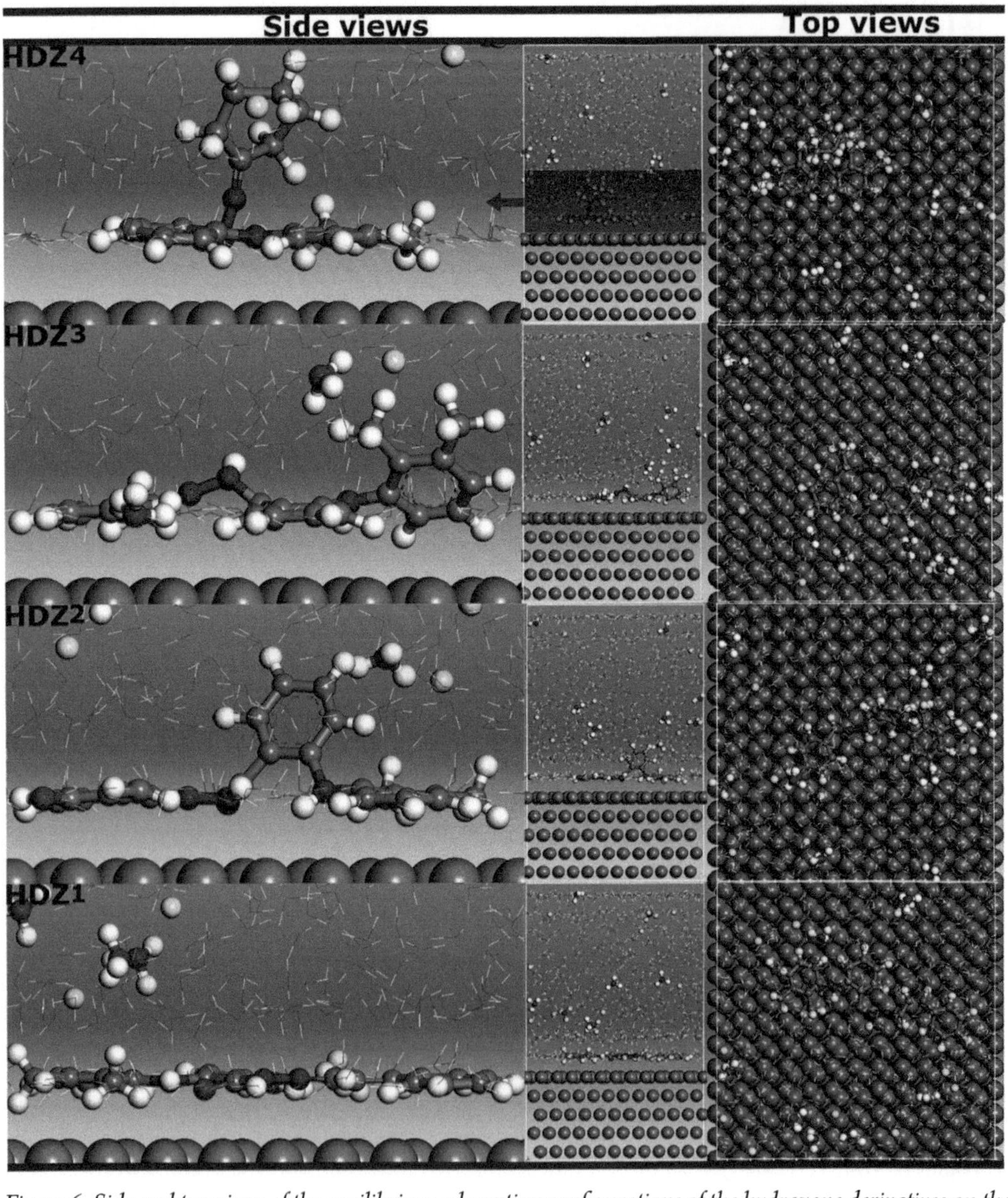

Figure 6. Side and top views of the equilibrium adsorption configurations of the hydrazone derivatives on the Fe(110) surface in solution. Reproduced with permission from reference (29). Copyright 2021 Elsevier.

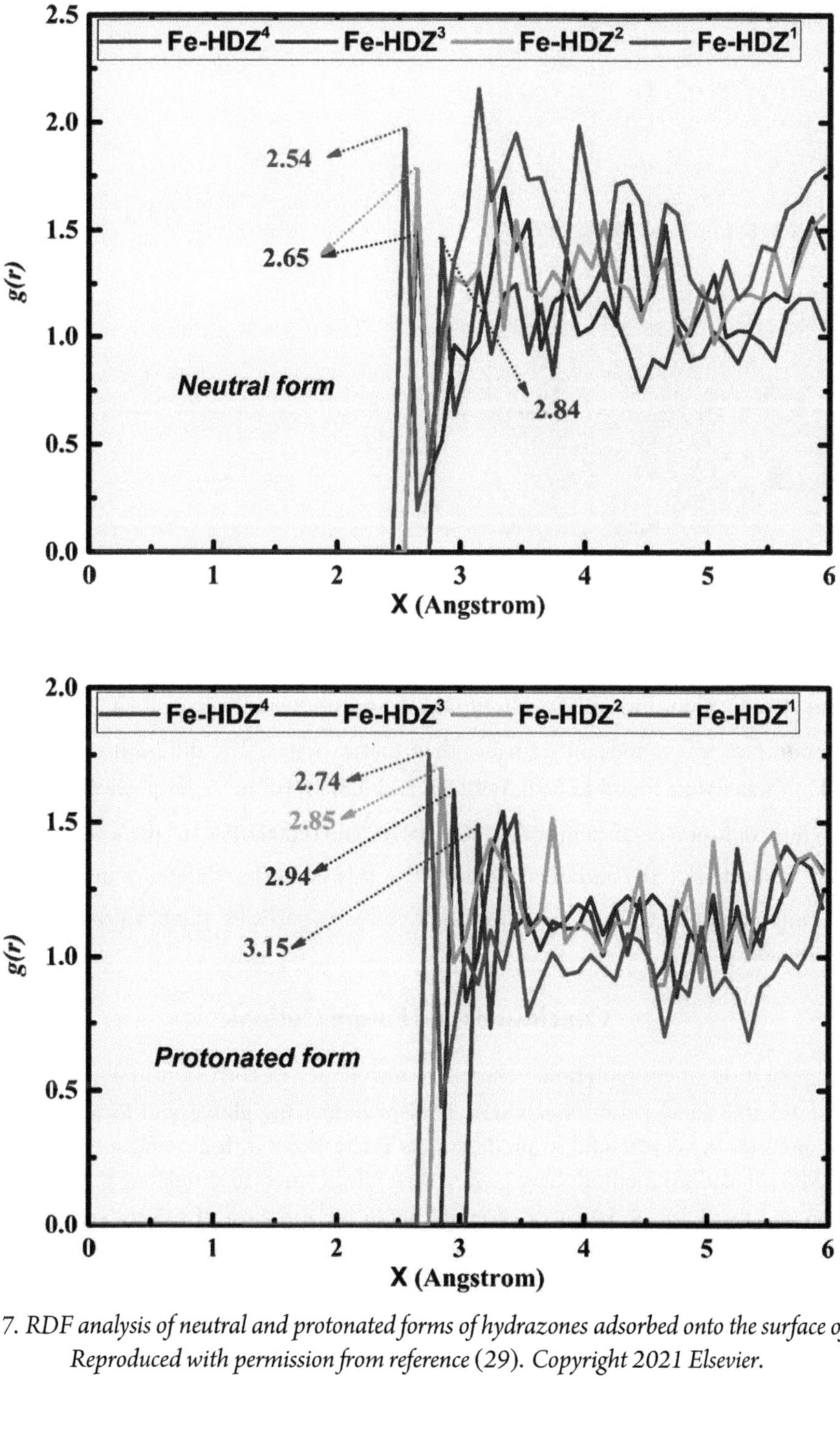

Figure 7. RDF analysis of neutral and protonated forms of hydrazones adsorbed onto the surface of iron. Reproduced with permission from reference (29). Copyright 2021 Elsevier.

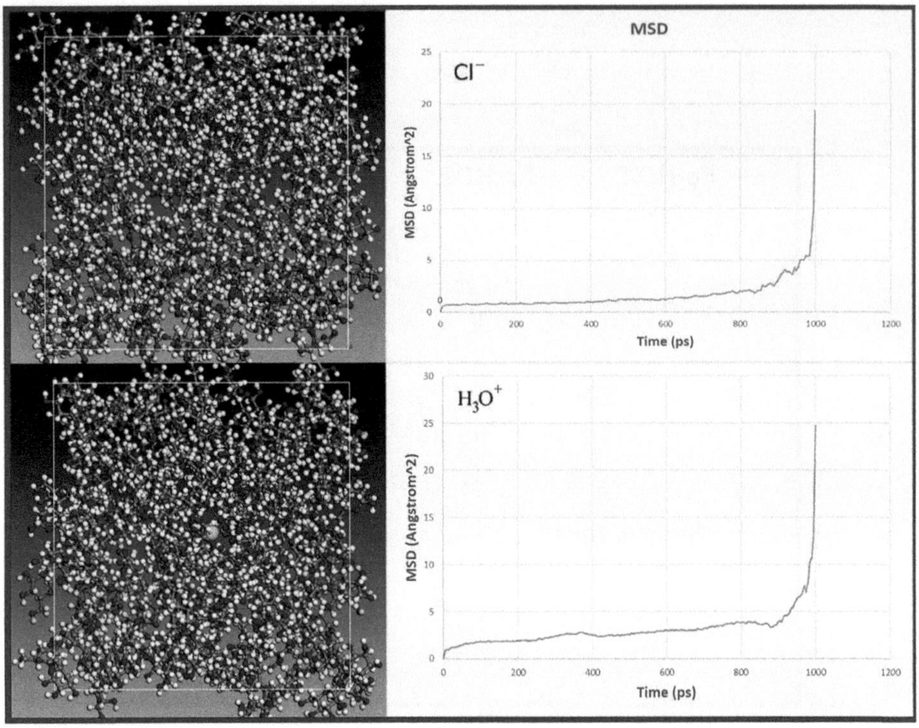

Figure 8. Diffusion model and MSD plots of corrosive particles in the inhibitor membrane. Reproduced with permission from reference (59). Copyright 2021 Elsevier.

In this investigation, the authors found that the diffusion coefficient of Cl^- and H_3O^+ in the FG inhibition film was significantly smaller than that in water. The diffusion coefficients of Cl^- and H_3O^+ in water were found to be 0.349×10^9 and 0.874×10^9 m^2/s, respectively. Inside the FG inhibition film, diffusion coefficients of 0.000376×10^9 and 0.000205×10^9 m^2/s were found for Cl^- and H_3O^+, respectively. The authors concluded that this significant difference indicates the ability of the FG inhibition film to limit the diffusion of corrosive particles, thereby protecting mild steel against corrosion.

Conclusions and Future Outlook

The application of computational chemistry techniques in corrosion inhibition studies have attracted (and still does) many researchers. Understanding the global and local reactivity of an inhibitor molecule is very helpful in predicting its interactions with a metal surface. In addition, MD and MC simulations methods have proved their effectiveness in simulating the metal–inhibitor interaction and the diffusion behavior of inhibitors in the presence of solvent species. While the DFT method focus on the inhibitor itself, atomic simulation methods tend to mimic the corrosion inhibition process, providing useful insight about the adsorption of inhibitors over metals. However, in spite of the considerable progress made in using these computational techniques, it remains challenging to theoretically predict the corrosion inhibition mechanism because of its complexity. The most notable limitation of reactivity descriptors derived from DFT calculations is their reliance on the molecular structure of the inhibitor without considering the metal surface, indicating that quantum chemical parameters cannot always predict the experimental corrosion inhibition performance. On the other hand, MD and MC simulations are unable to describe bond formation/breaking. In response to these limitations, several attempts were initiated involving the development of the reactive force fields by MD and ab initio DFT methods. Further research efforts should focus on developing reactive force fields for a wide range of adsorption systems and making ab initio DFT calculations possible for large-sized molecules.

Suggested Reading

Lewars, E. (2003). Computational chemistry. Introduction to the theory and applications of molecular and quantum mechanics. Kluwer Academic, 318.

Cramer, C. J. (2013). Essentials of computational chemistry: theories and models. John Wiley & Sons.

Satoh, A. (2010). Introduction to practice of molecular simulation: MD, MC, Brownian dynamics, Lattice Boltzmann and dissipative particle dynamics. Elsevier.

Rapaport, D. C. (2004). The art of MD simulation. Cambridge University Press.

Acknowledgments

This research was supported by a basic science research program through the National Research Foundation of Korea funded by the Ministry of Science, ICT, and Future Planning (No. 2015R1A5A1037548).

References

1. Obot, I.; Macdonald, D.; Gasem, Z. M. Density functional theory (DFT) as a powerful tool for designing new organic corrosion inhibitors. Part 1: an overview. *Corrosion Science* **2015**, *99*, 1–30.
2. Quraishi, M. A.; Chauhan, D. S.; Saji, V. S., 3 - Computational methods of inhibitor evaluation. In *Heterocyclic Organic Corrosion Inhibitors*, Quraishi, M. A.; Chauhan, D. S.; Saji, V. S., Eds. Elsevier: 2020; pp 59-86.
3. Sasikumar, Y.; Adekunle, A. S.; Olasunkanmi, L. O.; Bahadur, I.; Baskar, R.; Kabanda, M. M.; Obot, I. B.; Ebenso, E. E. Experimental, quantum chemical and Monte Carlo simulation studies on the corrosion inhibition of some alkyl imidazolium ionic liquids containing tetrafluoroborate anion on mild steel in acidic medium. *Journal of Molecular Liquids* **2015**, *211*, 105–118.
4. Shariatinia, Z.; Ahmadi-Ashtiani, A. Corrosion inhibition efficiency of some phosphoramide derivatives: DFT computations and MD simulations. *Journal of Molecular Liquids* **2019**, *292*.
5. Verma, C.; Lgaz, H.; Verma, D. K.; Ebenso, E. E.; Bahadur, I.; Quraishi, M. A. Molecular dynamics and Monte Carlo simulations as powerful tools for study of interfacial adsorption behavior of corrosion inhibitors in aqueous phase: A review. *Journal of Molecular Liquids* **2018**, *260*, 99–120.
6. Kokalj, A.; Peljhan, S.; Finsgar, M.; Milosev, I. What determines the inhibition effectiveness of ATA, BTAH, and BTAOH corrosion inhibitors on copper? *J. Am. Chem. Soc.* **2010**, *132* (46), 16657–16668.

7. Kozlica, D. K.; Kokalj, A.; Milošev, I. Synergistic effect of 2-mercaptobenzimidazole and octylphosphonic acid as corrosion inhibitors for copper and aluminium – An electrochemical, XPS, FTIR and DFT study. *Corrosion Science* **2020**, 109082.
8. Chafiq, M.; Chaouiki, A.; Lgaz, H.; Salghi, R.; Gaonkar, S. L.; Bhat, K. S.; Marzouki, R.; Ali, I. H.; Khan, M. I.; Shimizu, H.; Chung, I.-M. Synthesis and corrosion inhibition evaluation of a new schiff base hydrazone for mild steel corrosion in HCl medium: electrochemical, DFT, and molecular dynamics simulations studies. *Journal of Adhesion Science and Technology* **2020**, *34* (12), 1283–1314.
9. Gouron, A.; Le Mapihan, K.; Camperos, S.; Al Farra, A.; Lair, V.; Ringuedé, A.; Cassir, M.; Diawara, B. New insights in self-assembled monolayer of imidazolines on iron oxide investigated by DFT. *Applied Surface Science* **2018**, *456*, 437–444.
10. Belghiti, M. E.; Echihi, S.; Mahsoune, A.; Karzazi, Y.; Aboulmouhajir, A.; Dafali, A.; Bahadur, I. Piperine derivatives as green corrosion inhibitors on iron surface; DFT, Monte Carlo dynamics study and complexation modes. *Journal of Molecular Liquids* **2018**, *261*, 62–75.
11. Shechtman, D.; Blech, I.; Gratias, D.; Cahn, J. W. Metallic Phase with Long-Range Orientational Order and No Translational Symmetry. *Physical Review Letters* **1984**, *53* (20), 1951–1953.
12. Thomas, L. H. The calculation of atomic fields. *Mathematical Proceedings of the Cambridge Philosophical Society* **1927**, *23* (5), 542–548.
13. Kohn, W.; Sham, L. J. Self-Consistent Equations Including Exchange and Correlation Effects. *Physical Review* **1965**, *140* (4A), A1133–A1138.
14. Pearson, R. G. Hard and soft acids and bases. *J. Am. Chem. Soc.* **1963**, *85* (22), 3533–3539.
15. Gázquez, J. L. The hard and soft acids and bases principle. *The Journal of Physical Chemistry A* **1997**, *101* (26), 4657–4659.
16. Görling, A. Density-functional theory beyond the Hohenberg-Kohn theorem. *Phys. Rev. A* **1999**, *59* (5), 3359.
17. Hohenberg, P.; Kohn, W. Inhomogeneous electron gas. *Phys. Rev.* **1964**, *136* (3B), B864.
18. Kohn, W.; Sham, L. J. Quantum density oscillations in an inhomogeneous electron gas. *Phys. Rev.* **1965**, *137* (6A), A1697.
19. Slater, J. C. A simplification of the Hartree-Fock method. *Phys. Rev.* **1951**, *81* (3), 385.
20. Wu, Z.; Cohen, R. E. More accurate generalized gradient approximation for solids. *Phys. Rev. B* **2006**, *73* (23), 235116.
21. Lee, C.; Yang, W.; Parr, R. G. Development of the Colle-Salvetti correlation-energy formula into a functional of the electron density. *Phys. Rev. B* **1988**, *37* (2), 785.
22. Fukui, K. Role of frontier orbitals in chemical reactions. *Science* **1982**, *218* (4574), 747–754.
23. Koopmans, T. Über die Zuordnung von Wellenfunktionen und Eigenwerten zu den einzelnen Elektronen eines Atoms. *Physica* **1934**, *1* (1-6), 104–113.
24. Chaouiki, A.; Chafiq, M.; Rbaa, M.; Lgaz, H.; Salghi, R.; Lakhrissi, B.; Ali, I. H.; Masroor, S.; Cho, Y. New 8-Hydroxyquinoline-Bearing Quinoxaline Derivatives as Effective Corrosion Inhibitors for Mild Steel in HCl: Electrochemical and Computational Investigations. *Coatings* **2020**, *10* (9), 811.
25. Iczkowski, R. P.; Margrave, J. L. Electronegativity. *J. Am. Chem. Soc.* **1961**, *83* (17), 3547–3551.

26. Parr, R. G.; Pearson, R. G. Absolute hardness: companion parameter to absolute electronegativity. *J. Am. Chem. Soc.* **1983,** *105* (26), 7512–7516.
27. Morell, C.; Gázquez, J. L.; Vela, A.; Guégan, F.; Chermette, H. Revisiting electroaccepting and electrodonating powers: proposals for local electrophilicity and local nucleophilicity descriptors. *Phys. Chem. Chem. Phys.* **2014,** *16* (48), 26832–26842.
28. Lukovits, I.; Kalman, E.; Zucchi, F. Corrosion inhibitors—correlation between electronic structure and efficiency. *Corrosion* **2001,** *57* (1), 3–8.
29. Lgaz, H.; Salghi, R.; Masroor, S.; Kim, S.-H.; Kwon, C.; Kim, S. Y.; Yang, Y.-J.; Chung, I.-M. Assessing corrosion inhibition characteristics of hydrazone derivatives on mild steel in HCl: Insights from electronic-scale DFT and atomic-scale molecular dynamics. *Journal of Molecular Liquids* **2020,** *308*, 112998.
30. Meller, J. *Molecular dynamics.* **2001**.
31. Hansson, T.; Oostenbrink, C.; van Gunsteren, W. Molecular dynamics simulations. *Curr. Opin. Struct. Biol.* **2002,** *12* (2), 190–196.
32. Leimkuhler, B.; Matthews, C., *Molecular Dynamics*. Springer: 2016.
33. Hollingsworth, S. A.; Dror, R. O. Molecular Dynamics Simulation for All. *Neuron* **2018,** *99* (6), 1129–1143.
34. Mathews, D. H.; Sabina, J.; Zuker, M.; Turner, D. H. Expanded sequence dependence of thermodynamic parameters improves prediction of RNA secondary structure. *Journal of molecular biology* **1999,** *288* (5), 911–40.
35. Zuker, M. Calculating nucleic acid secondary structure. *Current opinion in structural biology* **2000,** *10* (3), 303–10.
36. Bahlakeh, G.; Ramezanzadeh, M.; Ramezanzadeh, B. Experimental and theoretical studies of the synergistic inhibition effects between the plant leaves extract (PLE) and zinc salt (ZS) in corrosion control of carbon steel in chloride solution. *Journal of Molecular Liquids* **2017**, *248*, 854–870.
37. Chafai, N.; Chafaa, S.; Benbouguerra, K.; Daoud, D.; Hellal, A.; Mehri, M. Synthesis, characterization and the inhibition activity of a new alpha-aminophosphonic derivative on the corrosion of XC48 carbon steel in 0.5 M H2SO4: Experimental and theoretical studies. *Journal of the Taiwan Institute of Chemical Engineers* **2017**, *70*, 331–344.
38. Fouda, A. S.; Ismail, M. A.; Abousalem, A. S.; Elewady, G. Y. Experimental and theoretical studies on corrosion inhibition of 4-amidinophenyl-2,2′-bifuran and its analogues in acidic media. *Rsc Advances* **2017**, 7 (73), 46414–46430.
39. Heakal, F. E.; Attia, S. K.; Rizk, S. A.; Abou Essa, M. A.; Elkholy, A. E. Synthesis, characterization and computational chemical study of novel pyrazole derivatives as anticorrosion and antiscalant agents. *Journal of Molecular Structure* **2017**, *1147*, 714–724.
40. Lgaz, H.; Salghi, R.; Jodeh, S.; Hammouti, B. Effect of clozapine on inhibition of mild steel corrosion in 1.0 M HCl medium. *Journal of Molecular Liquids* **2017**, *225*, 271–280.
41. Meng, Y.; Ning, W. B.; Xu, B.; Yang, W. Z.; Zhang, K. G.; Chen, Y.; Li, L. H.; Liu, X.; Zheng, J. H.; Zhang, Y. M. Inhibition of mild steel corrosion in hydrochloric acid using two novel pyridine Schiff base derivatives: a comparative study of experimental and theoretical results. *Rsc Advances* **2017**, 7 (68), 43014–43029.

42. Sanaei, Z.; Bahlakeh, G.; Ramezanzadeh, B. Active corrosion protection of mild steel by an epoxy ester coating reinforced with hybrid organic/inorganic green inhibitive pigment. *Journal of Alloys and Compounds* **2017**, *728*, 1289–1304.
43. Zhang, C.; Zhao, J. M. Synergistic inhibition effects of octadecylamine and tetradecyl trimethyl ammonium bromide on carbon steel corrosion in the H2S and CO2 brine solution. *Corrosion Science* **2017**, *126*, 247–254.
44. Ivani, I.; Dans, P. D.; Noy, A.; Pérez, A.; Faustino, I.; Hospital, A.; Walther, J.; Andrio, P.; Goñi, R.; Balaceanu, A. Parmbsc1: a refined force field for DNA simulations. *Nature Methods* **2016**, *13* (1), 55.
45. Case, D. A.; Cheatham, T. E., III; Darden, T.; Gohlke, H.; Luo, R.; Merz, K. M., Jr.; Onufriev, A.; Simmerling, C.; Wang, B.; Woods, R. J. *The Amber biomolecular simulation programs.* **2005**, *26* (16), 1668–1688.
46. Casewit, C. J.; Colwell, K. S.; Rappe, A. K. Application of a universal force field to main group compounds. *Journal of the American Chemical Society* **1992**, *114* (25), 10046–10053.
47. Rappe, A. K.; Casewit, C. J.; Colwell, K. S.; Goddard, W. A.; Skiff, W. M. UFF, a full periodic table force field for molecular mechanics and molecular dynamics simulations. *Journal of the American Chemical Society* **1992**, *114* (25), 10024–10035.
48. Hagler, A. T.; Huler, E.; Lifson, S. Energy functions for peptides and proteins. I. Derivation of a consistent force field including the hydrogen bond from amide crystals. *Journal of the American Chemical Society* **1974**, *96* (17), 5319–5327.
49. Hagler, A. T.; Lifson, S. Energy functions for peptides and proteins. II. Amide hydrogen bond and calculation of amide crystal properties. *Journal of the American Chemical Society* **1974**, *96* (17), 5327–5335.
50. Lifson, S.; Hagler, A. T.; Dauber, P. Consistent force field studies of intermolecular forces in hydrogen-bonded crystals. 1. Carboxylic acids, amides, and the C:O.cntdot..cntdot..cntdot.H- hydrogen bonds. *Journal of the American Chemical Society* **1979**, *101* (18), 5111–5121.
51. Sun, H. COMPASS: an ab initio force-field optimized for condensed-phase applications overview with details on alkane and benzene compounds. *The Journal of Physical Chemistry B* **1998**, *102* (38), 7338–7364.
52. Brooks, C. L. Computer simulation of liquids. *Journal of Solution Chemistry* **1989**, *18* (1), 99–99.
53. Hitchcock, D. B. A History of the Metropolis–Hastings Algorithm. *The American Statistician* **2003**, *57* (4), 254–257.
54. Rapaport, D. C., *The art of molecular dynamics simulation*. Cambridge university press: 2004.
55. Saha, S. K.; Dutta, A.; Ghosh, P.; Sukul, D.; Banerjee, P. Novel Schiff-base molecules as efficient corrosion inhibitors for mild steel surface in 1 M HCl medium: experimental and theoretical approach. *Physical Chemistry Chemical Physics* **2016**, *18* (27), 17898–17911.
56. Lgaz, H.; Bhat, K. S.; Salghi, R.; Shubhalaxmi; Jodeh, S.; Algarra, M.; Hammouti, B.; Ali, I. H.; Essamri, A. Insights into corrosion inhibition behavior of three chalcone derivatives for mild steel in hydrochloric acid solution. *Journal of Molecular Liquids* **2017**, *238*, 71–83.

57. Lgaz, H.; Salghi, R.; Bhat, K. S.; Chaouiki, A.; Shubhalaxmi; Jodeh, S. Correlated experimental and theoretical study on inhibition behavior of novel quinoline derivatives for the corrosion of mild steel in hydrochloric acid solution. *Journal of Molecular Liquids* **2017**, *244*, 154–168.
58. Chugh, B.; Singh, A. K.; Thakur, S.; Pan, B.; Pandey, A. K.; Lgaz, H.; Chung, I. M.; Ebenso, E. E. An Exploration about the Interaction of Mild Steel with Hydrochloric Acid in the Presence of N-(Benzo d thiazole-2-yl)-1-phenylethan-1-imines. *Journal of Physical Chemistry C* **2019**, *123* (37), 22897–22917.
59. Lgaz, H.; Chung, I. M.; Salghi, R.; Ali, I. H.; Chaouiki, A.; El Aoufir, Y.; Khan, M. I. On the understanding of the adsorption of Fenugreek gum on mild steel in an acidic medium: Insights from experimental and computational studies. *Applied Surface Science* **2019**, *463*, 647–658.
60. Khaled, K. F.; El-Maghraby, A. Experimental, Monte Carlo and molecular dynamics simulations to investigate corrosion inhibition of mild steel in hydrochloric acid solutions. *Arabian Journal of Chemistry* **2014**, 7 (3), 319–326.

Chapter 7

Corrosion Inhibitors for Acidic Environments

Manilal Murmu,[1,2] Sourav Kr. Saha,[3] Naresh Chandra Murmu,[1,2] and Priyabrata Banerjee[1,2,*]

[1]Surface Engineering & Tribology Group,
Council of Scientific and Industrial Research (CSIR)-Central Mechanical Engineering Research Institute, Mahatma Gandhi Avenue, Durgapur 713209, India

[2]Academy of Scientific and Innovative Research (AcSIR), AcSIR Headquarters, CSIR-Human Resource Development Centre Campus, Sector-19, Kamla Nehru Nagar, Ghaziabad 201002, India

[3]Department of Materials Science and Engineering, Pusan National University, Busan 46241, Republic of Korea

[*]Email: pr_banerjee@cmeri.res.in.

The cost of corrosion and its consequent catastrophes have adversely affected the economy of almost all nations across the globe, which is reflected in the fractional reduction in gross domestic products. The use of corrosion inhibitors derived from organic materials is the most practiced as well as most effective strategy for protecting metallic materials from corrosive acidic environments. The corrosion inhibitors possess several active sites capable of adsorbing onto susceptible metal surfaces exposed to corrosive environments by displacing the preadsorbed corrosive ingredients, and they act as protective layers upon adsorption, which inhibits the further ingress of corrosive ingredients toward the metal surface. The presence of planar geometry, aromatic moieties, π bonds, heteroatoms having a lone pair of electrons, and also sometimes enriching moieties contributes to the enhancement of their adsorption followed by the corrosion-inhibiting properties of the corrosion inhibitors. The chapter covers the most progress in the development of promising corrosion inhibitors analyzed for their applicability in monobasic acidic media, such as an aqueous solution of hydrochloric acid and nitric acid. The mechanism of corrosion and its mitigation have been highlighted based on experimental observations and correlated with insightful theoretical approaches.

Introduction

Equipment made of metals or its alloys is extensively utilized in several industrial activities, such as chemical processing (*1*), metal processing (*2*), petroleum refineries (*3*), marine industries (*4*), biomedical implants (*5*), construction work (*6*), and several other industries (*7, 8*). For the maintenance of such metallic materials, several industrial practices, such as acid cleansing, pickling, descaling, or other techniques, are executed by using acidic solutions mostly of hydrochloric acid, nitric acid, and sulfuric acid (*9, 10*). But it has been frequently reported that when the metallic materials (iron and others) are exposed to such solutions, corrosion takes place. For instance, corrosion in petroleum (*11*), marine (*12*), construction (*13*), and several other industries is frequently observed. Similarly, aluminum alloys are also widely utilized in several fields, including household appliances, military hardware, ship building, and automotive, aviation, and aerospace industries (*14–18*), because of their low density, desired mechanical strength, and appreciable corrosion-inhibiting properties (*19–21*). The mechanical stability and resistive behavior of an aluminum alloy are attributed to the dispersion of intermetallic phases formed due to the presence of elemental compositions of copper, magnesium, silicon, iron, and manganese in the alloy's bulk (*20, 21*). But when these alloy materials are exposed to aggressive media containing hydronium (H_3O^+) ions, chloride (Cl^-) ions, and others, they cause the intermetallic materials to corrode locally (*22–24*).

The use of corrosion-inhibiting materials as additives is the most practicable approach to combat the degradation of metallic materials from such corrosive environments. There have been a plenty of corrosion inhibitors based on organic molecules possessing several types of functional groups and geometrical structures, which have been developed and synthesized by various synthetic procedures for their use as corrosion inhibitors. In this work, the most importantly synthesized or explored organic molecule-based corrosion-inhibiting additives are collectively discussed with the progress of the determining techniques. This discussion of wet chemical and electrochemically determined corrosion-inhibiting properties exhibited in the studied corrosion-inhibiting molecules is correlated with theoretical insights. The corrosive dissolution of selective metallic materials is discussed. An overview of the corrosion-inhibiting nature of additives in monobasic corrosive acidic solutions at various ranges of temperature and various exposure times is essential for researchers involved in the corrosion-mitigating arena. The correlation of experimentally determined corrosion inhibition with theoretically extracted outcomes is very necessary. An overview of the significant literature, herein, dealing in the progressive exploration or development of the analysis or investigation in corrosion inhibitors and inhibition mechanism elucidation have been emphasized. This chapter also highlights trends in the exploration and elucidation of the intrinsic molecular properties determined from quantum chemical calculations of the used corrosion-inhibiting molecules. The simulation and visualization of the interacting interfaces between corrosion-inhibiting molecules and metals' surface atoms in the absence and presence of various corrosive species explored using several simulation techniques, and insights on protection capability are explained and discussed.

Corrosion in Acidic Environments

Aqueous solutions of hydrochloric acid (HCl) and nitric acid (HNO_3) are mostly used for acid cleansing and pickling of various metallic materials in several industries. HCl is a weak reducing acid, and the complexing ability of a chloride ion (Cl^-) makes it an excellent solvent for metal oxides. Likewise, HNO_3 is also an oxidizing acid, meaning it dissolves most metals by forming soluble metal nitrates. But after their cleansing actions, the nascent metal surfaces are vulnerable to corrosive dissolution, and metallic corrosion is subsequently initiated. It is well known that when metallic materials come in contact with the aggressive or corrosive environment, the dissolution of the exposed surface takes place with the extent of exposure (*25–29*). This phenomenon is referred to as corrosion. The mechanism of the corrosion of some selected metallic materials exposed into the corrosive acidic solution is discussed in the subsequent section.

Corrosion Mechanism in a Hydrochloric Acid Medium

In an aqueous hydrochloric acid (HCl) solution, metallic materials undergo oxidative dissolution. The following reactions may be expected during the electrochemical reactions at the anodic sites (*30, 31*):

$$Fe + Cl^- \rightarrow (FeCl^-)_{ads} \quad (1)$$

$$(FeCl^-)_{ads} \leftrightarrow (FeCl)_{ads} + e^- \quad (2)$$

$$(FeCl)_{ads} \leftrightarrow FeCl^+ + e^- \quad (3)$$

$$FeCl^+ \leftrightarrow Fe^{2+} + Cl^- \quad (4)$$

The evolution of hydrogen takes place in the cathode by the following reactions (*30, 31*):

$$Fe + H^+ \leftrightarrow (FeH^+)_{ads} \quad (5)$$

$$(FeH^+)_{ads} + e^- \rightarrow (FeH)_{ads} \quad (6)$$

$$Fe(H)_{ads} + H^+ + e^- \rightarrow Fe + H_2 \quad (7)$$

In the case of aluminum alloys, it is known that alumina gain stability in 4–9 pH ranges, while in acidic media—for instance, a 0.1 M HCl solution (i.e., pH = 1)—the superficial alumina of an AA6060 alloy has been reported to experience a chemical reaction with H^+ ions represented as:

$$Al_2O_3(s) + 6H^+(aq) \rightarrow Al^{3+}(aq) + 3H_2O(l) \quad (8)$$

Aluminum alloys are highly positively charged in strong acidic media. Because of this, aggressive anions such as chloride ions present in the corrosive media are prone to adsorb or deposit at the surface of the alloy (*32, 33*). These adsorbed chloride ions initiate localized corrosion at the intermetallic regions leading to pitting (*34*) and film breakdown (*35*). In this way, the aluminum matrix is barer and susceptible to corrosive attacks by H^+ and Cl^- ions present in the corrosive media. The anodic reaction of bare aluminum alloy leads to oxidation of Al to Al^{3+} as follows:

$$Al(s) \rightarrow Al^{3+}(aq) + 3e^- \quad (9)$$

Furthermore, chloride ions are chemically adsorbed on positively charged surfaces of metal. Thereafter it accelerates anodic dissolution of the aluminum alloy by forming $AlCl_2^+$ (*36, 37*):

$$Al(s) + Cl^-(aq) \rightarrow AlCl^-(ads) \quad (10)$$

$$AlCl^-(ads) + Cl^-(aq) \rightarrow AlCl_2^+(ads) + 3e^- \quad (11)$$

These formed $AlCl_2^+$ ions diffuse toward a bulk solution, leading the corrosion reaction to continue further.

Alternatively, the electrochemical reaction on the cathodic site continues simultaneously, accompanied by the evolution of hydrogen gas as follows:

$$H^+ + e^- \rightarrow H(ads) \quad (12)$$

$$H(ads) + H(ads) \rightarrow H_2 \quad (13)$$

Furthermore, a reduction of oxygen takes place at the cathode site and is shown as:

$$O_2 + 4H^+ + 4e^+ \rightarrow 2H_2O \quad (14)$$

Corrosion Mechanism in a Nitric Acid Medium

When exposed to a corrosive nitric acid (HNO_3) solution, the corrosive dissolution of metal substrates takes place. The HNO_3 remains completely ionized to H^+ and NO_3^- ions in an aqueous solution due to its strong acid and strong oxidation properties. The anodic dissolution of iron (Fe) undergoes the following reactions in an aqueous solution of 0.1 M HNO_3 (*37–39*):

$$Fe + NO_3^- \rightarrow Fe(NO_3^-)_{ads} \quad (15)$$

$$Fe(NO_3^-)_{ads} \rightarrow Fe(NO_3)_{ads} + e^- \quad (16)$$

$$Fe(NO_3)_{ads} \rightarrow Fe(NO_3^+) + e^- \quad (17)$$

$$Fe(NO_3^+) + \rightarrow Fe^{2+} + NO_3^- \quad (18)$$

The generated Fe^{2+} ions may be further oxidized to Fe^{3+} ions by NO_3^- and H^+ by the following reactions:

$$Fe^{2+} + 2H^+ + NO_3^- \rightarrow 2Fe^{3+} + 2H_2O + NO_2^- \quad (19)$$

$$3Fe^{2+} + 4H^+ + NO_3^- \rightarrow 3Fe^{3+} + 2H_2O + NO \uparrow \quad (20)$$

The Fe can be also oxidized to Fe^{2+} by Fe^{3+} as follows:

$$Fe + Fe^{3+} \rightarrow 3Fe^{2+} \quad (21)$$

Copper (Cu) or its alloys also undergo a dissolution reaction leading to corrosion. The dissolution mechanism of copper substrates exposed to an HNO_3 solution, as proposed by Turnbull et al., is shown in Figure 1 (*40*). It is reported that initially the oxidation of Cu takes place. Subsequently, the O_2 is reduced (process 1), and it activates NO_3^- as a cathodic reagent and solubilizes Cu^+ ions (process 2). The activated NO_3^- ions facilitate further oxidation of Cu^+ to Cu^{2+} (process 3) through their reduction to NO_2^- (process 4), and the reduced NO_2^- is released into the media (process 5). The rapid formation of NO_2^- and NO (processes 5 and 6) increases the complete dissolution of Cu (process 7). It is noteworthy to mention that in the absence of NO_3^- there is a possibility of Cu_2O passive layer formation by the deposition of Cu^+ at the surface (process 8), which is also capable of preventing the metallic dissolution of copper at the surface. But if the formation of the Cu_2O passive layer is slow, it leads to an increase in the corrosion of copper. Again, if both NO_3^- and NO_2^- are present in the media, then it competes for its adsorption onto copper's surface and interferes with protective oxide layer formation. Consequently, it stimulates the impending occurrence of corrosion phenomenon severely.

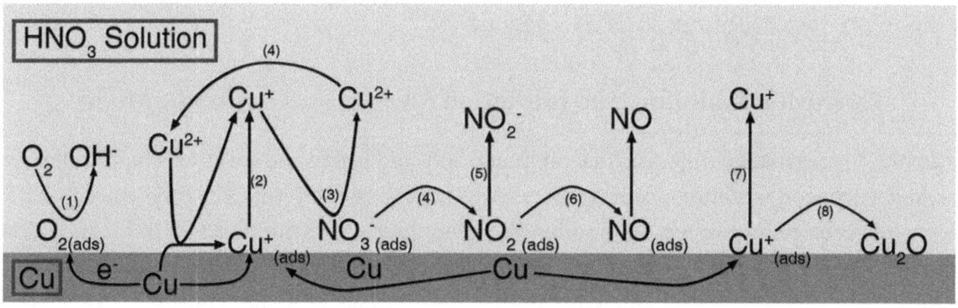

Figure 1. Mechanism of copper corrosion in an HNO_3 medium. (Reproduced with permission from ref (40). Copyright 2020 Elsevier Ltd.)

According to Turnbull et al., in the environment containing O_2 and HNO_3, there is an immense possibility for the passivating layer formation of copper oxides/nitrates (i.e., Cu_2O, CuO, and $Cu_2NO_3(OH)_3$ layer formation), which sometimes protect the substrates from further dissolution (*40, 41*).

The aluminum alloy is easily oxidized by O_2 to form its oxide, Al_2O_3, when exposed to an aqueous HNO_3 medium.

$$Al + \frac{3}{2}O_2 \rightarrow Al_2O_3 \quad (22)$$

This Al_2O_3 layer protects the metallic aluminum, but when exposed to an HNO_3 solution, it is damaged as follows:

$$Al_2O_3 + 3H^+ \rightarrow Al^{3+} + \frac{3}{2}H_2O \quad (23)$$

It was reported that bubbles came out when aluminum alloy was exposed to an HNO_3 solution. This was due to the breakdown of the Al_2O_3 layer and the liberation of H_2, NO, and NO_2 gases produced in the chemical reactions (*42*).

With the increase in the concentration of HNO_3, there is greater oxidative dissolution of the aluminum alloy through the following chemical reactions (*42*):

$$Al + H_2O \leftrightarrow (AlOH)_{ads} + H^+ + e^- \quad (24)$$

$$(AlOH)_{ads} + 5H_2O + H^+ \rightarrow Al^{3+} + 6H_2O + 2e^- \quad (25)$$

$$Al^{3+} + H_2O \rightarrow [AlOH]^{2+} + H^+ \quad (26)$$

The hydrated cation $[AlOH]^{2+}$ complexes with NO_3^- anions, producing a water-soluble complex, $[AlOHNO_3]^+$, by the following reaction (*42*):

$$[AlOH]^{2+} + NO_3^- \rightarrow [AlOHNO_3]^+ \quad (27)$$

The formation of the soluble complexes leads to the uncontrolled dissolution of the aluminum alloy exposed to the concentrated HNO_3 solution.

Corrosion Inhibitors and Inhibition Mechanisms for Acidic Media

Mostly, the corrosion inhibitors that are practically applicable and easily and highly effective are the organic corrosion inhibitors. Among the various corrosion inhibitors, the molecules that possess nitrogen, oxygen, phosphorous, and sulfur as heteroatoms, unsaturated π bonds, and aromatic moieties in their molecular structure are highly capable of adsorbing onto metallic surfaces that encumber corrosive attacks of aggressive media on exposed metallic substrates. The corrosion inhibitors applied in the corrosive electrolytes adsorb onto the surface of the metallic substrate either through physisorption (i.e., electrostatic or van der Waals interactions between corrosion-inhibiting molecules and targeted atoms at the surface of the metal) or through the chemisorption pathway (i.e., electron sharing or coordinate bond formation by unshared electrons with unfilled d orbitals of the metal substrate, or the synergistic back-bonding formation due to the electron retrodonated from the excess electrons from the metal centers to the inhibitor molecules); sometimes both phenomena may take place together (*43–47*).

The molecular adsorption of the inhibitor molecules can be represented as follows (*48*):

$$\text{Inhibitor} + \text{metal surface} \rightarrow \text{Inhibitor–metal surface} \quad (28)$$

Furthermore, the inhibitor molecule may possibly adsorb onto the metal surface with the release of hydrogen atoms (i.e., through the dissociative adsorption pathway), which is shown as follows (*48*):

$$\text{Inhibitor} + \text{metal surface} \rightarrow \text{Inhibitor} - \text{metal surface} + \frac{1}{2}H_2 \quad (29)$$

This chapter focuses on outlining the use of organic compounds as corrosion-inhibiting additives in HCl and HNO_3 media. It is overviewed in subsequent sections.

Corrosion Inhibitors Used in a Hydrochloric Acid Medium

Chloride (Cl^-) ions possess a small degree of hydration. These anions are capable of adsorbing over metal surfaces with positive charges. Owing to this specific adsorption, they expose excess negative charges accumulated on surfaces to the corrosive solution. Alternatively, this phenomenon favors the adsorption of positively charged corrosion-inhibiting molecules onto metal surfaces (*30*, *49*).

In such an acidic solution, these applied corrosion-inhibiting molecules may be protonated as shown in equation (30).

$$\text{Inhibitor} + H^+ \leftrightarrow \text{InhibitorH}^+ \quad (30)$$

The positively charged corrosion-inhibiting molecules, InhibitorH+, interact electrostatically with preadsorbed chloride anions (which are negatively charged) on metal surface atoms as follows (*50, 51*):

$$(\text{FeCl}^-)_{ads} + \text{InhibitorH}^+ \rightarrow (\text{FeCl}^-\text{InhibitorH}^+)_{ads} \quad (31)$$

Additionally, the lone pairs of electrons present in heteroatoms or π electrons of organic molecules undergo interaction with available d orbitals of the metal's surface atom, forming the coordinate bonds that lead to chemisorption. The adsorption of inhibitor molecules also affects cathodic reactions by competing with hydrogen adsorption and reduces the hydrogen evolution reactions at the cathode.

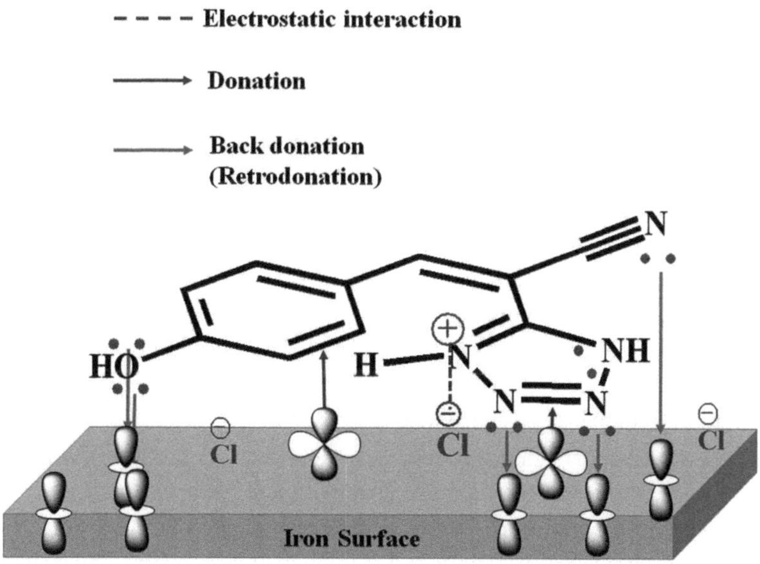

Figure 2. Diagrammatic representation of adsorption of a tetrazole derivative, HTA, on mild steel surface atoms.

There are several organic molecules, such as monomeric/Gemini surfactants (*52–61*), heterocyclic compounds (*62–72*), ionic liquids (*73–82*), and more, that have been studied for their corrosion-inhibiting properties explored in acidic solutions, as summarized in Table 1. The corrosion-inhibiting behaviors have been extensively explored, and their experimentally determined corrosion-inhibiting efficacy has been validated through insightful theoretical studies such as density functional theory (DFT) and molecular dynamics (MD) simulations (*83, 84*). A selection of the most important literature is overviewed briefly. Verma et al. (*85*) explored and reported on substituted tetrazole derivatives, namely (E)-3-phenyl-2-(1H-tetrazole-5-yl)acrylonitrile (PTA), (E)-3-(4-nitrophenyl)-2-(1H-tetrazole-5-yl)acrylonitrile (NTA), and (E)-3-(4-hydroxyphenyl)-2-(1H-tetrazole-5-yl)acrylonitrile (HTA), as efficient inhibitors to impending corrosion of mild steel exposed to 1 M HCl. The percentages of corrosion-inhibiting efficacy determined by gravimetric measurement of tetrazole derivatives are found in the order HTA > PTA > NTA. The determined efficacy increased when the concentration increased and resulted in a decreased rate of corrosion for the mild steel specimens owing to the inhibitors' adsorption over the metal surface. But there was also a decrease in corrosion-inhibiting capabilities of the tetrazole-based derivatives with a variation of temperature from 305 to 338 K, which is attributed to the desorption of adsorbed tetrazole derivatives at elevated temperatures. From the adsorption isotherm analysis, it was found that these tetrazole derivatives follow the Langmuir adsorption isotherm, meaning adsorption of these molecules is governed by a monolayer protective film of tetrazole derivatives adsorbed onto metal surfaces. Furthermore, the equilibrium adsorption–desorption constant was high, which suggests the strong adsorption capability of the tetrazole derivatives in the order HTA > PTA > NTA and is consistent with experimentally obtained corrosion-inhibiting efficacies. The free energy of adsorption, an additional significant parameter, helped categorize adsorption mechanisms of surface-adsorbing corrosion inhibitors following physisorption (i.e., electrostatic interactions between oppositely charged inhibitor species and surface metal) or chemisorption (i.e., electron sharing between unshared electrons or π electrons present in the corrosion-inhibiting molecules and available d orbitals of metal surface atoms leading to the formation of a metal–inhibitor coordinate bond). It was reported that if the free energy of adsorption is −20 kJ/mol or less, then physisorption is favored, but if it is −40 kJ/mol or highly negative, then chemisorption is favored. These tetrazole derivatives were found to undergo both the physisorption as well as chemisorption pathway during their adsorption as reflected from the determined free energy of adsorption in the intermediate range. The potentiodynamic polarization (PDP) result also affirmed the adsorption of these tetrazole derivatives. The corrosion potential determined from the PDP study established that these tetrazole derivatives acted as mixed-type corrosion inhibitors with a predominantly cathodic nature; that is, they were capable of retarding both cathodic and anodic corrosion of metallic materials simultaneously, predominantly retarding the cathodic reaction. The observations of double-layer capacitance determined from the Nyquist plot were found to be reduced, revealing their adsorption capability on mild steel surfaces. Furthermore, the surface heterogeneity factor determined from the Nyquist plot confirmed the development of smoother surfaces by the inhibitors' adsorption. This resulted in a deviation from the ideal capacitance as reflected from the semicircle loop of the Nyquist plot. There was also an increase in charge transfer resistance with the addition of inhibitor molecules on the corrosive media, and furthermore, it escalated with increasing concentration. The Bode plot also confirmed that there was an increase in the phase angle with the addition of tetrazole derivatives, indicating a decrease in surface imperfections owing to their adsorption onto mild steel surfaces. It is noteworthy to mention that the obtained trends of improvement in the electrochemical properties, implying the retardation of corrosion-inhibiting properties, is HTA > PTA > NTA. Additionally, DFT was utilized to justify probable reasons behind wet chemically determined corrosion-inhibiting efficacy trends. The greater adsorption of these tetrazole derivatives was attributed to the electron density distribution in the highest occupied molecular orbital (HOMO) and the lowest unoccupied molecular orbital (LUMO) of the molecules. The presence of the hydroxyl (−OH) group acted as a ring activator in the phenyl moiety, which enhanced its adsorption over metal surfaces, while the ring-deactivating nitro ($-NO_2$) group in the phenyl moiety resulted in the comparatively lower adsorption capability of NTA compared to that of other derivatives. The determined quantum chemical parameters, such as HOMO energy, LUMO energy, HOMO–LUMO energy gap, dipole moment, and softness of neutral and protonated tetrazole derivatives, also followed the same trend as the outcomes of corrosion-inhibiting properties determined from wet chemical experimentations. From the analysis of the wet chemical results and the DFT results, it could be predicted that there is a possibility of chemisorption followed by physisorption (i.e., physiochemisorption by the tetrazole derivative). The physisorption phenomenon was accomplished through interactions between oppositely charged corrosion-inhibiting additives and metal surface atoms electrostatically, while chemisorption was accomplished through electron sharing due to electron donating and accepting between the lone pairs of electrons present in the heteroatoms (namely, N, O) of the tetrazole derivatives, π electrons of the aromatic moiety, and the polar functional group with the metal d orbital of the metal surface atoms. Back bonding was also possible due to the retrodonation of excess electrons from the metal centers to the suitable orbital of the corrosion-inhibiting molecules, as shown schematically for HTA adsorption onto an iron surface in Figure 2. The strengthened adsorption of these molecules protected those metals exposed to corrosive electrolytes.

Table 1. Some Explored Gemini Surfactants, Heterocyclic Compounds, and Ionic Liquid-Based Corrosion-Inhibiting Molecules for Protection from Corrosive Hydrochloric Acid Solutions

Sl. no.	*Molecules*	*Metal/ Acidic solution*	*Studies*	*Results*	*Ref*
1.	N,N′-((cyclohexane-1,4-diylbis(oxy))bis(2-oxoethane-2,1-diyl))bis(N,N-dimethyloctan-1-aminium) methyl sulfate (CHOGS-8), N,N′-((cyclohexane-1,4-diylbis(oxy))bis(2-oxoethane-2,1-diyl))bis(N,N-dimethyldodecan-1-aminium) methyl sulfate (CHOGS-12), N,N′-((cyclohexane-1,4-diylbis(oxy))bis(2-oxoethane-2,1-diyl))bis(N,N-dimethylhexadecan-1-aminium) methyl sulfate (CHOGS-16)	Carbon steel/ 15% HCl	WL, PDP, EIS, SEM, EDXS, DFT	It showed 97.51%, 98.73%, and 99.61% corrosion-inhibiting efficacy for CHOGS-8, CHOGS-12, and CHOGS-16 at 200 ppm.#	(52)
2.	4-(1-hydroxy-13-oxo-3,6,9,12-tetraoxapentadecan-15-yl)phenyldodecanoate (HTOPD), 4-(1-hydroxy-13-oxo-3,6,9,12-tetraoxapentadecan-15-yl)phenyltetradecanoate (HTOPT), 4-(1-hydroxy-13-oxo-3,6,9,12-tetraoxapentadecan-15-yl)phenylhexadecanoate (HTOPH)	Mild steel/ 0.5 M HCl	PDP, EIS, AFM, antimicrobial activity	It showed 91.75%, 94.53%, and 95.01% corrosion-inhibiting efficacy for HTOPD, HTOPT, and HTOPH at 35 ppm.#	(53)
3.	Imidazolium Gemini surfactant ([C14-4-C_{14}im]Br_2), its monomer ([C_{14}mim]Br)	A3 carbon steel/ 0.1 mol L^{-1} HCl	WL, PDP, EIS, DFT	It showed 97.08% and 96.62% corrosion-inhibiting efficacy for [C14-4-C_{14}im]Br_2 and [C_{14}mim]Br at 100 × 10^{-6} mol L^{-1}.#	(54)
4.	N^1,N^1,N^1,N^2,N^2,N^2-hexadodecylethane-1,2-diaminium bromide (IL2), N^1,N^1,N^1,N^2,N^2,N^2-hexadodecylpropane-1,3-diaminium bromide (IL3), N^1,N^1,N^1,N^2,N^2,N^2-hexadodecylhexane-1,6-diaminium bromide (IL3),	Mild steel/1 M HCl	WL, PDP, EIS	It showed 90.55%, 92.41%, and 94.09% corrosion-inhibiting efficacy for IL-2, IL-3, and IL6 at 5 × 10^{-3} M.#	(55)

Table 1. (Continued). Some Explored Gemini Surfactants, Heterocyclic Compounds, and Ionic Liquid-Based Corrosion-Inhibiting Molecules for Protection from Corrosive Hydrochloric Acid Solutions

Sl. no.	*Molecules*	*Metal/ Acidic solution*	*Studies*	*Results*	*Ref*
5.	3-azamethyl-1,5-pentamethylene-bis(N-dodecyl-N,N-dimethylammonium bromide) (3N-12), 3-azamethyl-1,5-pentamethylene-bis(N-octadecyl-N,N-dimethylammonium bromide) (3N-18)	Stainless steel/3 M HCl	PDP, EIS, SEM	The corrosion-inhibiting efficacy for 3N-12 reached 95% at 1 Mm.#	(56)
6.	3-oxa-1,5-pentamethylene-bis(N-alkyl-N, N-dimethylammonium chloride) (12-O-12, R= $C_{12}H_{25}$; 18-O-18, R= $C_{18}H_{37}$), 3-oxa-1,5-pentamethylene-bis(N-alkyl-N-hydroxyethyl-N-methylammonium chloride) (12-MOH-O-MOH-12, R= $C_{12}H_{25}$; 18-MOH-O-MOH-18, R= $C_{18}H_{37}$)	Stainless steel/3 M HCl	PDP, EIS, SEM	The compound with twelve carbon atoms in the alkyl chains and hydroxyethyl groups linked to the positively charged nitrogen atoms appeared to be the most effective inhibitors, showing 98% at concentration 3 Mm.#	(57)
7.	N,N′-didodecyl-N,N,N′,N′-tetramethyl-1,5,9-triazonane diodide (TMTN[1]2I), N,N′-didodecyl-N,N,N′,N′-tetramethyl-1,5,9-triazonane dibromide (TMTN[1]2Br)	Stainless steel AISI 304/3 M HCl	PDP, EIS, SEM, CLSM	It showed 96% and 91% corrosion-inhibiting efficacy for TMTN[1]2I and TMTN[1]2Br at 3 mM.#	(58)
8.	Ethane-1,2-diyl bis(N,N-dimethyl-N-alkylammonium acetoxy)dichloride (m = 12, 14)	Mild steel/1 M HCl	WL, PDP, EIS, DFT	The corrosion-inhibiting efficacy was found to be 97.71% and 95.79%, for m = 14 and 12, at 50 °C, at 1×10^{-2}, respectively.#,*	(59)

Table 1. (Continued). Some Explored Gemini Surfactants, Heterocyclic Compounds, and Ionic Liquid-Based Corrosion-Inhibiting Molecules for Protection from Corrosive Hydrochloric Acid Solutions

Sl. no.	*Molecules*	*Metal/ Acidic solution*	*Studies*	*Results*	*Ref*
9.	Ethane-1,2-dyl bis (N, N-dimethyl-N-dodecyl ammoniumacetoxy) dichloride (12-E2-12), Ethane-1,2-dyl bis (N, N-dimethyl-N-tetradecyl ammoniumacetoxy) dichloride (14-E2-14), Ethane-1,2-dyl bis (N, N-dimethyl-N-hexadecyl ammoniumacetoxy) dichloride (16-E2-16), Sodium sulfate of carboxymethyl cellulose (NaCMC)	Mild steel/ 1 M HCl	WL, PDP, EIS, MD	Mixed inhibitor of 500 ppm of NaCMC and 1 ppm of 12-E2-12, 14-E2-14, and 16-E2-16 showed corrosion-inhibiting efficacy of 97% , 84% and 62%, respectively.#, *	*(60)*
10.	N,N-dimethyl-N-(3-(3-(4-(trifluoromethyl)phenyl) ureido)propyl)octan-1-aminium iodide (FUS-8), N,N-dimethyl-N-(3-(3-(4-(trifluoromethyl)phenyl) ureido)propyl)decan-1-aminium iodide (FUS-10), N,N-dimethyl-N-(3-(3-(4-(trifluoromethyl)phenyl)ureido)propyl)dodecan-1-aminium iodide (FUS-12)	Carbon steel/1 M HCl	WL (at 293–323 K), PDP, EIS, FESEM, EDXS, DFT, MC	The corrosion-protection capacities at an optimum concentration (0.5 mM) are 93.2% (FUS-8), 96.3% (FUS-10), and 97.6% (FUS-12) by PDP, comparable to the findings achieved by EIS.#,*	*(61)*
11.	Thiazole-4-carboxylic acid (TCA), 2-methyl-1,3-thiazole- 4-carboxylic acid (MTCA)	Mild steel/1 mol L^{-1} HCl	PDP, EIS, DFT, MC	The corrosion-inhibiting efficacy of MTCA and TCA are ~90% at an optimum concentration (150 ppm).#,*	*(62)*
12.	1-((8-hydroxyquinolin-5-yl)methyl)-3-methylthiourea (HQMT), 5-isothiocyanatomethyl-8-hydroxyquinoline (TCHQ)	Carbon steel/1 M HCl	WL, OCP, PDP, EIS, SEM, DFT, MC	The corrosion-inhibiting efficacy of HQMT and TCHQ are 91% and 89% at an optimum concentration (10^{-3} M).#,*	*(63)*

Table 1. (Continued). Some Explored Gemini Surfactants, Heterocyclic Compounds, and Ionic Liquid-Based Corrosion-Inhibiting Molecules for Protection from Corrosive Hydrochloric Acid Solutions

Sl. no.	*Molecules*	*Metal/ Acidic solution*	*Studies*	*Results*	*Ref*
13.	2-(1,4,5-triphenyl-1H-imidazol-2-yl)phenol (P1), 3-methoxy-4-(1,4,5-triphenyl-1H-imidazol-2-yl)phenol (P2)	Mild steel/1 M HCl	WL, PDP, EIS, SEM, EDXS, DFT, MC	The corrosion-inhibiting efficacy of P1 and TP2 are 94.4% and 96.2% at an optimum concentration (150 ppm).#,*	(*64*)
14.	2-thioureidobenzoxazole (BOT), 2-thioureidobenzimidazole (BIT), 2-thioureidobenzothiazole (BTT)	CS 1018 mild steel/2 M HCl	WL, PDP, EIS, SEM, DFT, MC	The corrosion-inhibiting efficacy of BOT, BIT, and BTT are 92.58%, 98.02%, and 98.81% at 2 mM. BTT showed a maximum inhibition efficacy of 95.6%, which is attributed to its better adsorption through two nucleophilic S atoms, which donates a lone pair of electrons to the empty d orbitals of the Fe atom.#,*	(*65*)
15.	5-((2-bromoethoxy) methyl) quinolin-8-ol (QC2-Br), 5-((3-bromopropoxy) methyl) quinolin-8-ol (QC3-Br)	CS 1018 mild steel/1 M HCl	WL, PDP, EIS, SEM, EDXS, DFT, MC, MD	Both QC2-Br and QC3-Br achieved a corrosion-inhibiting efficacy of 96% at 1 $\times 10^{-3}$ M.#,*	(*66*)
16.	Methyl (E)-5-(4-chlorophenyl)-2-(((E)-4-methylbenzylidene)hydrazono)-2,3-dihydrothiazole-4-carboxylate (T1), methyl (E)-5-(4-chlorophenyl)-2-(((E)-4-nitrobenzylidene)hydrazono)-2,3-dihydrothiazole-4-carboxylate (T2), methyl (E)-2-(((E)-4-chlorobenzylidene)hydrazono)-5-(4-chlorophenyl)-2,3-dihydrothiazole-4-carboxylate (T3)	Mild steel/1 mol L^{-1} HCl	WL, PDP, EIS, SEM, EDXS, DFT, MD	The corrosion inhibitors T1, T2, and T3 showed efficacy of 77.27%, 87.54%, and 92.11% at 5×10^{-4} mol L^{-1}.#,*	(*67*)

Table 1. (Continued). Some Explored Gemini Surfactants, Heterocyclic Compounds, and Ionic Liquid-Based Corrosion-Inhibiting Molecules for Protection from Corrosive Hydrochloric Acid Solutions

Sl. no.	*Molecules*	*Metal/ Acidic solution*	*Studies*	*Results*	*Ref*
17.	5-(((R)-2-((3aR,5R,6S,6aR)-6-(hexyloxy)-2,2-dimethyltetrahydrofuro[2,3-d][1,3]dioxol-5-yl)-2-hydroxyethoxy)methyl)quinolin-8-ol (GQ-6), 5-(((R)-2-((3aR,5R,6S,6aR)-6-(dodecyloxy)-2,2-dimethyltetrahydrofuro[2,3-d][1,3]dioxol-5-yl)-2-hydroxyethoxy)methyl) quinolin-8-ol (GQ-12)	Mild steel/1 M HCl	WL, PDP, EIS, SEM, EDXS, DFT, MC	At optimal concentration (1×10^{-3} M), it exhibited corrosion-inhibiting efficacies of 96.5% for compound QG-12 and 85.9% for compound QG-6.#,*	*(68)*
18.	(Z)-4-(4-hydroxy-3-methoxybenzylidene)-3-methylisoxazol-5(4H)-one (IOD)	Mild steel/1 M HCl	WL, PDP, EIS, SEM, EDXS, AFM	At optimal concentration (300 ppm/1.2×10^{-3} M), it exhibited corrosion-inhibiting efficacy of 96.22%.#,*	*(69)*
19.	2-nitroacridone	Low carbon steel/1 M HCl	WL, PDP, EIS, SEM, EDXS, AFM, DFT	At optimal concentration (10 mM), it exhibited corrosion-inhibiting efficacy of 92%.#,*	*(70)*
20.	4,4′-(1,4-phenelyne)bis(6-amino-3-methyl-2,4-dihydropyrano[2,3-c]pyrazole-5-carbonitrile) (BP-1), 4,4′-(1,4-phenelyne)bis(6-amino-3-methyl-1-phenyl-1,4-dihydropyrano[2,3-c]pyrazole-5-carbonitrile) (BP-2)	Mild steel/1 M HCl	WL, PDP, EIS, SEM, EDXS, AFM, XPS, Contact angle, DFT	At optimal concentration (1.72×10^{-4} mol), BP-1 and BP-2 exhibited corrosion-inhibiting efficacy of 94.96% and 97.80%.#,*	*(71)*
21	N-(quinolin-8-yl) quinoline-2-carboxamide (Hqcq), N-(quinolin-7-yl) pyrazine-2-carboxamide) (Hqpzc)	Mild steel/1 mol L^{-1} HCl	EIS, SEM	The corrosion-inhibiting efficacy for Hqcq was 94.34% and for Hqpzc was 92.18% at 100 ppm.#	(72)
22.	1-ethyl-3-methylimidazoliumthiocyanate	API 5L X52 steel/ H_2SO_4 and HCl	PDP, EIS, FESEM, EDXS, AFM, XPS	Corrosion-inhibiting efficacy: 82.9% and 77.4% for H_2SO_4 (75 ppm) and HCl (100 ppm) solutions.#	(73)

Table 1. (Continued). Some Explored Gemini Surfactants, Heterocyclic Compounds, and Ionic Liquid-Based Corrosion-Inhibiting Molecules for Protection from Corrosive Hydrochloric Acid Solutions

Sl. no.	*Molecules*	*Metal/ Acidic solution*	*Studies*	*Results*	*Ref*
23.	Poly-2-acrylamido-2-methylpropane sulfonic acid triethanolamine derivative (P1), poly-2-acrylamido-2-methylpropane sulfonic acid triethylamine derivative (P2), poly-2-acrylamido-2-methylpropane sulfonic acid trimethylamine derivative (P3)	Carbon steel/1 mol L^{-1} HCl	PDP, EIS, SEM	Corrosion-inhibiting efficacy: 83.1% (P1), 80.7% (P2), and 60.0% (P3) (250 ppm addition). These are mixed-type inhibitors.	(74)
24.	1-(6-ethoxy-6-oxohexyl)pyridazin-1-ium bromide (S1), 1-(2-bromoacetyl) pyridazinium bromide (S2)	Carbon steel/1 M HCl	EIS, MC	Inhibition efficacy increased with the concentration of S1 and S2, and reached 84% for S1 and 82% for S2 at 10−3 M (303 K).#	(75)
25.	1-methyl-3-propylimidazolium iodide (MPIMI), 1-butyl-3-methylimidazolium iodide (BMIMI), 1-hexyl-3-methylimidazolium iodide (HMIMI)	Mild steel/1 M HCl	PDP, EFM, EIS, SEM, AFM	Maximum inhibitions of 93.1%, 87.8%, and 80.4% for HMIMI, BMIMI, and MPIMI was found using 5 × 10−3 M.#	(76)
26.	Benzyltributylammonium tetrachloroaluminate	Carbon steel/2 N HCl	ENA, EIS, SEM, EDXS, MC	Maximum inhibition efficacy of 97% was found using 400 ppm.#	(77)
27.	5-(trifluoromethyl) dibenzothiophenium tetrafluoroborate (TDTB), 5-(trifluoromethyl) dibenzothiophenium trifluoromethanesulphonate (TDTM), 1-ethyl-3-methylimidazolium ethyl sulfate ((EMIM)(ESO_4))	Mild steel/1 M HCl	WL, PDP, EIS, SEM, EDXS, DFT, MC	Maximum inhibition efficacy found for TDTB, TDTM, and (EMIM)(ESO_4) at 8.0×10^{-2} M was 90.08%, 90.08%, and 83.36%, respectively.#,*	(78)

Table 1. (Continued). Some Explored Gemini Surfactants, Heterocyclic Compounds, and Ionic Liquid-Based Corrosion-Inhibiting Molecules for Protection from Corrosive Hydrochloric Acid Solutions

Sl. no.	*Molecules*	*Metal/ Acidic solution*	*Studies*	*Results*	*Ref*
28.	L-phenylalanine methyl ester saccharinate ([PheME][Sac]), L-leucine methyl ester saccharinate ([LeuME][Sac]), L-alanine methyl ester saccharinate ([AlaME][Sac])	Mild steel/1 M HCl	WL, PDP, EIS, SEM, EDXS, Contact angle, DFT, MC	Maximum inhibition efficacy found for [PheME][Sac], [LeuME][Sac], and [AlaME][Sac] was 79.90% (100 ppm/ 2.7×10^{-4} M), 77.78% (100 ppm/ 2.7×10^{-4} M), and 69.87% (100 ppm/ 2.7×10^{-4} M), respectively.#,*	(*79*)
29.	3-butyl-1-vinylimidazolium bromide (PIL)	Mild steel/1 M HCl	PDP, EIS, SEM, DFT	Maximum inhibition efficacy found for PIL at 400 ppm was 92%.$,*	(*80*)
30.	1-hexylpyridinium bromide (NR3)	Carbon steel/1 M HCl	WL, LPR, PDP, EIS, SEM	Maximum inhibition efficacy found for PIL at 400 ppm was 92.%.#,*	(*81*)
31.	Benzyltributylammonium tetrachloroferrate ($[BTBA]^+[FeCl_4]^-$)	Mild steel/1 N HCl	EIS, SECM, SEM, DFT, MC	Maximum inhibition efficacy found for PIL at 400 ppm was 99.5%.#,*	(*82*)

These molecules performed by simultaneously retarding cathodic and anodic reactions. Their adsorption was governed by the Langmuir adsorption isotherm. * Theoretical finding supported experimental results. $ These molecules performed by simultaneously retarding cathodic and anodic reactions. Their adsorption was governed by the Frumkin adsorption isotherm. [AFM = atomic force microscopy; CLSM = confocal laser scanning microscopy; DFT = density functional theory; EDXS = energy-dispersive X-ray spectroscopy; EFM = electrochemical frequency modulation; EIS = electrochemical impedance spectroscopy; ENA = electrochemical noise analysis; FESEM = field emission scanning electron microscopy; LPR = linear polarization resistance; MC = Monte Carlo simulation approach; MD = molecular dynamics simulation; OCP = open circuit potential; SECM = scanning electrochemical microscopy; SEM = scanning electron microscopy; WL = weight loss measurement; XPS = X-ray photoelectron spectroscopy].

Gutiérrez et al. employed multiple linear regression methods by using appropriate descriptors to find the structure–property relationship model to foretell the corrosion-inhibiting efficacy of the designed corrosion inhibitors quantitatively in the MobyDigs program. The relation is shown in Figure 3 (*86*).

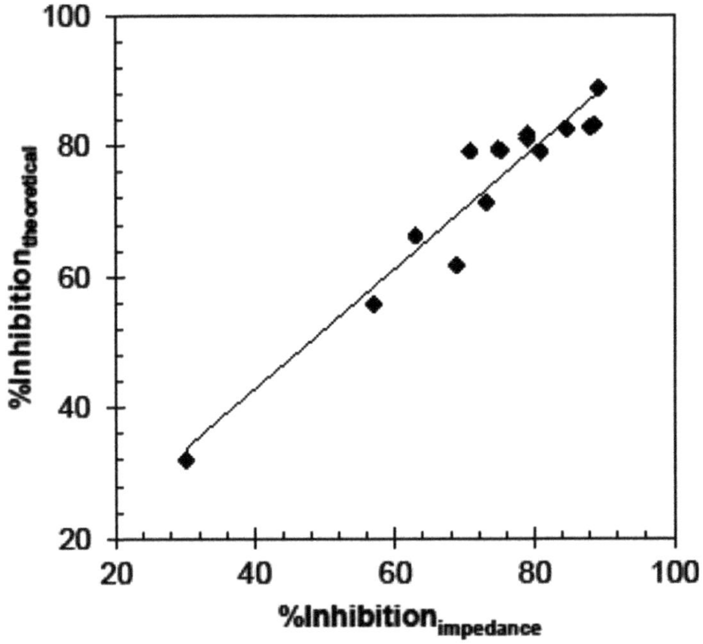

Figure 3. Comparison of theoretically and experimentally determined corrosion-inhibiting efficacy (percentage). (Reproduced with permission from ref (86). Copyright 2016 Elsevier Ltd.)

The corrosion inhibition might be predicted from equation (32).

$$\% \text{ Inhibition} = 513.95 - 32.03\chi + 5334\text{bq}^{\text{ISO}} + 0.37V + 1433.78q_{N1} \quad (32)$$

χ is the electronegativity of the corrosion-inhibiting molecule, bq^{ISO} refers to the aromaticity index, V refers to the volume, and q_{N1} refers to the charge on the nitrogen atom $N1$.

Herein a series of imidazole derivatives have been used for theoretical investigation. The open circuit potential (OCP), PDP, and electrochemical impedance measurement were employed to experimentally determine the corrosion-inhibiting efficacy of 2-methylimidazole, 4-(imidazol-1-yl)phenol, 2-(aminomethyl)benzimidazole, and 6-bromo-1H-benzimidazole as representative corrosion inhibitors. The test molecules whose electrochemical parameters were determined were 2-chloro-1H-imidazole, 2-Methylimidazole, 4′-(Imidazol-1-yl)acetophenone, 4-(1H-Imidazol-1-yl)aniline, 4′-(Imidazol-1-yl)benzaldehyde, 1-(4-bromophenyl)-1H-imidazole, 1-(4-chlorophenyl)imidazole, 4-(midazol-1-yl)phenol, 1-(3-chlorophenyl)imidazole, 1-(2-chlorophenyl)imidazole, 2-(chloromethyl)benzimidazole, 6-bromo-1H-benzimidazole, 1H-benzimidazol-5-amine, 2-(aminomethyl)benzimidazole, and 1H-benzimidazol-5-ylmethanol. A good approximation between the experimentally determined values and the theoretically predicted values was established for imidazole and benzimidazole. The experimentally determined percentages of corrosion-inhibiting efficacy for imidazole were 27.6% and 26.1%, respectively, while for benzimidazole it was 62.4% and 56.7%, respectively, under the same experimental conditions. It was summarized that the presence of heteroatoms, lone pairs of electrons, and π electrons played a crucial function in their interactions with metallic surface atoms leading to the adsorption of corrosion-inhibiting molecules. Additionally, the presence of halogen, an electron-donating or withdrawing group, an aromatic ring, different volume, as well as a steric factor altered the corrosion-inhibiting properties of the organic molecules (*86*). Similarly, Olasunkanmi et al. established the quantitative structure–activity relationship model in proposing a good correlation of molecular quantum chemical descriptors with experimentally determined corrosion-inhibiting efficacies for N-{n-[1-R-5-(quinoxalin-6-yl)-4,5-dihydropyrazol-3-yl]phenyl}methanesulfonamides on mild steel in 1 M HCl (*87*).

Yilmaz et al. (*88*) reported the corrosion-inhibiting behavior of 2-((thiazole-2-ylimino)methyl)phenol (THYMP) in 2 M HCl for mild steel by weight loss and electrochemical techniques. The mechanism of adsorption of THYMP and its subsequent corrosion-inhibiting property was explained by the potential of zero charge (E_{pzc}). A surface charge of a metal surface under investigation was determined from the electrochemical impedance spectroscopy (EIS) study indirectly because its direct measurement is experimentally difficult. When the surface polarization resistance was at its maximum or the capacitance was at its minimum, it indicated that at this potential the surface of the electrode carried zero charge, and the ionic double layer was absent. At this potential of zero charge, the electrode surface was more activated to adsorb substances present in the electrolyte (*89–91*). The plot of polarization resistance with respect to the potential for 1 hour of exposure in the presence of 0.01 M of THYMP in 2 M HCl is shown in Figure 4. The surface charge of the electrode is determined by the following relation:

$$E_r = E_{ocp} - E_{(q=0)} \tag{33}$$

$E_{(q=0)}$ is the E_{pzc}, E_{ocp} refers to Antropov's rational corrosion potential (*92, 93*), and E_{ocp} symbolizes OCP. It is said that the negative value of E_r is due to the negative charge accumulation on the electrode surface and vice versa. In such cases, the adsorption of cations is favored over negatively charged electrode surfaces, so the adsorption of anions takes place on positively charged electrode surfaces. The computation of E_{pzc} is important for studying the adsorption of corrosion-inhibiting molecules. The value of E E_{pzc} is more negative than that of E_{ocp}. Furthermore, it has been reported that the value of E_r for the organic molecule is positive with respect to E_{pzc}. This suggests that the metal surface was positively charged. In this situation, the chloride ions were initially adsorbed onto the metal surface, which was positive due to oxidation. Subsequently, THYMP molecules were adsorbed over preadsorbed chloride ions through electrostatic interactions because the THYMP molecule remained protonated in highly acidic hydrochloric acid. In this way, the chloride ions acted as interconnecting bridges between the positively charged molecule and the positively charged metal electrode surface. Additionally, there could be electron sharing between free electrons of the heteroatom (N, O, S), π electrons present in inhibitor molecules, and unoccupied d orbitals of iron atoms, leading to chemical adsorption. Although it may be expected that the protonation and involvement of the lone pair in the resonance may somewhat reduce the occurrence of chemisorption, the additional lone pair of electrons at the sulfur atom could interact with empty d orbitals of iron atoms, leading to coordinate bond formation. THYMP, which is comprised of an oxygen-like heteroatom, an azomethine linkage, and an aromatic benzene ring, facilitated coordinate bonding. Simultaneously, the available molecular orbitals were capable of accommodating excess electrons coming from transition metal d orbitals, giving rise to back bonding. In this way, it stabilized the existing metal–inhibitor bond. In addition to the electronic properties of this molecule, its molecular structure covered a greater surface area, isolating the exposed metals from the corrosive electrolyte.

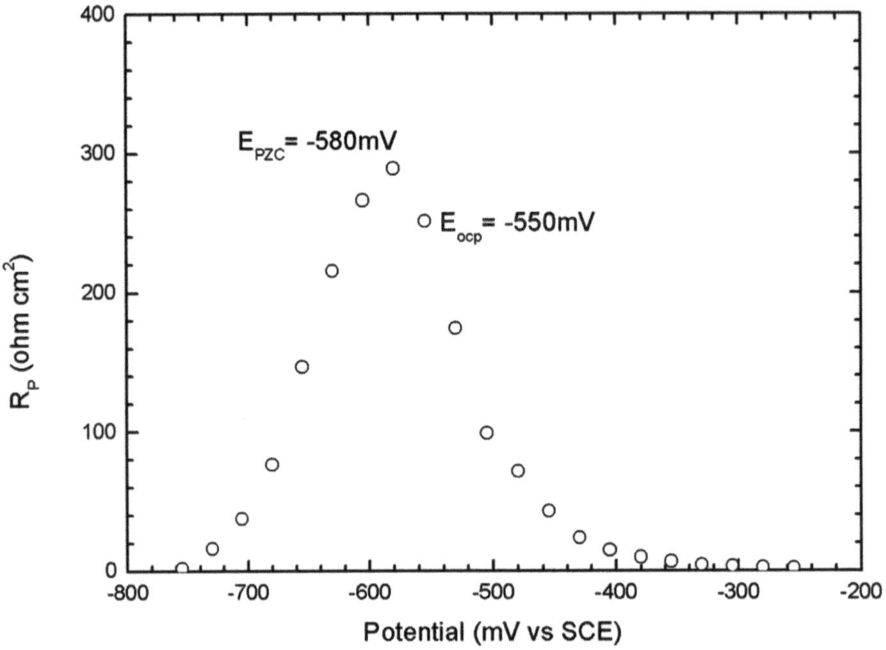

Figure 4. The plot of polarization resistance with respect to the potential presence of 0.01 M of THYMP. (Reproduced with permission from ref (88). Copyright 2016 Elsevier Ltd.)

The electron density distributions present in frontier molecular orbitals, that is, HOMOs and LUMOs of the molecule, were analyzed using DFT and are presented in Figure 5. It is clear that electron density in both the HOMO and LUMO was completely distributed throughout the whole structure of the molecule. Because of this, almost all the sites were reactive in nature, facilitating electron sharing between HOMOs of these inhibitor molecules with empty d orbitals present in the metal center to make coordinate bonds; at the same time, the LUMO of the molecule accommodated the electron coming toward it for back bonding. The electron sharing between the inhibitor molecules and the metal center led to its adsorption over it, and a protective layer was formed, which could inhibit the oxidative degradation of metals exposed to harsh media (*88*). The corrosion-inhibiting properties of these molecules were also found to be enhanced with the addition of another inhibitor molecule, which may synergistically improve the property to the desired level.

Ramya et al. (*94*) explored corrosion inhibition enhancement through synergistic hydrogen-bonding interactions of alkyl benimidazoles with urea in hydrochloric acid media. 2-methylbenzimidazole (MBI), 2-ethylbenzimidazole (EBI), and 2-propylbenzimidazole (PBI) derivatives were synthesized. The PDP and EIS revealed their high corrosion-inhibiting properties in 1 M HCl when combined with urea. The optimized combined form (150 ppm benzimidazole derivatives (MBI, EBI, PBI) and 50 ppm urea) was more effective at the temperature range of 303–313 K. These combined forms acted as mixed-type corrosion inhibitors. These inhibitor mixtures worked through a coadsorption process at elevated temperatures, but at low temperatures, the competitive adsorption was inferred from the obtained synergism parameters. It was also found that within 303–308 K, the inhibitor followed the Langmuir adsorption isotherm, while at 313 K, the Temkin adsorption isotherm was witnessed. The temperature-dependent inhibition efficiency was found for this set of corrosion inhibitors. Additionally, the synergistic increase in the corrosion-inhibiting effect was validated theoretically from quantum chemical calculations. The stability of the optimized geometry of the combined form of the benzimidazole derivatives with urea was established by increasing intermolecular hydrogen-bonding interactions, which was revealed by the closest distance of targeted atoms of MBI, EBI, and PBI from urea being ~1.8 Å, as shown in Figure 6. Hydrogen bonding was expected between protons of –NH in benzimidazole derivatives and oxygen in urea molecules. The adsorption of such molecular associations covered greater surface area, which enhanced the protection of the metallic materials from corrosion. The HOMO and LUMO determined from the optimized form of associated inhibitor molecules were mostly confined to the alkyl benzimidazole derivative only. It may be expected that major electron sharing leads to the adsorption of alkyl benzimidazole moiety.

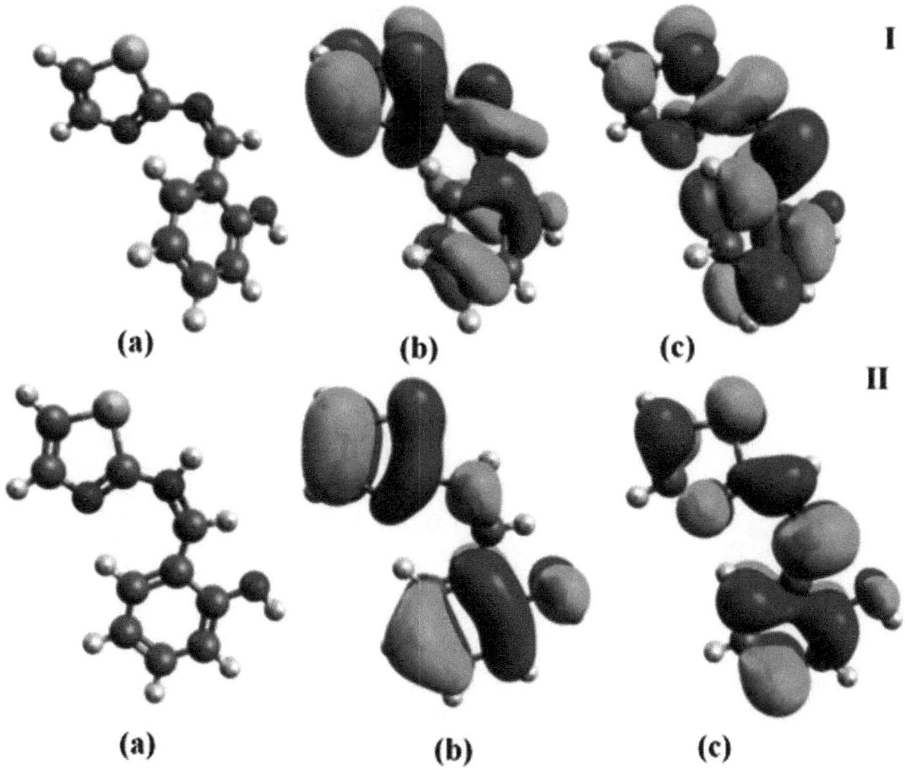

Figure 5. The optimized structure of (I)(a) neutral THYMP in a solvent and (II)(a) protonated THYMP in a solvent, and their corresponding (b) HOMOs as well as (c) LUMOs. (Reproduced with permission from ref (88). Copyright 2016 Elsevier Ltd.)

Similarly, with the progress of the exploration, Ramya et al. (95) reported the synergistic effect comprising alkyl benzimidazoles, namely, MBI, EBI, and PBI, with semicarbazide (SC) in inhibiting the corrosion of mild steel in 1 M HCl studied by both electrochemical and theoretical methods (95). From the electrochemical studies it was reported that the alkyl benzimidazole derivatives worked as corrosion inhibitors with the variation of temperatures. The specific mixture of 150 ppm of alkyl benzimidazole derivatives and 50 ppm of SC exhibited good corrosion-inhibiting properties at 303, 308, and 313 K. It was revealed from the PDP results that these combinations (alkyl benzimidazole derivatives and SC) worked as mixed-type inhibitors. The corrosion-inhibiting molecules obeyed the Langmuir adsorption isotherm, and their adsorption blocked available reactive regions. The corrosive species attacking it were decreased when an alkyl benzimidazole–SC mixture was added because it existed as a synergistic pair covering a greater surface area. The synergistic effect in the enhancement of the electronic property of inhibitors was investigated using DFT. The DFT-optimized combined structures of these inhibitor molecules facilitated H-bonding with bond lengths up to 1.9 Å which were formed by the alkyl benzimidazoles interacting with SC. The combined adsorption of the molecules covered greater surface area and protected the suspected metals from corrosion damages.

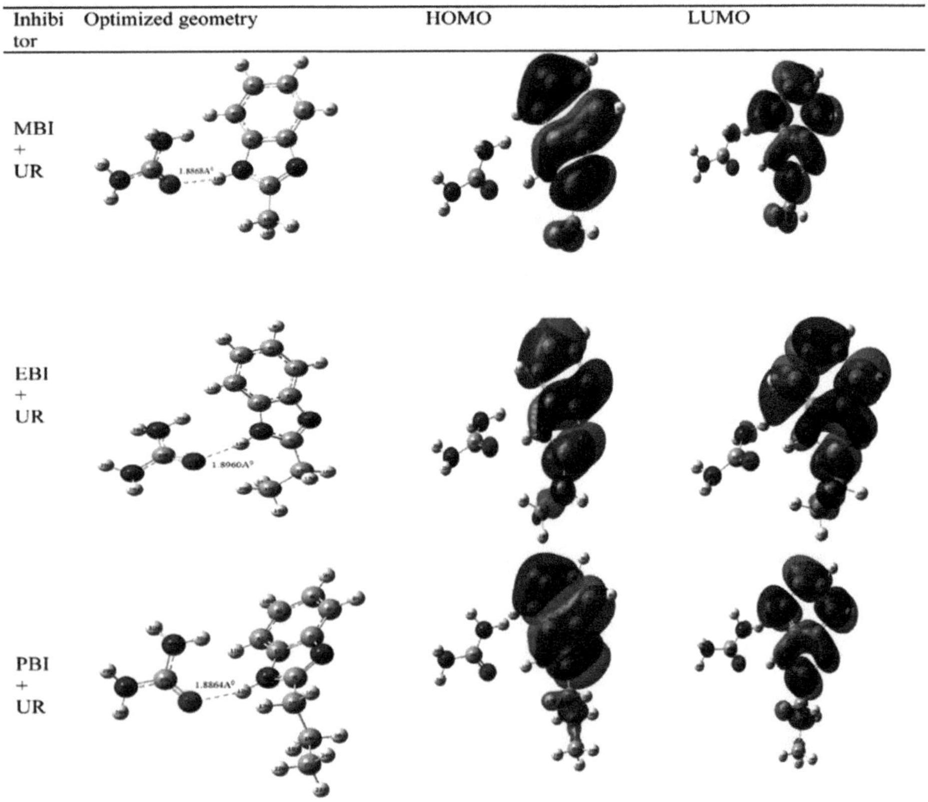

Figure 6. The optimized geometries, HOMO plot, and LUMO plot of synergistic benzimidazole and urea pairs. UR = urea. (Reproduced from ref (94). Copyright 2016 Elsevier Ltd.)

Furthermore, 1-(4-nitrophenyloimino)-1-(phenylhydrazono)-propan-2-one (NO2AM) was also explored to establish it as a good corrosion-inhibiting additive in a 1 M HCl solution for protecting mild steels through gravimetric and electrochemical studies along with quantum chemical studies (96). It was revealed that NO2AM adsorbed onto a mild steel surface through physisorption, chemisorption, and retrodonation. The adsorption was established to be spontaneous in nature, and it followed the Langmuir adsorption isotherm. It worked like a mixed-type corrosion inhibitor. The outcomes derived from DFT-based quantum chemical parameters corroborated with experimental findings. Furthermore, the DFT approach was used to explain the molecular electrostatic potential because it was related to the electron density of corrosion-inhibiting additives. Furthermore, Mulliken population analysis was widely utilized to estimate distributions of atomic charges in the entire molecule and to predict the sites responsible for adsorption. The nitrogen, oxygen, and some carbon atoms carried negative charges, while the rest of the carbon atoms carried positive charges.

The atoms carrying negative charges acted as electron donors and could form a coordinate bond with the Fe atoms, while the atoms carrying positive charges could form a back bond. Furthermore, the natural bonding orbital analysis provided information regarding intra or intermolecular H-bonding and interactions among bonds, and helped explore charge transfer and conjugative interactions in the molecular systems (97). The negative charges concentrated in the nitrogen and oxygen, as well as in a few carbon, atoms of the benzene ring of NO2AM were the most favorable sites for donating their electrons, and facilitated coordinate bonding with metal d orbitals of mild steel surfaces.

Sites responsible for nucleophilic as well as electrophilic attacks were also well explained based on molecular electrostatic potential (98). The regions having rich electron density were mostly situated around heteroatoms and conjugated double bonds. These were capable of forming coordinate bonds between π electrons of benzene moiety and unfilled d orbitals of iron atoms, facilitating chemisorption. The mild steel surface atoms acted as electrophiles, and these atoms could facilitate the attraction of the negatively charged sites present in the inhibitors. Normally, the nucleophilic centers of inhibitors are heteroatoms having freely available lone pairs of electrons, electronegative groups, and π electrons. These are easily accessible for bond forming and facilitate its adsorption (96).

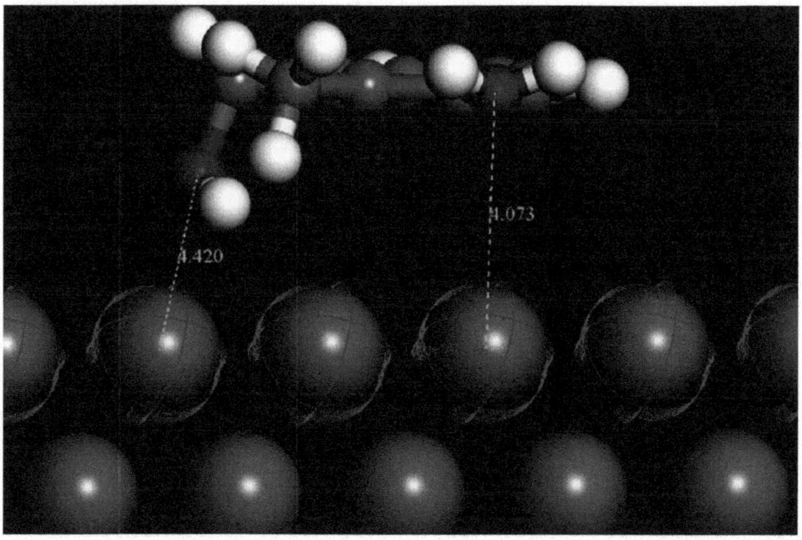

Figure 7. The stable adsorbed corrosion inhibitor AMBA on an iron surface obtained from an MC simulation approach. (Reproduced with permission from ref (99). Copyright 2016 Elsevier Ltd.)

It was reported that 3-amino-2-methylbenzylalcohol (AMBA) can also work for corrosion inhibition in the protection of mild steel in an acidic solution (99). The anticorrosive property increased with increasing concentration and was found to be effective up to 24 hours exposure in the acidic solutions. The addition of iodide anions enhanced the corrosion-inhibiting property of AMBA. The DFT study revealed the insight of corrosion inhibition mechanisms. The nitrogen heteroatom and the benzene ring of AMBA played a crucial role that involved complexation with metal surface atoms leading to its adsorption. A Monte Carlo (MC) simulation was performed to gain in-depth understanding of the stability and adsorption strength of AMBA. The adsorption of AMBA with the lowest energy revealed that AMBA was adsorbed onto iron surface in a horizontal manner, implying its adsorption onto the iron surface, as shown in Figure 7. Accompanied by electron donation from the π electron of the benzene moiety, the unshared electrons of nitrogen heteroatoms of AMBA molecules interacted with the vacant d orbitals of the iron. The negative magnitude of obtained adsorption energy of AMBA molecules suggested stable and strong adsorption capability.

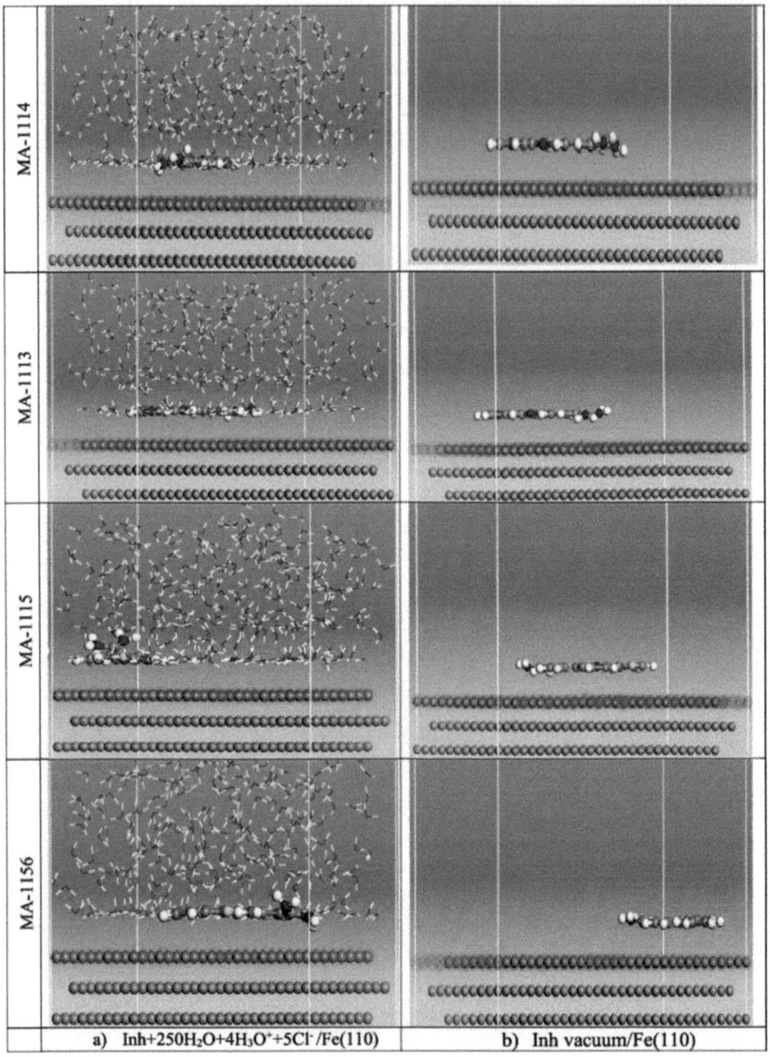

Figure 8. Configuration of bichalcophene derivatives after interaction with an Fe (110) surface in (a) an HCl solution and (b) a vacuum. (Reproduced with permission from ref (100). Copyright 2018 Elsevier Ltd.)

Abousalem et al. reported the eco-friendly fluoropheny-2,2′-bichalcophenes as potential inhibitors to protect carbon steel from acidic environments through electrochemical techniques. These fluorinated bichalcophenes designated as M1113, M1114, M1115, and M1156, are biocide agents. Out of these, M1156 acted as the best corrosion inhibitor, exhibiting 98% corrosion-inhibiting efficacy at 18 μM concentration (*100*). In order to validate the experimental findings from electrochemical measurements, DFT, Fukui indices, and Mulliken charge analyses were performed. From the DFT-derived parameters, it was found that among the other bichalcophene derivatives, the thiophene derivative (i.e., M1156) acted as the best corrosion inhibitor, which was reflected from the lowest HOMO–LUMO energy gap and the highest global softness value. The Fukui indices analysis also showed that the thiophene-containing bichalcophene derivatives were mostly reactive due to the presence of sulfur heteroatoms, which facilitated strong interaction with the iron surface atom and promoted its adsorption. From the Mulliken atomic charge analysis it was revealed that in thiophene derivative M1156, the number of atoms acting as nucleophilic sites was higher than those acting as electrophilic sites. Additionally, the π electrons in the bichalcophene moiety and the sulfur and nitrogen heteroatoms participated in both electrostatic as well as chemical adsorption. The adsorption of these bichalcophene derivatives onto iron surfaces was compared in three different media, namely, a vacuum, simulated water, and a corrosive solution using an MC simulation approach. The adsorption configuration is shown in Figure 8, and the comparative chart showing the adsorption energy is graphically presented in Figure 9. The bichalcophene derivatives were found to be disposed nearly horizontally over targeted iron surfaces in all three media. The obtained adsorption energy was large and negative, suggesting the spontaneous nature of adsorption of the inhibitors. It was largest for the simulated corrosive solution (i.e., 1 M HCl), followed by the water medium, and lastly the vacuum. This implies that the presence of chloride and hydronium anions facilitated adsorption through electrostatic interaction. The adsorption energy of sulfur-containing bichalcophene derivatives was greater and highly negative compared to the bifuran-based bichalcophene derivatives, suggesting that the more stable coordinate bonds were formed through the sulfur atom from the bithiophene moiety and through the nitrogen from the amido group with the iron surface atom. These findings validated the results obtained from the DFT-based quantum chemical calculation approach.

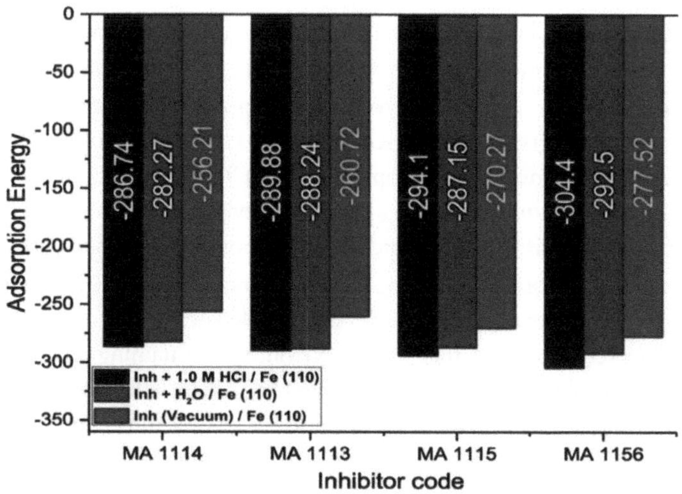

Figure 9. Configuration of bichalcophenes after interaction with an Fe (110) surface in vacuum, water, and acidic media. (Reproduced with permission from ref (100). Copyright 2018 Elsevier Ltd.)

The benzimidazole derivatives were also explored as anticorrosive additives for mild steels in acidic solutions (*101*, *102*). The anticorrosive property of 2-phenylimidazo [1,2a]pyrimidine-3-carbaldehyde (IPM) was explored to analyze its corrosion-inhibiting effect on carbon steel in a 1.0 M HCl medium through chemical, electrochemical, and theoretical procedures (*103*). The corrosion-inhibiting property of IPM was high at higher concentrations. It behaved as a mixed-type inhibitor that was cathodic in nature as obtained from PDP. The increase in charge transfer resistance and decrease in double-layer capacitance ascertained the adsorption of IPM onto metal surfaces. The Langmuir adsorption isotherm was followed during its adsorption. The kinetic and thermodynamic studies for IPM suggested the occurrence of chemisorption and physisorption mechanisms during its adsorption onto carbon steel. The quantum chemical parameters were correlated with experimental findings. The MC simulation results simulating the adsorption of IPM onto an alpha-Fe_2O_3 (111) surface considering water molecules and HCl showed that it could be adsorbed onto alpha-Fe_2O_3 (111), as shown in Figure 10. The horizontally disposed, well-adsorbed IPM with highly negative values suggested the spontaneity and stability of adsorbed configurations of IPM onto alpha-Fe_2O_3 (111) facets. This implied that IPM was capable of displacing the corrosive species present near the oxide surface and being adsorbed through the interaction of the lone pairs of electrons in the heteroatoms of IPM and the alpha-Fe_2O_3 (111) surface.

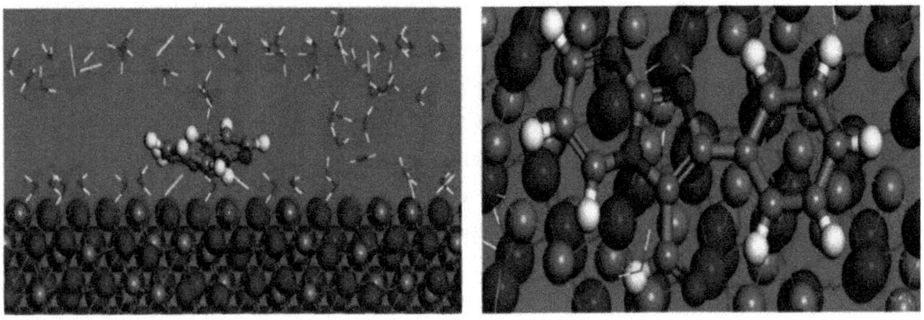

Figure 10. Side and top view of the stable, adsorbed corrosion inhibitor IPM onto α-Fe_2O_3 (111) in the presence H_2O and HCl. (Reproduced with permission from ref (103). Copyright 2017 Elsevier Ltd.)

Verma et al. (*104*) explored the influence and size of the ring on corrosion-inhibiting properties on mild steels by three 3-amino alkylated indoles, namely, N-((1H-indol-3-yl)(phenyl)methyl)-N-ethylethanamine (AAI-1), 3-(phenyl(pyrrolidin-1-yl)methyl)-1H-indole (AAI-2), and 3-(phenyl(piperidin-1-yl)methyl)-1H-indole (AAI-3), in a 1 M HCl solution by means of weight loss, electrochemical, surface morphology/topography, DFT, and MD simulation methods. It was found that the cyclic amino group containing 3-amino alkylated indoles showed better corrosion-inhibiting capability in comparison with an open chain comprising modified derivatives of an amino group. Furthermore, the corrosion-inhibiting capability increased with the increase of the cyclic amino group's ring size. This piperidine-based 3-amino alkylated indole exhibited high corrosion-inhibiting performance compared to that of the pyrrolidine-containing analogue. Moreover, the corrosion-inhibiting capability increased as the inhibitor content added to the corrosive solution increased. It was found that at a maximum concentration (0.862 mM), these inhibitor molecules showed 94.34%, 96.08%, and 96.95% corrosion-inhibiting efficacy for AAI-1, AAI-2, and AAI-3, respectively. The adsorption of these corrosion inhibitors onto a steel surface was revealed by EIS measurements. The adsorption obeyed the Langmuir adsorption isotherm. Polarization studies unveiled cathodic-type corrosion-inhibiting properties. The surface morphology and topography analyses also affirmed the adsorption of 3-amino alkylated indole derivatives onto the steel surface. The HOMO–LUMO energy and electronegativity derived from quantum chemical calculations supported the inhibitors' probable adsorption. The MD simulation study revealed AAI-1, AAI-2, and AAI-3 adsorbed onto an Fe (110) surface, attaining equilibrium energy, as shown in Figure 11. The computed binding energies supported the experimentally determined inhibition efficacies. As shown in Figure 11, all of these 3-amino alkylated indole derivatives tended to move toward the iron surface and be disposed with almost flat orientation. This flat disposition was a result of the interaction between the unshared electrons of nitrogen atoms and the π electrons of the indole, and the phenyl moieties provided its electrons to vacant d orbitals of iron atoms at the surface of metals; this helped to form a stable coordinate bond. The determined adsorption energy was found to be negative, suggesting the spontaneous interaction of the inhibitors with iron surface atoms. The order of the binding of the studied inhibitors was AAI-1 < AAI-2 < AAI-3. These findings were also corroborated with experimentally determined inhibition efficacies. The resulting interaction energy with an extremely negative magnitude for AAI-3 implied its high adsorption capability onto a metal surface; because of this, it reflected the maximum corrosion-inhibiting efficacy.

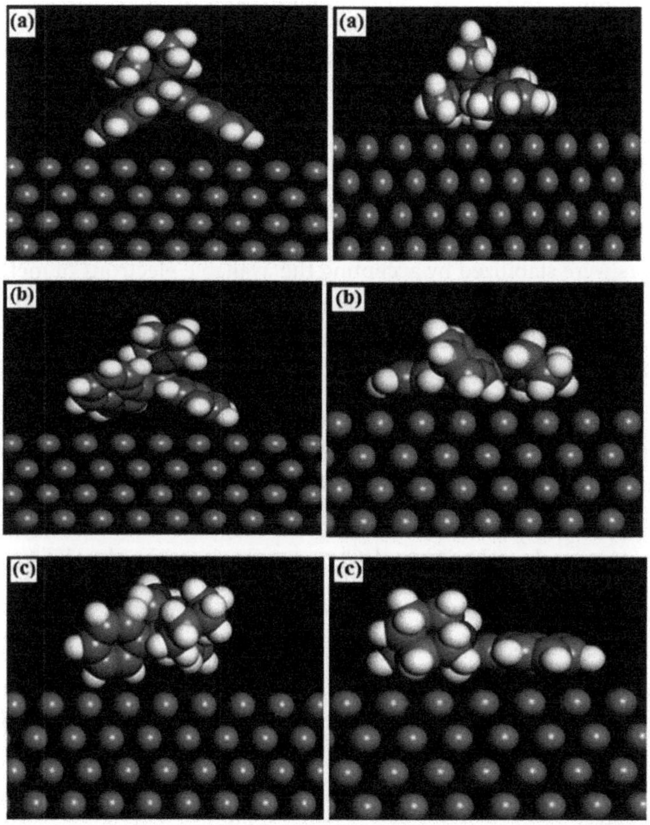

Figure 11. Side view of before (left) and after (right) equilibrium adsorption of inhibitors on an Fe (110) surface. (Reproduced with permission from ref (104). Copyright 2016 Elsevier Ltd.)

Dohare et al. (*105*) synthesized three pyrazolo-pyridine derivatives, namely, 4-(4-methoyphenyl)3,5-dimethyl-1,4,7,8-tetrahydrodipyrazolopyridine (PP-1), 3,5dimethyl-4-phenyl-1,4,7,8-tetrahydrodipyrazolopyridine (PP-2), and 3,5-dimethyl-4-(3-nitrophenyl)-1,4,7,8-tetrahyddrodipyrazolopyridine (PP-3), with different substituents using ultrasound-mediated, one-step multicomponent reactions, which are a green strategy of chemical synthesis. These pyrazolo-pyridine derivatives were found to be efficient corrosion inhibitors through gravimetric and electrochemical measurements. The adsorption of these pyrazolo-pyridine derivatives was confirmed by the inspection of surface morphology and topography. The Langmuir adsorption isotherm model fit well with experimentally determined values, suggesting the formation of a monolayer film on the inhibitor molecules without intermolecular interaction. The PDP measurement revealed its effective reduction capability on the rate of corrosion of the inhibitor molecules, and it possessed a mixed-type corrosion-inhibiting property. The EIS measurement exhibited high values of polarization resistance, suggesting high corrosion-inhibiting capability. Furthermore, the corrosion-inhibiting efficacies predicted from theoretically determined quantum chemical parameters were well supported. The corrosion inhibition order of these three inhibitor molecules was found to follow the trend PP-1 > PP-2 > PP-3. The higher corrosion-inhibiting property of PP-1 was mainly due to its parallel adsorption onto targeted surfaces despite the presence of corrosive species, and it covered a greater surface area in the MD simulation results, as shown in Figure 12. It can be said that the electron-releasing $-OCH_3$ group present in PP-1 increased the electron density on the molecular skeleton and facilitated its adherence onto surface metal atoms (*106*), while the electron-withdrawing $-NO_2$ group present in PP-3 reduced the skeletal electron density, which decreased its interaction propensity with metal atoms (*107*). Furthermore, the greater interaction energy obtained by the adsorption of the protonated inhibitor suggested greater adsorption in its protonated form. It suggested that the protonated inhibitors better adsorb over metal surfaces. It was proposed that initially the Cl^- ions were adsorbed onto the iron surface, which caused the negative charges to accumulate over the metal surface, and then the positively charged corrosion-inhibiting molecules were adsorbed over it through electrostatic interactions. The protonated inhibitors were stabilized by accepting the available electrons of metal surface atoms. In addition, the lone pairs of the molecules were shared with the empty d orbitals of surface metal atoms and eased coordinate bonding. Furthermore, the adsorption of the inhibitors was strengthened by the formation of back bonds due to the retrodonation of electrons by metal surface atoms to the corrosion-inhibiting molecules adsorbed over it. In this way, the adsorbed corrosion inhibitor formed a thin protective film, which was capable of inhibiting the penetration of aggressive ingredients toward the metal surface, thereby protecting the exposed metal from further dissolution. In this regard, mean square displacement (MSD) was also employed, and the transport activity of the corrosive species, such as H_3O^+ and Cl^-, were explored by the diffusion models with the corresponding MSD plot shown in Figure 13. The diffusion coefficient of H_3O^+ and Cl^- was highly reduced owing to a high diffusion-inhibition capability for PP-1 compared to the other analogues. PP-1 showed a higher corrosion-inhibiting efficacy than that of other inhibitors.

Zhang et al. (*108*) unveiled the enhancement in corrosion mitigation of carbon steel through the addition of a combination of two organic molecules, namely, phytic acid and benzyltrimethylammonium bromide, into 0.5 M HCl, which was explored using polarization and EIS measurements. The combination of corrosion inhibitors acted as a mixed-type corrosion inhibitor. The corrosion-inhibiting capability changed with concentration variation, as well as the molar ratio of the combination. It exhibited maximum corrosion inhibition at a 1:2 molar ratio of phytic acid and benzyltrimethylammonium bromide. The DFT and MD simulation revealed that these inhibitor molecules efficiently adsorb over metal surfaces through a chemisorption pathway. Furthermore, the combination of phytic acid and benzyltrimethylammonium bromide adsorbed more closely than their individual adsorption, as revealed through MD simulation.

Similarly, Han et al. (*109*) designed and fabricated time-dependent self-strengthening sodiumdodecyl sulfate (SDS) and 4-mercaptopyridine (4MP) as corrosion inhibitors for protecting air-hardening tool steel (A3) exposed in 0.5 M HCl for prolonged submersion. Initially, it may be presumed that the corrosion-inhibiting film of mixture of 4MP and SDS (MP–SDS) formed a monolayer, and with the extension of the immersion period, an additional layer was formed over the previous layer, giving rise to a bilayer formation. This bilayer of corrosion-inhibiting molecules efficiently minimized corrosion after recuperating the deadsorption of the previously adsorbed layer of inhibitors (*109*). It was found in the PDP study that the order of corrosion-inhibiting efficacy, as revealed through reducing corrosion current density, was SDS < MP < MP–SDS, while the individual corrosion-inhibiting efficacy increased with the increasing concentrations of these molecules. From EIS results it was found that initially there was a formation of a single depressed capacitive loop with and without the addition of SDS. It indicated single-layered SDS was adsorbed over the metal surface. With the addition of MP and SDS inhibitors into the corrosive medium and with the elapse of time, two depressed capacitive semicircles as well as two time constants were seen in Nyquist and Bode plots obtained from the EIS, which suggested the appearance of two organic layers on the electrode's surface. Initially, with the MP–SDS addition, there was a decline in anticorrosive behavior, but increases were apparent within the range of 4 to 16 hours, suggesting the increasing synergistic effect. Thus, a remarkable adsorption durability was achieved. Reduction of corrosion-inhibiting properties was encouraged by removing previously adsorbed organic film; after the prolonged immersion time, it was recovered owing to its significant property of double-layered inhibitor formation. From a weight loss measurement, it was observed that by adding 100 mg L^{-1} of an inhibitor into the test solution, the corrosion rate of A3 steel was reduced in the following trend: MP–SDS $<$ MP $<$ SDS. Alternatively, the corrosion-inhibiting efficacy was enhanced as SDS $<$ MP $<$ MP–SDS. This suggested that the mixing of MP and SDS synergistically increased corrosion-inhibiting efficacy.

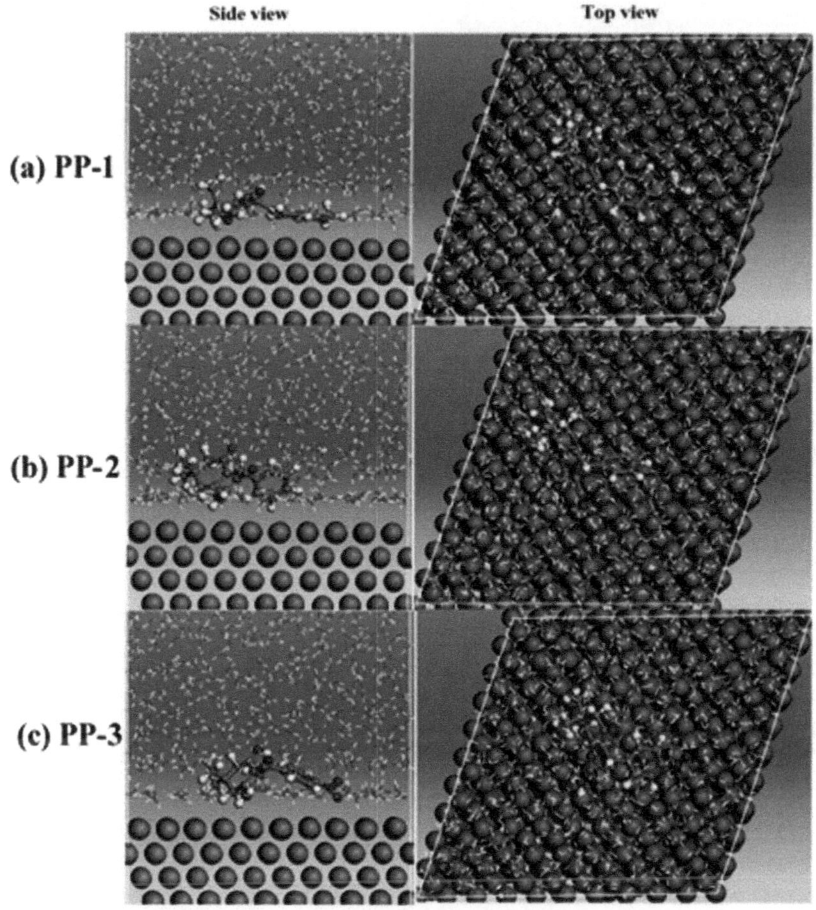

Figure 12. Equilibrium adsorbed configuration of (a) PP-1, (b) PP-2, and (c) PP-3 on an Fe (110) surface. Images displayed on the left- and right-hand sides refer to side and top views of adsorbed configuration, respectively. (Reproduced from ref (105). Copyright 2019 Creative Commons CC BY license.)

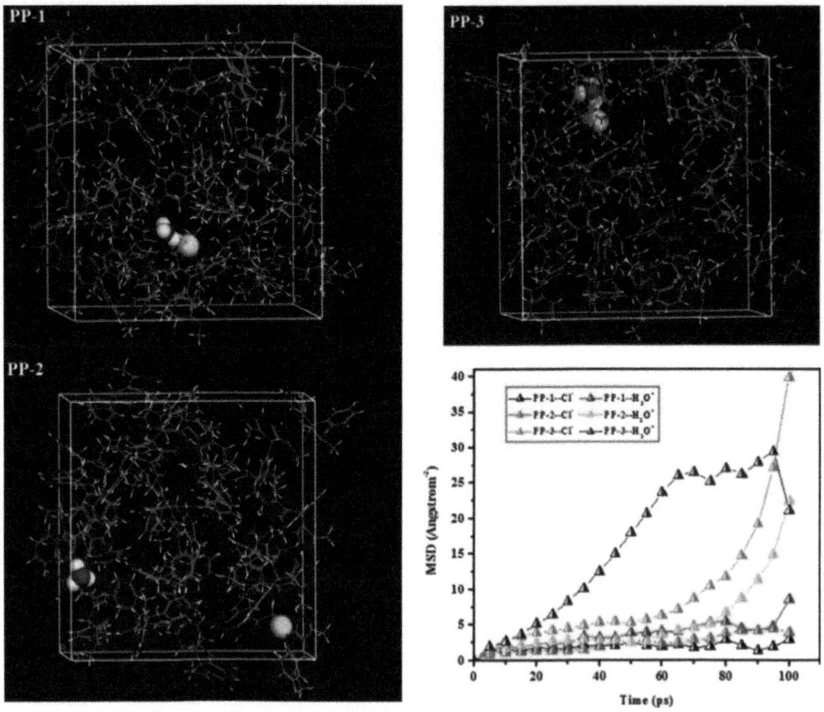

Figure 13. Model of diffusion and MSD curves for Cl^- and H_3O^+ determined in the film of inhibitors. (Reproduced from ref (105). Copyright 2019 Creative Commons CC BY license.)

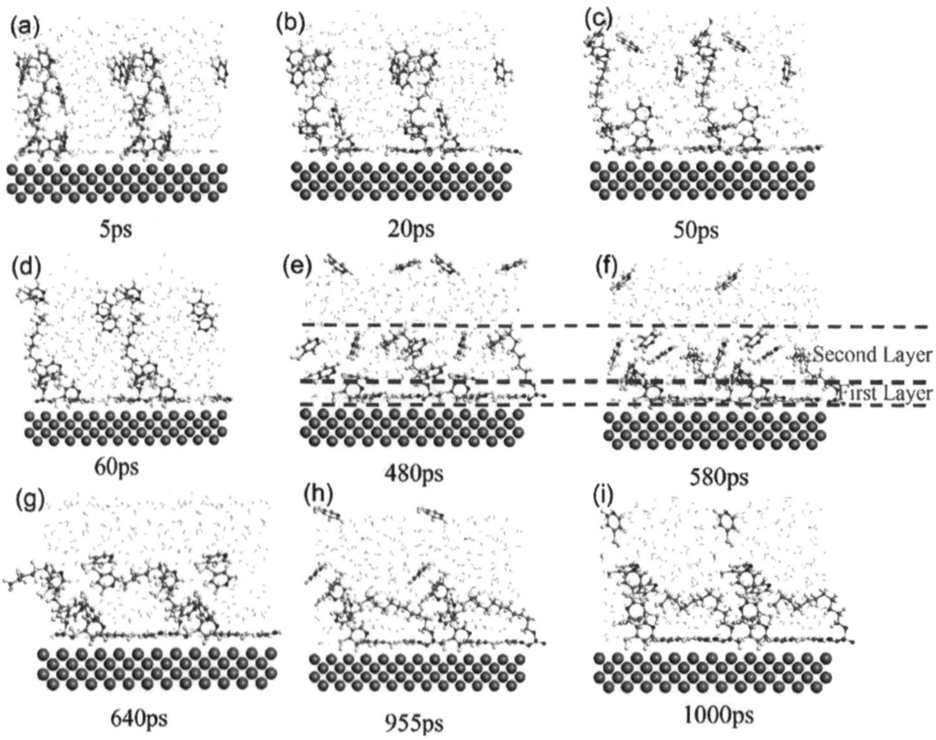

Figure 14. Snapshots of the dynamic adsorption and coadsorption of corrosion-inhibiting molecules with the progress of time. (Reproduced with permission from ref (109). Copyright 2018 Elsevier Ltd.)

From the MD simulation study, the observation of the dynamic frames of interactions also clearly helped visualize the bilayer formation due to the interaction of MP and SDS on the Fe (001) surface. The relative snapshots showed that the MP adsorbed onto the iron surface in a horizontal fashion. In the case of the SDS molecule, the hydrophilic group adsorbed onto the iron surface with the aliphatic chain disposed away from the iron surface when stretched within the solution phase (see Figure 14(a–d)). In this way the coadsorption of MP and SDS formed a protective monolayer and protected the Fe atoms from further dissolution. With the passage of time, the adsorption led to the formation of a bilayer with some porosity caused by MP interacting with the tails of SDS molecules (see Figure 14(e)). Afterward, the compactness of the bilayer was improved as the exposure time increased with the prolonged increase in the immersion time (see Figure 14(f–i)), consequently enhancing corrosion inhibition. So it can be said that the alkyl chain of SDS facilitated synergistic interaction with the MP molecules. Formation of such an inhibitor film displaced the water molecules present near the Fe surface, which hindered the migration of the corrosive species and helped reduce the corrosive dissolution.

For better understanding, Figure 15 shows the adsorption of MP (a) and the subsequent (b, c, d) bilayer formation by the interaction of MP and SDS, leading to a dense and compact film with the progress of time.

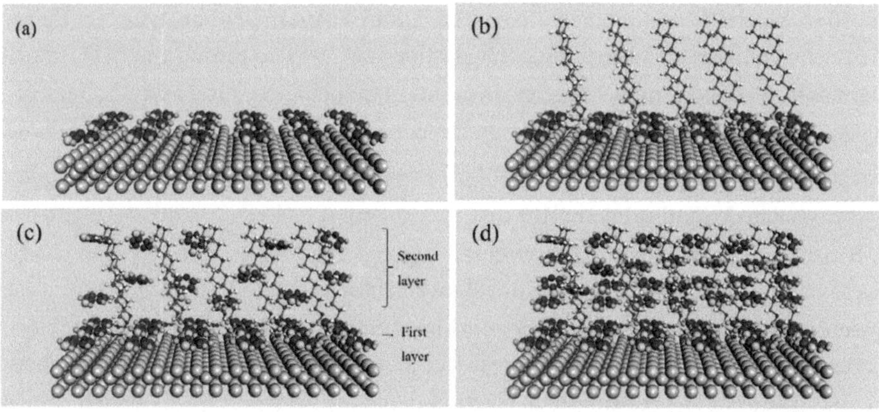

Figure 15. Schematic diagram showing the adsorption of (a) MP and (b, c, d) the subsequent bilayer formation by the interaction of MP and SDS, leading to a dense and compact film with the progress of time. (Reproduced with permission from ref (109). Copyright 2018 Elsevier Ltd.)

Some derivatives of chalcone (i.e., 3-(4-(dimethylamino)phenyl)-1-(4-hydroxyphenyl)prop-2-en-1-one, 3-(3-hydroxyphenyl)-1-(4-hydroxyphenyl)prop-2-en-1-one, and 1-(4-hydroxyphenyl)-3-phenylprop-2-en-1-one) were reported to be efficient corrosion-inhibiting additives explored in HCl media for protecting carbon steel, and their corresponding corrosion-inhibitng efficacies, analyzed using electrochemical measurement, were 91.3%, 87.0%, and 85.9%, respectively (*110*). The probable reason behind the adsorbing capability of chalcone derivatives on carbon steel surfaces was analyzed from the density functional-based tight binding (DFTB) calculation to understand its adsorption and subsequent corrosion-inhibiting property by Guo et al. (*111*). The chalcone derivatives were found to adsorb onto the Fe (110) plane in a horizontal fashion. The interaction of these chalcone derivatives with iron surface atoms led to charge redistribution. The differences arising in the charge density were due to adsorbing chalcone derivatives onto Fe (110). Furthermore, the chalcone derivatives were capable of adsorbing onto the Fe (110) surface, as revealed from quantitatively analyzing Mulliken atomic charges of the heteroatoms of the free and adsorbed chalcone derivatives, which showed significant changes. Again, the significant variation of projected density of states (PDOS) before and after chalcones adsorbed onto the Fe (110) surface showed that there were discrete peaks of PDOS for the free chalcone derivatives, but on adsorption, the molecular PDOS were unsaturated nearby Fermi energy for the three horizontally disposed configurations with a significant decrease in intensity, suggesting the occurrence of a strong hybridization of orbitals of reactive sites of the chalcone derivatives over metal surface atoms. Moreover, the electron density distribution analysis in frontier molecular orbitals, herein HOMO-1, HOMO, as well as LUMO of free and adsorbed chalcone derivatives on metal surfaces, revealed electron sharing between the inhibitor and metallic surface. This was further ascertained by a decrease in the ΔE of the adsorbed chalcone derivatives compared to the isolated one, implying the stabilization of the inhibitor molecules on adsorption (*112*).

Saha et al. (*113*, *114*) and Murmu et al. (*115*, *116*) also revealed potential corrosion-inhibiting properties of some Schiff base molecules as efficient corrosion inhibitors in 1 M HCl media for protecting mild steels explored using weight loss and electrochemical measurements. The corrosion-inhibiting property was due to appreciable adsorption of azomethine-functionalized organic compounds onto mild steel surfaces. Though the adsorption was validated using surface characterization through field emission scanning electron microscopy (FESEM) and energy-dispersive X-ray spectrometry (EDXS), the insight behind the intrinsic adsorption property and the adsorption capability was predicted using DFT and Fukui indices analysis. Furthermore, the competitive and equilibrium adsorption configuration was revealed through an MD simulation in a model containing various corrosive species. In addition to such theoretical analysis, Saha et al. (*113*) also employed DFTB calculations to envisage the occurrence of electron transference between the corrosion-inhibiting molecules, namely, 2-((2-hydroxyethylimino)methyl)-6-methoxyphenol (L^1), 2-((1-hydroxybutan-2-ylimino)methyl)-6-methoxyphenol (L^2), and 2-((2-hydroxy-3-methoxybenzylideneimino)phenol (L^3), over the targeted metal facets reflected from charge density difference, as shown in Figure 16. The regions shown in blue in Figure 16 portray the accumulation of electron density, whereas the red regions denote loss of electrons. The interaction of the electron-rich centers present in the Schiff bases with iron surface atoms was affected by an electron-dense group, leading to its adsorption. It is obvious that electron depletion is highest for L^3, which is attributed to its greater electron-sharing capability with the iron surface atoms as compared to others (L^1 and L^2). Because of this, the DFTB technique can be employed to effectively visualize the consequences of an inhibitor–metal interaction leading to its equilibrium adsorption.

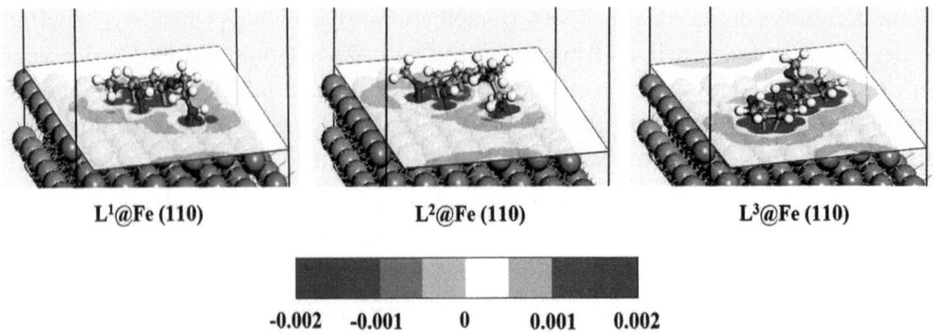

Figure 16. The difference in charge density due to adsorbed organic molecules on an iron surface. (Reproduced with permission from ref (113). Copyright 2018 Royal Society of Chemistry.)

Furthermore, Murmu et al. (*115*) also explored the effect of the stereochemical configuration of 4-(4-((Pyridin-2-yl)methyleneamino)phenoxy)-N-((pyridin-2-yl) methylene)benzenamine (PMB) and 4-(4-(4-((Pyridin-2-yl)methyleneamino)phenoxy)phenoxy)-N-((pyridin-2-yl)methylene)benzenamine (PPMB) on their adsorption, and the surface coverage of these inhibitors helped increase the corrosion-inhibiting property as determined from the experimental techniques. The experimental findings revealed that the PPMB showed greater corrosion-inhibiting efficacy (93%) compared to PMB (~89%), which was due to high surface coverage by PPMB adsorption onto the metal surface atoms. Both these Schiff bases acted as a mixed-type corrosion inhibitor capable of inhibiting the cathodic as well as anodic reactions simultaneously and followed the Langmuir adsorption isotherm. The adsorption of these inhibitors was confirmed from FESEM and EDXS studies. Furthermore, the surface topography change was ascertained from the atomic force microscopy study. It showed that the mild steel samples taken out from such aggressive solution exposure were found to be smoother and to have lower average surface roughness values, compared to the uninhibited solution. It indicated the adsorption of the added inhibitors. This adsorption also modified the surface of mild steels toward hydrphobicity as revealed from the water contact angle measurement. The adsorption capability of both PMB and PPMB were explored at an electronic as well as molecular level. The almost horizontal geometrical configuration of PPMB made it a better surface-covering adsorbed corrosion inhibitor than PMB. The obtained quantum chemical parameters supported the efficient and better adsorption of PPMB than of PMB. The presence of sufficient reactive centers in the structure of the inhibitors additionally facilitated their adsorption. The adsorption of these inhibitors was also visualized from the charge density difference arising due to the inhibitors' interaction with targeted metal surface atoms, which was confirmed from DFTB calculations simulating the adsorption of reactive molecular fragments of inhibitors with targeted metal surfaces. Additionally, the MD simulation study clearly helped visualize the spontaneous adsorption of the inhibitor molecules, even in the presence of corrosive species.

Again, it was also reported that azomethine-functionalized triazole derivatives, namely, 5-((furan-2-yl)methyleneamino)-2H-1,2,4-triazole-3-thiol (FMT) and 5-((thiophen-2-yl)methyleneamino)-2H-1,2,4-triazole-3-thiol (TMT) inhibitor molecules, interacted competitively with other corrosive species present in an acidic environment, thereby being adsorbed over metal surface atoms. After their adsorption, they protected the substrate from further corrosion damages (*116*). The efficient corrosion-inhibiting property was revealed from electrochemical analysis. Additionally, the outcomes derived from the DFT, Fukui indices, and MD study were corroborated with experimental findings. The presence of the most polarizable sulfur atoms in TMT molecules was attributed to high adsorption compared to FMT. The adsorption of these molecules were also elucidated from the HOMO and LUMO plot extracted from the DFT calculation before and after the simulated adsorption study performed by MD simulation revealed a remarkable change (*116*). The interaction or electron sharing leading to adsorption was furthermore ascertained from the changes unveiled in the electron density distribution of these inhibitors adsorbed onto metal surface atoms, as shown in Figure 17. It confirmed the inhibitors' adsorption onto an iron surface. Similar observations were also reported elsewhere in the literature (*111*).

Amino acids also worked as corrosion inhibitors. It was reported that amino acids are efficient as corrosion inhibitors for metallic materials in corrosive media (*117–120*), but deeper theoretical insights were not explored. Kumar et al. revealed more insight on adsorption of the four amino acids, namely, cysteine (CYS), glutamic acid (GLA), glycine (GLY), and glutathione derivative (GLT), on a copper surface through DFT (*121*). Quantum chemical descriptors, namely, E_{HOMO}, E_{LUMO}, and energy gap, established a worthy relationship between these descriptors with experimentally obtained corrosion-inhibiting efficacy. A comprehensible study to explore the adsorption of these amino acids were carried out for Cu (111), Cu (100), and Cu (110) surfaces.

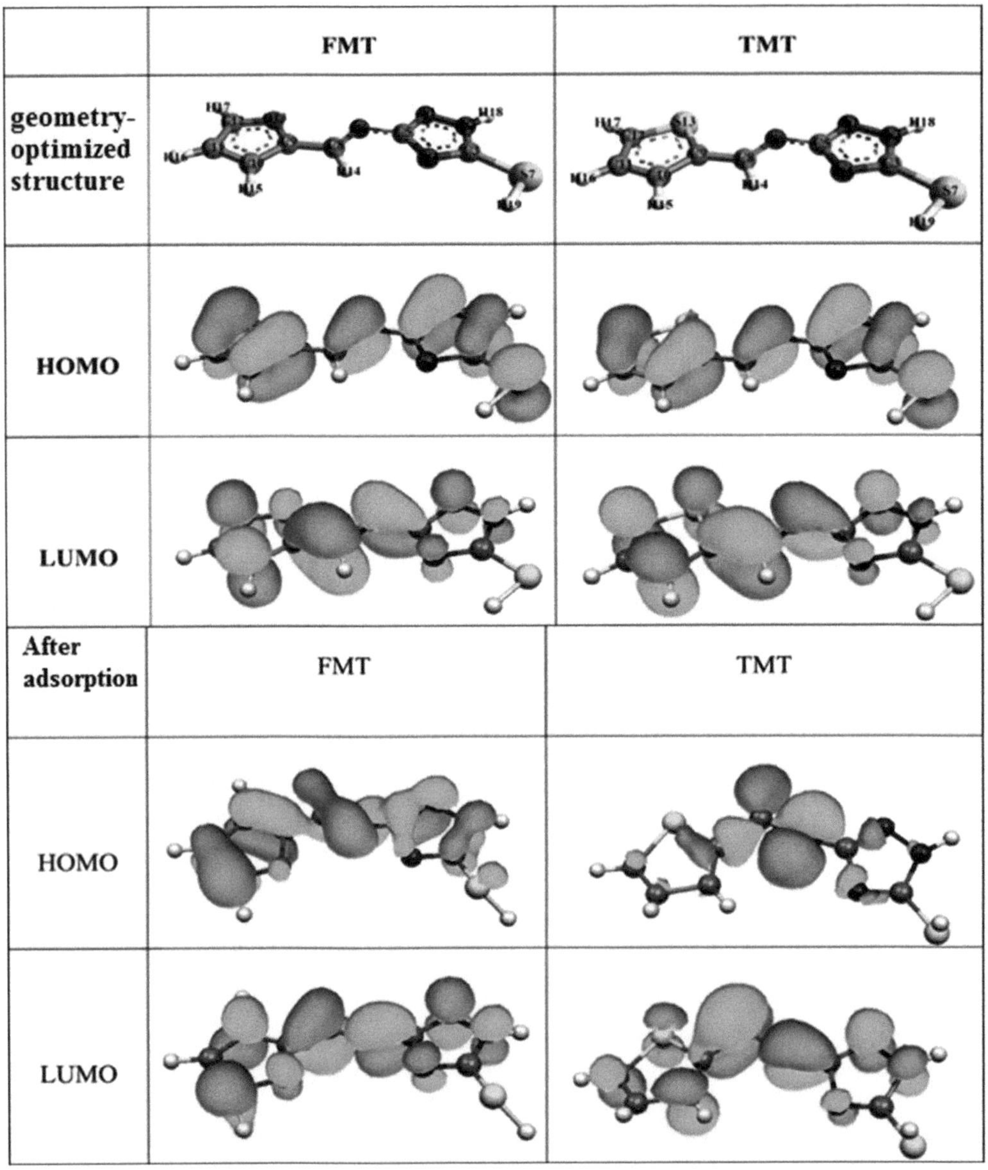

Figure 17. The HOMO and LUMO plot for free and adsorbed forms of inhibitors. (Reproduced with permission from ref (116). Copyright 2018 Elsevier Ltd.)

From the optimized structure, it was found that chemical bonds formed between nitrogen, oxygen, and sulfur atoms of these molecules and the Cu (111) surface, which was ascertained from the changes found in the density of electrons, as shown in Figure 18, and PDOS plots, as shown in Figure 19, respectively. The region colored in blue shows the depletion of electrons, and the region in reddish brown denotes the accretion of electrons. From Figure 18, it is obvious that the formation of a coordinate bond between a heteroatom of amino acids and a copper atom, namely, N-Cu, O-Cu, and S-Cu, was facilitated and well visible from the charge redistribution. Comparing the geometrically optimized adsorption configuration of GLT, CYS, GLA, and GLY onto a Cu

(111) surface, it was found that the GLT–Cu (111) system exhibited greater adsorption, which was attributed to the higher localization of charges amid GLT and the Cu (111) plane, resulting in the ensuing adsorption strength trend GLT > CYS > GLA ~ GLY.

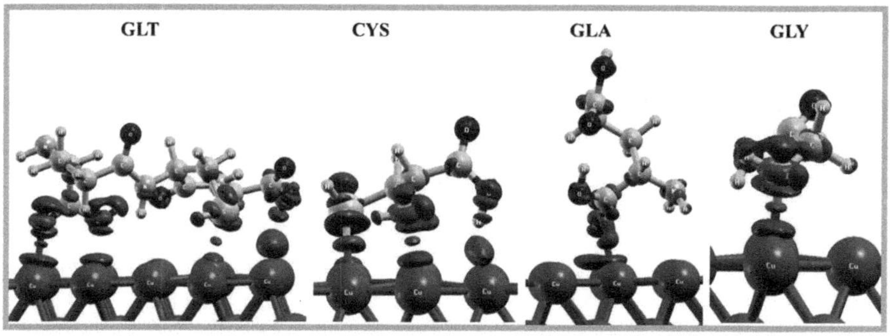

Figure 18. The optimized geometry of GLT, CYS, GLA, and GLY adsorbed onto a copper surface showing the electron density differences. (Reproduced with permission from ref (121). Copyright 2020 Elsevier Ltd.)

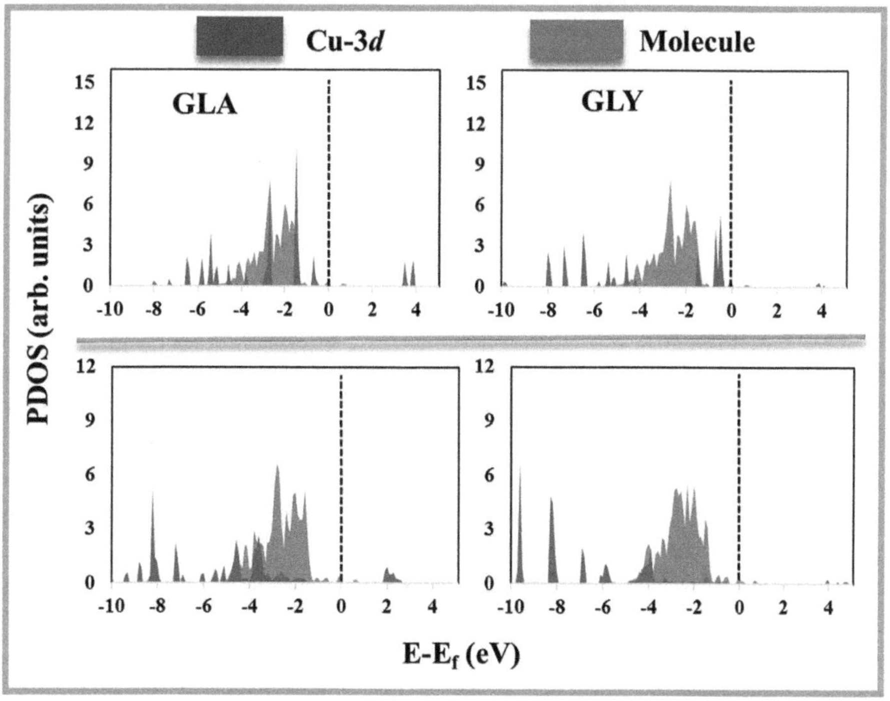

Figure 19. PDOS for GLA and GLY before and after adsorption on a Cu (111) surface. (Reproduced with permission from ref (121). Copyright 2020 Elsevier Ltd.)

Moreover, from the PDOS of bonded atoms, some of the studied amino acids on copper surface atoms shown in Figure 19 revealed that numerous molecular states existed in the copper 3*d* band, and hybridization of molecular orbitals of amino acids with the copper surface atom takes place. It is evident from the figure that prior to adsorption, the molecular peaks were sharp and discrete, but after the adsorption of the amino acid onto the copper surface atom, significant broadening of the peaks occurred, implying that the strong interaction occurred between the sorbate and substrate. A careful observation revealed that the molecular states existing in the 3*d* band copper atoms varied as GLT > CYS > GLA > GLY. This indicates that maximum hybridization occurred for the GLT–Cu

(111) system and gave rise to maximum adsorption. The analysis of the PDOS validated the adsorption energy and corrosion-inhibiting efficacy trends. The effective inhibiting property was found when the corrosion-inhibiting molecules adsorbed onto a metal surface, developing a shielding layer that was dense, compact, as well as free of defects. This was elucidated theoretically by increasing surface coverage by corrosion-inhibiting molecules, as shown in Figure 20 for two and three molecules. It was found that with the increase of the surface coverage, there was no change in the adsorption energy. However, an intermolecular interaction could occur, primarily H-bonding, as shown in Figure 20. In addition to that, the calculated adsorption energy remained unaltered when the number of molecules was increased from one to three on the Cu (111) plane.

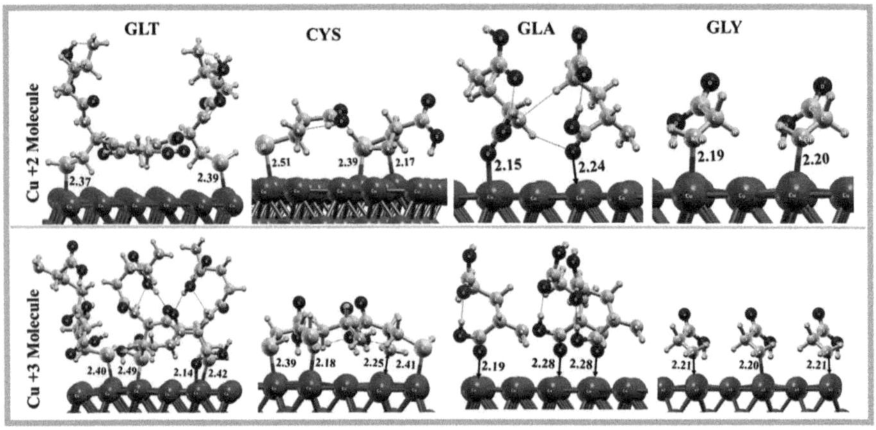

Figure 20. Energetically minimized adsorbed configuration amino acids with an increase in number from one to three on a Cu (111) surface. (Reproduced with permission from ref (121). Copyright 2020 Elsevier Ltd.)

Boutouil et al. (*122*) reported on the corrosion-inhibiting efficacy of heterocycle 1,2,3-triazole, namely, (1-p-tolyl-1H-1,2,3-triazol-4-yl)methanol (TTM), in a 1 M HCl medium by PDP within temperature ranges of 298 to 333 K. The addition of the TTM inhibitor into the test solution decreased both cathodic as well as anodic current density. Thus, it behaved like a mixed-type corrosion inhibitor. An EIS study revealed a remarkable reduction in corrosion inhibition by the TTM molecule, showing 90% efficacy after 60 minutes of immersion. The TTM molecule reacted spontaneously with surface atoms and adsorbed onto mild steel following the Langmuir adsorption isotherm. However, at elevated temperatures, a deviation of slope from unity was found, which suggested there might be additional sites for adsorption at the metal–solution interface when the TTM molecule was adsorbed onto it. Upon fitting the results into the El Ewady thermodynamic kinetic model, it was found that at high temperatures the result was well fitted. It suggested that each TTM molecule was coordinated over more active sites present in targeted electrodes at elevated temperatures (*123*). The quantum chemical parameters, such as electron density distribution in the HOMO–LUMO of TTM, revealed that the studied inhibitor molecule had sufficient electron distribution in the HOMO of the molecules so it could be donated to the vacant d orbital of the Fe atom from the entire structure of the protonated species, and the protonated TTM molecule could participate in interaction through the aromatic and heterocyclic ring. Alternatively, the electron acceptance of TTM could be accomplished by accepting the incoming electron from the metal centers toward the inhibitor molecule, mostly on the aromatic and heterocyclic moiety in the neutral molecule, and mostly on the aromatic moiety on the protonated moiety in the TTM molecule. The electron transfer from TTM to metal surface atoms was more feasible as evidenced from the positive values of the calculated fraction of electron transfer. The local reactivity analysis showed the presence of both electrophilic as well as nucleophilic centers on TTM molecules, which were susceptible to interacting with metal atoms to adsorb. Similarly, molecular electrostatic potential mapping showed that the most negative region, where the electron density was more concentrated, was located in the triazole moiety of TTM, suggesting that the nitrogen and oxygen heteroatoms present in TTM molecules participate in adsorption by electron donation to an electron-deficient site in metals. Furthermore, the topology analysis of the electron localization function through quantum chemical methods provided significant information on the electron density distribution. The determined electron localization function values revealed that the electrons were mostly localized, showing its predominant organometallic bond formation behavior. This property helped the TTM molecule adsorb onto the metal surface atom.

Moreover, MD simulation was very helpful in exploring theoretical insights on the interaction between the inhibitor molecule TTM and the corroding species like H_2O, H_3O^+, and Cl^-. It revealed that the protective layer moved toward the metal surface atoms and was stabilized upon interacting with the metal surface. From the radial distribution function, it was revealed that the TTM inhibitor adsorbed onto the Fe atom through nitrogen, oxygen, and chlorine heteroatoms of TTM molecules because the closest distance between a heteroatom and a surface Fe atom was less than 3.5 Å, and the adsorption was considered to be chemisorption. The adsorbed organic film isolated the iron surface from the harsh electrolyte, thereby inhibiting the ingress of the corrosive species. The coefficient of diffusion calculated from the mean square displacement of the penetration of the corrosive species H_2O, H_3O^+, and Cl^- revealed that the displacement of H_2O is more pronounced than that of H_3O^+ and Cl^-, which implies that the adsorbed layer of TTM on the Fe surface can highly inhibit the penetration of H_3O^+ and Cl^- ions. The diffusion coefficient order of H_2O, H_3O^+, and Cl^- at 303 K was $Cl^- < H_3O^+ < H_2O$. From the fraction of free volume (FFV), it was also confirmed that the smaller the FFV, the harder it is to penetrate the organic layer. The computed FFVs are 13.8% for Cl^- and 13% for H_3O^+ ions, as shown in Figure 21, suggesting that this formed protective layer highly constrained the diffusion of the highly aggressive species present in the corrosive solution.

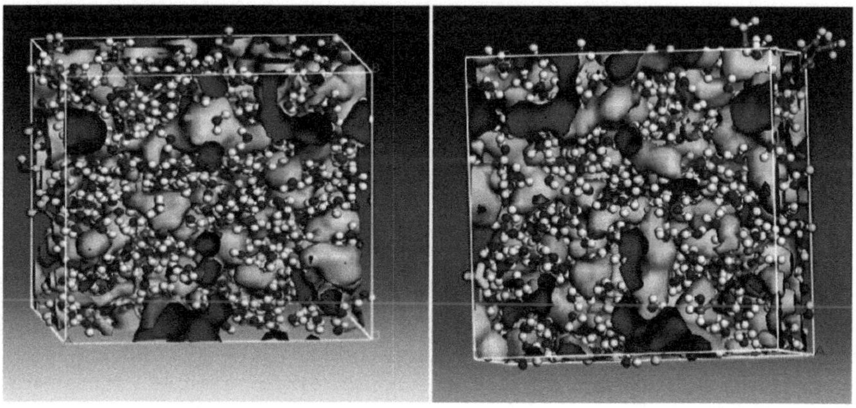

Figure 21. FFV on TTM film probed using Cl^- (left-hand side) as well as H_3O^+ (right-hand side), respectively. (Reproduced with permission from ref (122). Copyright 2016 Elsevier Ltd.)

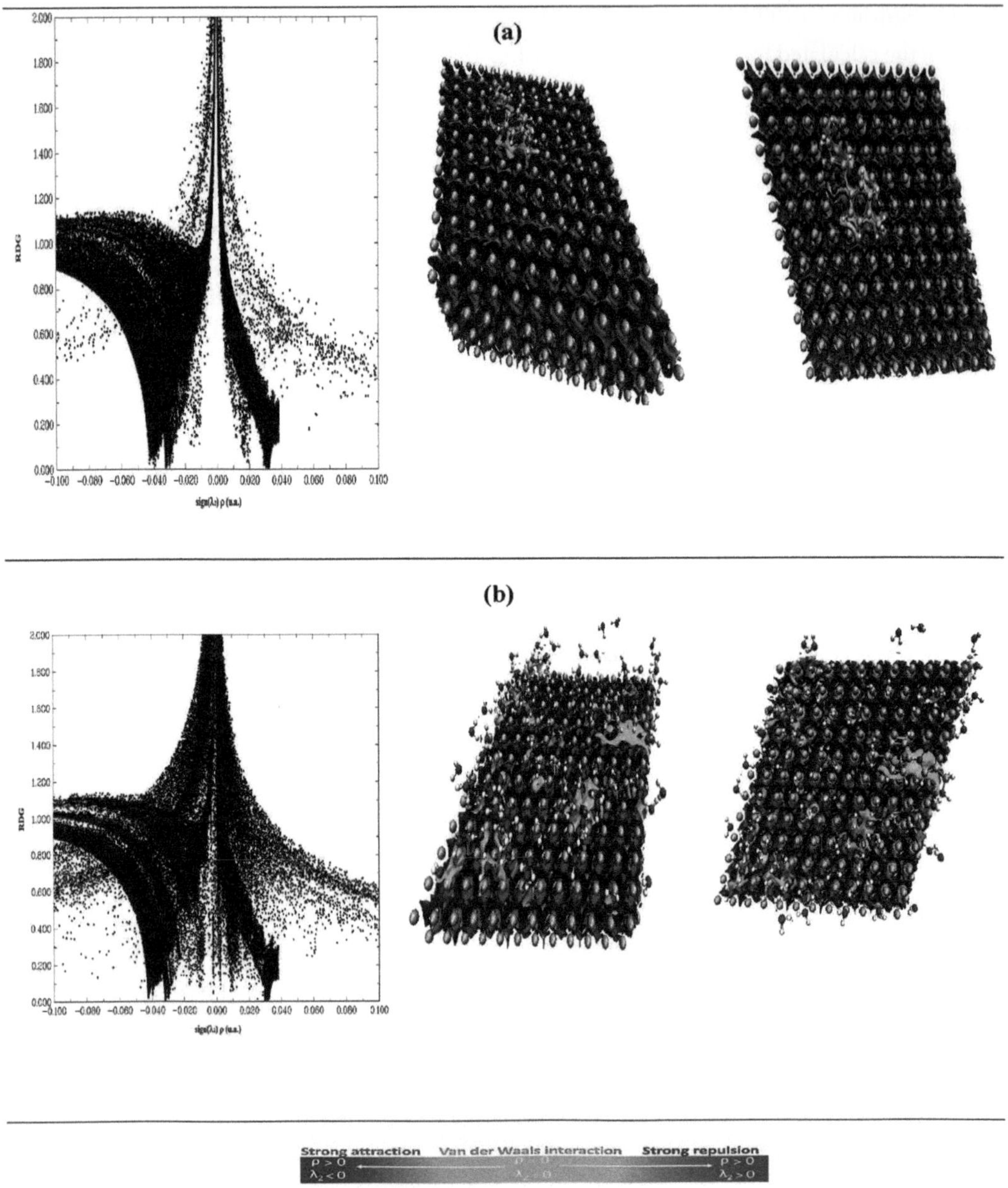

Figure 22. Reduced density gradient (RDG) vs. effective density (sign ($\lambda 2$) × ρ) plot for an Fe (110) surface in (a) the gaseous phase and (b) an acidic medium. (Reproduced with permission from ref (122). Copyright 2016 Elsevier Ltd.)

Furthermore, an analysis of the noncovalent interaction (NCI) based on a reduced density gradient (RDG) was a convenient method for mapping as well as analyzing weak interactions in the three-dimensional space (*124*). The interpretations of the RDG were visualized using a specific isosurface where the coloring effect depended on effective density (sign ($\lambda 2$) × ρ). The negative value of effective density indicated the attractive interactions (like H-bonding), whereas the positive values and the values close to zero indicated the occurrence of nonbonding interactions and van der Waals interactions, respectively. Even though such sorts of interactions (like H-bonding, London dispersion, and steric hindrances) were weak compared to the covalent bonding, when these interactions acted collectively, their effect could be indispensable. The variation of RDG with respect to effective density (i.e., sign (λ2) × ρ) was plotted for the considered system, as shown in Figure 22. The three types of interactions were graphically seen from the obtained NCI isosurface, where the blue, green, and red regions were associated respectively to H-bonding, steric hindrances, and London forces (see Figure 22). It is evident from Figure 22 that the van der Waals interactions play a key role in bridging the inhibitor molecule TTM with the electrode surface atoms through cooperative interaction, namely, π ... π, C–H ... π, and X– π (where, X = N, O, Cl atoms present in TTM). However, a steric repulsive interaction existed, as shown by the red region, which was due to the interaction of phenyl aromatic and triazole moiety. Thus, the adsorption of corrosion-inhibiting molecules over metallic atoms were correlated with the NCI results.

The corrosion inhibition of an organic molecule was also enhanced by the addition of a coadditive, which could synergistically enhance its adsorption over the metal substrate and result in increased surface coverage. According to Aramaki and Hackerman, the synergistic parameter (S_i) was determined using the following relation (*125*):

$$S_i = \frac{1-\theta_{1+2}}{1-\theta'_{1+2}} \qquad (34)$$

$\theta_{1+2} = (\theta_1 - \theta_2) - (\theta_1\theta_2)$, and θ_1 and θ_2 refer to the surface coverage due to the additive and coadditive, respectively, and θ'_{1+2} refers to the combined surface coverage of the additive and coadditive.

The preferred synergism was obtained when the determined $S_i > 1$, and the antagonistic effect was inferred if $S_i < 1$, while $S_i = 1$ indicated no interaction between the additive and the coadditive (*125*).

Generally, potassium iodide is added as a coadditive to enhance and stabilize the adsorption of inhibitors over metal surfaces through the coadsorption of these iodide ions (*126*). Aquino-Torres et al. (*127*) unveiled that corrosion-inhibiting efficacy of 0.1 mM cinnamic, p-coumaric, and caffeic acids increased from 70 to 80% with the addition of 2 mM potassium iodide in a 1 M HCl solution obtained from electrochemical measurements; this was attributed to the coadsorption of iodide ions. The presence of potassium iodide induced dipolar sites over metal surfaces that improved the adsorption of the inhibitors. The iodide ions played a crucial role in facilitating inhibitors' adsorption onto the metal surface atoms, which was inferred from the outcomes of simulated models, namely, (a) Fe (110) and (b) iodide monolayer adsorbed Fe (110) surfaces analyzed using DFT (see Figure 23). As shown in Figure 23(a), all the corrosion-inhibiting molecules adsorbed onto the metal through six sites, with four from the aromatic carbon of the ring and the other two through carbon and oxygen of the carboxylic group of the corrosion-inhibiting molecules (highlighted in green), which promoted a symmetric disposition of the aromatic moiety on the metal surface atoms. Furthermore, upon adsorption of these organic molecules onto surfaces, the hydroxyl group (–OH) and the hydrogen atoms were raised above the plane. The strong interaction of these atoms was confirmed from the obtained bond distance between the Fe and C atoms, and was more or less near the sum of the covalent radii (i.e., 1.97 Å). The second model, as shown in Figure 23(b), revealed equilibrium adsorption of these inhibitor molecules through seven sites, in which four centers coordinated through aromatic carbon of the ring, while the remaining three coordinated through three carbon atoms from the carboxylic functional groups (highlighted in blue). The configuration of adsorbed molecules over an iodide/Fe (110) surface possessed a planar and symmetrical distribution of the ring on the targeted surface. Additionally, the hydrogen atoms of the aromatic rings remained in plane with the ring with no distortion. In this case, the average Fe–C bond distance was 3.7 Å, while the distances between the iodide layer and the cinnamic, p-coumaric, and caffeic acids were 3.58 Å, 3.78 Å, and 3.83 Å, respectively. The order of the obtained adsorption energy trends corroborated with the corrosion-inhibiting efficacy trend of the molecules. The magnetic moment was condensed upon adsorption of these molecules, establishing cinnamic acid as the best corrosion inhibitor.

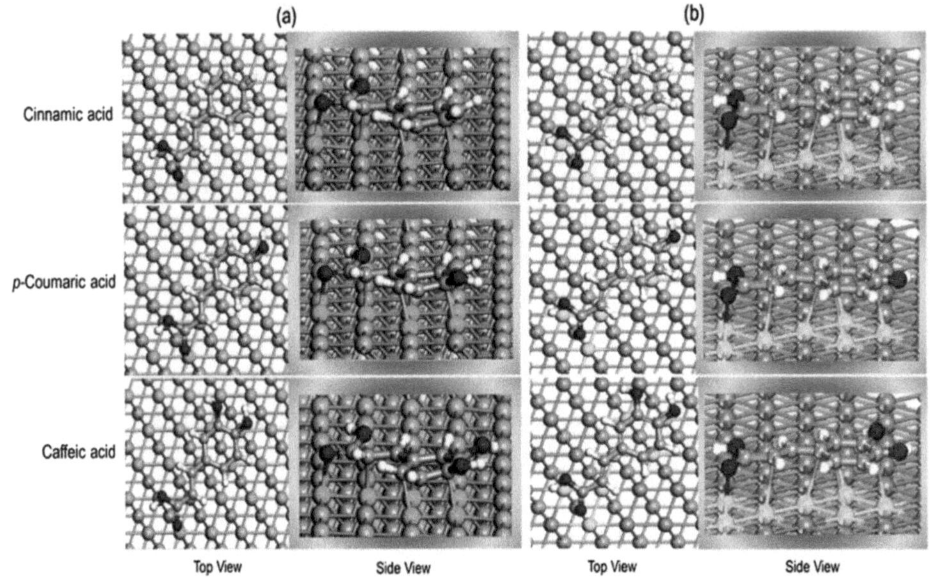

Figure 23. Adsorbed cinnamic, p-coumaric, and caffeic acid on (a) Fe (110) and (b) iodide preadsorbed Fe (110) planes. (Reproduced with permission from ref (127). Copyright 2016 Elsevier Ltd.)

Corrosion Inhibitors Used in a Nitric Acid Medium

The corrosion-inhibiting properties in nitric acid were also explored by the exploitation of several organic molecules. Li et al. (38) revealed the efficient corrosion-inhibiting property of 2-mercaptopyrimidine (2MP) derivatives (namely, thiol-MP, thione-MP, p-thiol-MP, and p-thione-MP) and its inhibition mechanism in corrosive HNO_3 for cold-rolled steel. 2-Mercaptopyrimidine was found to be highly efficient in retarding corrosion of cold-rolled steel compared to its thione, p-thiol, and p-thione derivatives, which was analyzed through weight loss measurements and electrochemical techniques. Based on the chemical structure of the substituted 2-mercaptopyrimidine derivatives, corrosion-inhibiting efficacy was found in the order: –SH > –Br > –NH2 > –OH > –Cl. Three main modes of adsorption of 2-mercaptopyrimidine derivatives were proposed. First, in an aqueous solution of HNO_3, the ionized NO_3^- could initially be adsorbed onto the cold-rolled steel surface, consequently creating a negatively charged surface. In such circumstances, the protonated 2-mercaptopyrimidine derivatives were adsorbed, thereby interfering with the successive dissolution of iron exposed to an HNO_3 solution.

Second, as shown in Figure 24, regardless of the protonation and tautomeric equilibriums between each form, the sulfur atoms could adsorb over metal surfaces. Additionally, unshared pairs of electrons as well as the π electrons present in the pyridine moiety also interacted with the suitable orbital of the metal, facilitating its entire molecular adsorption. The adsorption of inhibitor molecules over a greater surface area acted as barrier and retarded the corrosion reactions.

The inhibitor interacted with surface active atoms, which led to the formation of an inhibitor…metal complex (*128, 129*). The sulfur heteroatom present in the inhibitor molecules skeleton as a major reactive site bonded with the metal surface atom, making it easy to adsorb. The strong interaction of the 3*p* orbital of the sulfur heteroatom and the 3*d* orbital of the iron atom revealed by a Mayer bond order produced the most stable inhibitor…metal complex among the other derivative…metal complexes obtained from DFT. The theoretical calculations for the inhibitor…metal complex provided further insight on inhibitor–metal surface atom interaction. The optimized geometrical structures of 2MP complexed with Fe, namely, (a) thiol-MP-Fe, (b) thione-MP-Fe, (c) p-thiol-MP-Fe, and (d) p-thione-MP-Fe, as shown in Figure 25. All the four inhibitors…metal complexes were only formed easily when the sulfur atom bonded directly with the iron atom, indicating that the sulfur atom was the major active site for interacting with the Fe atom. The shorter S…Fe distance was close to 2.3–2.5 Å and a larger Mayor bond order ((a) 0.4068; (b) 0.4677; (c) 0.4127; and (d) 0.4736), implying the strong interaction of the sulfur atom with the Fe atom.

Third, the 2-mercaptopyrimidine possessed strong coordination capability due to the presence of sulfur and nitrogen heteroatoms on its pyridine moiety with the Fe^{2+} and Fe^{3+} ions forming an inhibitor–metal coordination complex. This acted as a shielding barrier separating the metal from corrosive damage in an HNO_3 solution. All these inhibitors functioned like mixed-type inhibitors primarily retarding the anodic reaction. The adsorption and corrosion inhibition property of 2-mercaptopyrimidine derivatives determined from wet chemical experimentations were found in the following order: thiol-MP < thione-MP < p-thiol-MP < p-thione-MP.

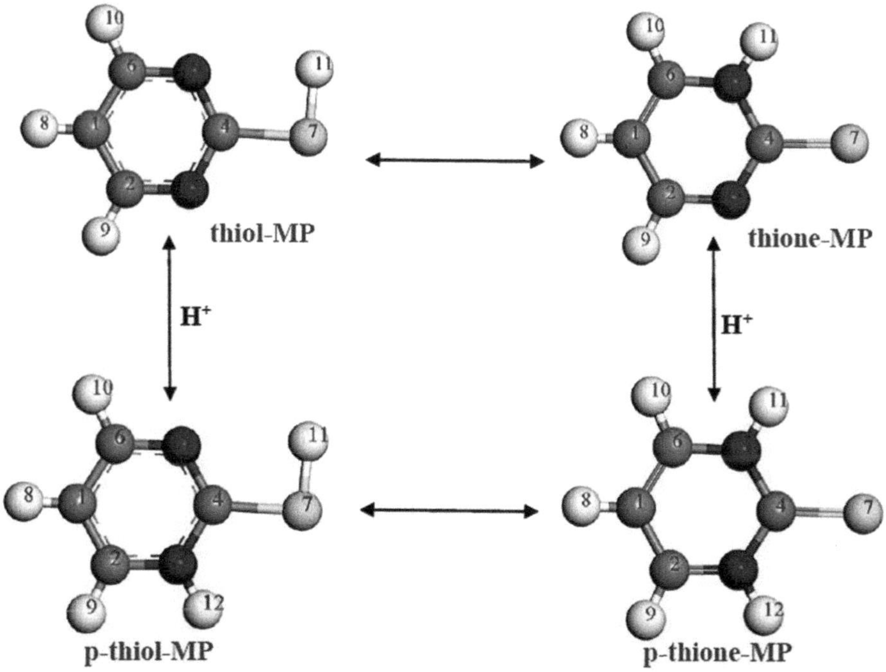

Figure 24. Optimized geometries of 2MP derivatives showing their protonation and tautomeric forms in equilibrium. (Reproduced with permission from ref (38). Copyright 2017 Elsevier Ltd.)

The insight of the adsorption of these derivatives was also explored and validated by MD simulation. It helped visualize the inhibitor's interaction with the iron atoms of the steel alloy as reflected from the large negative values of adsorption energies. Such integrations gave rise to horizontal adsorption leading to greater surface coverages, and the obtained order of adsorption was p-thione-MP > p-thiol-MP > thione-MP > thiol-MP, which corroborated well with the experimentally obtained results.

Similarly, other organic molecules, such as spiroprazole derivatives for inhibiting the impending corrosion of copper in 2 M HNO_3, was established by both experimentation as well as theoretical investigation. Such types of inhibitor molecules simultaneously retarded cathodic and anodic reactions after their adsorption onto the copper surface through physical adsorption. They followed the Freundlich adsorption isotherm explored with the variation of temperature. The corrosion-inhibiting capability increased as the concentration increased, and it decreased with temperature elevation. The decrease in the double-layer capacitance with the increase of concentration suggested the possession of a charge transfer minimizing effect, as reflected from the EIS (*130*).

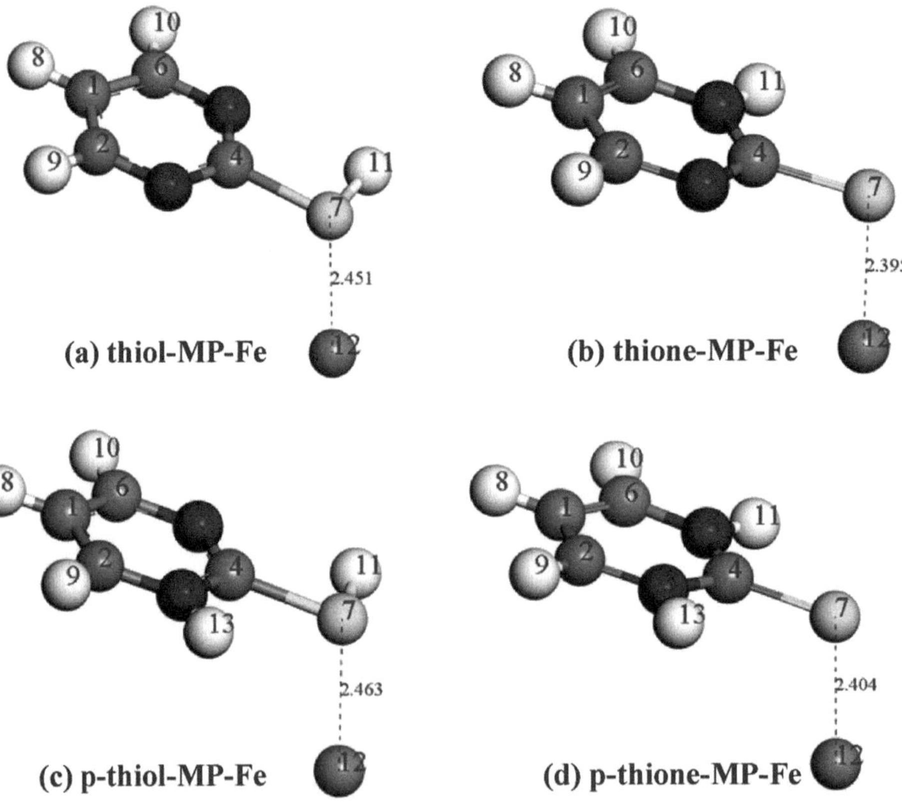

Figure 25. Optimized geometrical structures of 2MP complexing with Fe, namely, (a) thiol-MP-Fe; (b) thione-MP-Fe; (c) p-thiol-MP-Fe; and (d) p-thione-MP-Fe. (Reproduced with permission from ref (38). Copyright 2017 Elsevier Ltd.)

In addition to the organic molecules, drugs were also reported as efficient corrosion-inhibiting mechanisms in nitric acid media. Karthik and Sundaravaduvelu reported the efficient corrosion-inhibiting efficacy of a levetiracetam drug explored in a nitric acid solution for a copper substrate. Levetiracetam is the commercial name of (S)-2-(2-oxopyrrolidin-1-yl)butanamide, which possesses a heterocyclic ring with nitrogen and oxygen as heteroatoms, and a π bond in the skeleton of the molecule. The corrosion-inhibiting efficacy was found to be highest with approximately 95% efficacy at 300 ppm concentration added into a 0.5 M HNO_3 solution established by weight loss and electrochemical techniques. It was established to possess spontaneous adsorption undergone through both the physical as well as chemical pathway, acting as a mixed-type corrosion inhibitor, and it obeyed the Langmuir adsorption isotherm. The distribution of electron density in frontier molecular orbitals of the studied drug in its entire skeleton might be responsible for its interactions with copper surface atoms. Theoretically, the levetiracetam is expected to be highly reactive in nature. The presence of two nitrogen atoms and two oxygen atoms at the carbonyl center of the molecule carries more positive charges, which was susceptible to the formation of a coordinate bond using unoccupied orbitals of copper surface atoms. It led to its adsorption on its surface and inhibited the dissolution of copper ions toward the corrosive solution (*131*).

Fouda et al. (*132*) reported an appreciable corrosion-inhibiting property of azonitril derivatives, namely, 3-phenylazo-2-amino-1,1,3-tricyano propene, 3-p-tolylazo-2-amino-1,1,3-tricyanopropene, 3-p-anisylazo-2-amino-1,1,3-tricyanopropene, 3-p-nitrophenylazo-2-amino-1,1,3-tricyano propene, and 3-p-phenyl carboxylic azo-2-amino-1,1,3 tricyano propene, utilizing mass loss and electrochemical measurements in a 2 M HNO_3 solution. The corrosion-inhibiting efficacy increased with increased concentration of the inhibitor additives and decreased with the elevation of temperature. These azonitril derivatives worked as mixed-type inhibitors by adsorbing over copper surfaces following the Tempkin isotherm. The inhibitor molecules' adsorption was facilitated by the presence of polar units having nitrogen, heterocyclic, as well as aromatic moieties. It was expected that the mutual electron sharing amid added inhibitors and d orbitals of copper surface atoms would lead to the formation of chelate complexes, (i.e., giving rise to a firmly adsorbed inhibitor over the metal surface, which thereby acts as protective layer). The incorporation of nucleophilic methyl and methoxy groups enhanced the corrosion-inhibiting efficacy due to its electron density-enhancing capacity at the surface reactive sites, increasing the adsorption capability of the molecules. Alternatively, the incorporation of electrophilic carboxyl and nitro groups decreased the electron density, leading to poor adsorption of these inhibitors. The order of corrosion inhibition was well corroborated with the theoretically determined quantum chemical parameters (*132*).

Furthermore, plant extract has also been analyzed for exploring the electronic or molecular property of the components of the extract and the prediction/justification of the exhibited efficient corrosion inhibition. Li et al. (*42*) explored the possession of efficient corrosion-inhibiting properties of cassava starch. The cassava starch graft copolymer (CSGC) possessed efficient corrosion-inhibiting properties on aluminum alloy. It was attributed to the fact that the CSGC can exist in neutral and protonated forms in the nitric acid solution. The neutral CSGC possessed nitrogen and oxygen heteroatoms whose lone pair could be shared with the empty p orbital of the aluminum atom, forming coordinate bonds, and simultaneously, it was also capable of accepting electrons to form back bonding. Alternatively, the protonated CSGC (i.e., $CSGCH_x^{x+}$) can interact with NO_3^{2-} electrostatistically and thereby inhibit the reactions forming the soluble $[AlOHNO_3]^+$. In addition to it, both CSGC and $CSGCH_x^{x+}$ might form a coordinate bond with freshly generated Al^{3+} cations at the aluminum–solution interface. In this way, the coordinate bonding led to the formation of a protective inhibitor's film isolating the exposed metal from the corrosive solution and protecting it from further dissolution (*93*). The quantum chemical parameters, for instance, dipole moment, energy of HOMO and LUMO, or energy gap, explained the increase of adsorption as well as corrosion inhibition with the acrylamide grafting, and its subsequent enhancement property with the increase of an acrylamide unit in the grafting. The MD simulation results revealed the spontaneous adsorption capability with grafting and its subsequent enhancement property with the increase of an acrylamide unit in the grafting. All the grafted cassava starch exhibited stable adsorption configuration with flat disposition leading to greater surface coverage over the Al (110) surface, which was expected to protect vulnerable surfaces from corrosive attacks. The increase in the acrylamide unit in the grafting obviously increased the surface coverage. Because the inhibitor was furthermore hydrophobic in nature, it worked effectively as a corrosion inhibitor.

Summary and Outlook

There have been plenty of corrosion inhibitors based on organic molecules possessing several types of functional groups in addition to geometrical structures developed, as well as synthesized molecules produced by various synthetic procedures and used as corrosion inhibitor for decades. This chapter covers the progress in the development of promising corrosion inhibitors analyzed for their applicability in monobasic acidic media such as aqueous solutions of hydrochloric acid and nitric acid. The use of corrosion inhibitors derived from organic materials is the most practiced as well as most effective strategy for protecting the metallic materials from corrosive acidic environments. In this work, the most recently synthesized or explored organic molecule-based corrosion-inhibiting additives have been collectively discussed, along with the progress of determining and correlating the wet chemical and electrochemically determined corrosion-inhibiting properties exhibited in monobasic corrosive acidic solutions at various ranges of temperature and various exposure times. Some of the notable corrosion inhibitors such as amino acids or their derivatives, imidazoles, surfactants, heterocyclic compounds, ionic liquids, Schiff bases, and more, possess several active sites capable of adsorbing onto susceptible metal surfaces exposed to corrosive environments by displacing the preadsorbed corrosive ingredients. These corrosion-inhibiting additives form a protective layer after being adsorbed onto the targeted metal surfaces, which act as a barrier and inhibit further ingress of aggressive species toward metal surfaces. The presence of planar geometry, aromatic moieties, π bonds, and heteroatoms having a lone pair of electrons contribute to the enhancement of their adsorption followed by corrosion-inhibiting properties of the corrosion inhibitors, while the presence of long alkyl chains or other hydrophobic moieties also acts as a barrier protecting the metal surfaces.

In addition to the weight loss measurement, OCP, linear polarization potential, EIS, electrochemical noise analyses, electrochemical frequency modulation, and PDP have been mostly explored. Additionally, surface property and characterization have been performed mostly using scanning electron microscopy, FESEM, confocal laser scanning microscopy, atomic force microscopy, contact angle measurement, EDXS, and X-ray photoelectron spectroscopy. The theoretical techniques employed for elucidating and validating the experimentally obtained results were done using the quantitative structure–activity relationship method, with quantum chemical parameters determined based on DFT, DFTB, MC, and MD simulation methodologies. The mechanism of corrosion inhibition is facilitated by physisorption, chemisorption, or both pathways, leading to protective layer formation. Organic inhibitors are capable of protecting cathodic, anodic, or both reactions while simultaneously acting as mixed-type inhibitors. The adsorption may be of the Langmuir, Frumkin, Temkin, Freundlich, or more isotherms, depending on the reactive nature of the applied inhibitors.

The development of biobenign sustainable corrosion inhibitors with highly tunable functionality that can work efficiently in diverse fields and versatile environments is of immense need. The exploration of the corrosion-inhibiting properties of the inhibitors in hydrodynamic conditions can be of interest in near future. Theoretical prediction based on more authenticated electronic and molecular simulations in designing versatile corrosion inhibitors prior to wet chemical experimentation is required from an economic perspective prior to wet chemical experimentations. Development of cost-effective corrosion inhibitors, corrosion sensors, and the on-demand working of corrosion inhibitors is also required in the present scenario.

Acknowledgments

PB is thankful to the Department of Higher Education, Science & Technology and Biotechnology, Govt. of West Bengal, India, for providing financial assistance to carry out this research work (see sanction order no. 78 (Sanc.)/ST/P/S&T/6G-1/2018 dated January 31, 2019, and project no. GAP-225612). MM would like to acknowledge the Ministry of Tribal Affairs, New Delhi, India, for his National Fellowship for Higher Education of Scheduled Tribes candidates (Erstwhile Rajiv Gandhi National Fellowship for Scheduled Tribes, University Grants Commission, New Delhi, Govt. India) (see award letter no. F1-17.1/2014-15/RGNF-2014-15-ST-JHA-71559).

References

1. Elayaperumal, K.; Raja, V. S. Materials of Construction for Chemical Process Industries. In *Corrosion Failures: Theory, Case Studies, and Solutions*; John Wiley & Sons, Inc.: New Jersey, 2015; pp 75–110.
2. Walsh, R. A. *McGraw-Hill Machining and Metalworking Handbook*, 3rd ed; McGraw-Hill Companies Inc.: New York, 2006.
3. Harston, J. D.; Ropital, F. *Corrosion in Refineries*; European Federation of Corrosion (EFC) Series; Woodhead Publishing: Abington, Cambridge, 2007.
4. Kappenthuler, S.; Seeger, S. Holistic Evaluation of the Suitability of Metal Alloys for Sustainable Marine Construction from a Technical, Economic and Availability Perspective. *Ocean Eng.* **2021**, *219*, 108378.
5. Zhang, E.; Zhao, X.; Hu, J.; Wang, R.; Fu, S.; Qin, G. Antibacterial Metals and Alloys for Potential Biomedical Implants. *Bioact. Mater.* **2021**, 6, 2569–2612.
6. Khatib, J. M. *Sustainability of Construction Materials*, 2nd ed.; Woodhead Publishing Series in Civil and Structural Engineering; P. Lambert: Abington, Cambridge, 2016; pp 105–128.
7. Benvenuto, M. A. *Metals and Alloys Industrial Applications*; De Gruyter Textbook Series; Publisher-De Gruyter: Berlin, 2016.
8. Varshney, D.; Kumar, K. Application and Use of Different Aluminium Alloys with Respect to Workability, Strength and Welding Parameter Optimization. *Ain Shams Engineering Journal* **2021**, *12*, 1143–1152.
9. Dutta, A.; Saha, S. K.; Adhikari, U.; Banerjee, P.; Sukul, D. Effect of Substitution on Corrosion Inhibition Properties of 2-(substituted phenyl) benzimidazole Derivatives on Mild Steel in 1 M HCl Solution: A Combined Experimental and Theoretical Approach. *Corros. Sci.* **2017**, *123*, 256–266.
10. Goyal, M.; Kumar, S.; Bahadur, I.; Verma, C.; Ebenso, E. E. Organic Corrosion Inhibitors for Industrial Cleaning of Ferrous and Non-Ferrous Metals in Acidic Solutions: A Review. *J. Mol. Liq.* **2018**, *256*, 565–573.
11. Cramer, S. D.; Covino Jr., B. S. *ASM Handbook Volume 13C: Corrosion: Environments and Industries*; ASM International: Materials Park, OH, 2006.

12. Francis, R. *Corrosion Performance of Metals for the Marine Environment, EFC 63: A Basic Guide*; Maney Publishing: United Kingdom, 2012; p 108.
13. Gonzalez, A.; Schorr, M.; Valdez, B.; Mungaray, A. Bridges: Structures and Materials, Ancient and Modern. In *Infrastructure Management and Construction*; Sepasgozar, S.; Tahmasebinia, F.; Shirowzhan, S., Eds.; Intech Open: 2020; DOI: 10.5772/intechopen.90718.
14. Yuan, Q.; Cheng, R.; Zou, S.; Ding, C.; Liu, H.; Wang, Y.; Yang, D.; Xiao, X.; Jiang, Q.; Tang, R.; Chen, J. Isatin Thiosemicarbazone Derivatives as Inhibitors against Corrosion of AA6060 Aluminium Alloy in Acidic Chloride Medium: Substituent Effects. *J. Mater. Res. Technol.* **2020**, *9*, 11935–11947.
15. Abdel-Gawad, S. A.; Osman, W. M.; Fekry, A. M. Characterization and Corrosion Behavior of Anodized Aluminum Alloys for Military Industries Applications in Artificial Seawater. *Surf. Interfaces* **2019**, *14*, 314–323.
16. Jebaraj, A. V.; Aditya, K. V. V.; Kumar, T. S.; Ajaykumar, L.; Deepak, C. R. Mechanical and Corrosion Behaviour of Aluminum Alloy 5083 and Its Weldment for Marine Applications. *Mater. Today: Proc.* **2020**, *22*, 1470–1478.
17. Dhindaw, B. K.; Aditya, G. S. L.; Mandal, A. Recycling and Downstream Processing of Aluminium Alloys for Automotive Applications. In Saleem Hashmi and Imtiaz Ahmed Choudhury (Eds.). *Encyclopedia of Renewable and Sustainable Materials* **2020**, *3*, 154–161.
18. Wang, G. Q.; Zhao, Y. H.; Hao, Y. F. Friction Stir Welding of High-Strength Aerospace Aluminum Alloy and Application in Rocket Tank Manufacturing. *J. Mater. Sci. Technol. (Shenyang, China)* **2018**, *34*, 73–91.
19. Zhang, P.; Li, Y. Y.; Liu, Y.; Zhang, Y.; Liu, J. X. Analysis of the Microhardness, Mechanical Properties and Electrical Conductivity of 7055 Aluminum Alloy. *Vacuum* **2020**, *171*, 109005.
20. Otani, Y.; Sasaki, S. Effects of the Addition of Silicon to 7075 Aluminum Alloy on Microstructure, Mechanical Properties, and Selective Laser Melting Processability. *Materials Science and Engineering: A* **2020**, *777*, 139079.
21. Arif, M. A. M.; Omar, M. Z.; Sajuri, Z.; Salleh, M. S. Effects of Cu and Mg on Thixoformability and Mechanical Properties of Aluminium Alloy 2014. *Trans. Nonferrous Met. Soc. China* **2020**, *30*, 275–287.
22. Pinto, G. M.; Nayak, J.; Shetty, A. N. Corrosion Inhibition of 6061 Al–15 Vol. Pct. SiC(p) Composite and Its Base Alloy in a Mixture of Sulphuric Acid and Hydrochloric Acid by 4-(N,N-dimethyl amino) Benzaldehyde Thiosemicarbazone. *Mater. Chem. Phys.* **2011**, *125*, 628–640.
23. Prakashaiah, B. G.; Kumara, D. V.; Pandith, A. A.; Shetty, A. N.; Rani, B. E. A. Corrosion Inhibition of 2024-T3 Aluminum Alloy in 3.5% NaCl by Thiosemicarbazone Derivatives. *Corros. Sci.* **2018**, *136*, 326–338.
24. Liao, B.; Cen, H.; Chen, Z.; Guo, X. Corrosion Behavior of Sn-3.0Ag-0.5Cu Alloy under Chlorine-containing Thin Electrolyte Layers. *Corros. Sci.* **2018**, *143*, 347–361.
25. Quraishi, M. A.; Chauhan, D. S.; Saji, V. S. *Heterocyclic Organic Corrosion Inhibitors: Principles and Applications*; Amsterdam: Elsevier, 2020.
26. Hao, J.; Wang, Y.; Wu, Yu.; Guo, F. Metal Recovery from Waste Printed Circuit Boards: A Review for Current Status and Perspectives. *Resour., Conserv. Recycl.* **2020**, *157*, 104787.

27. Deyab, M. A. Understanding the Anti-Corrosion Mechanism and Performance of Ionic Liquids in Desalination, Petroleum, Pickling, De-Scaling, and Acid Cleaning Applications. *J. Mol. Liq.* **2020**, *309*, 113107.
28. Finšgar, M.; Jackson, J. Application of Corrosion Inhibitors for Steels in Acidic Media for the Oil and Gas Industry: A Review. *Corros. Sci.* **2014**, *86*, 17–41.
29. Gueccia, R.; Winter, D.; Randazzo, S.; Cipollina, A.; Kowski, J. K.; Micale, G. D. M. An Integrated Approach for the HCl and Metals Recovery from Waste Pickling Solutions: Pilot Plant Design Operations. *Chem. Eng. Res. Des.* **2021**, *168*, 383–396.
30. Yüce, A. O.; Telli, E.; Mert, B. D.; Kardaş, G.; Yazıcı, B. Experimental and Quantum Chemical Studies on Corrosion Inhibition Effect of 5,5 diphenyl 2-thiohydantoin on Mild Steel in HCl Solution. *J. Mol. Liq.* **2016**, *218*, 384–392.
31. Ma, X.; Jiang, X.; Xia, S.; Shan, M.; Li, X.; Yu, L.; Tang, Q. New Corrosion Inhibitor Acrylamide Methyl Ether for Mild Steel in 1 M HCl. *Appl. Surf. Sci.* **2016**, *371*, 248–257.
32. Birbilis, N.; Buchheit, R. G. Investigation and Discussion of Characteristics for Intermetallic Phases Common to Aluminium Alloys as a Function of Solution pH. *J. Electrochem. Soc.* **2008**, *155*, 117–126.
33. Zhang, T. B.; Jiang, W. F.; Wang, H. L.; Zhang, S. F. Synthesis and Localized Inhibition Behaviour of New Triazine-Methionine Corrosion Inhibitor in 1 M HCl for 2024-T3 Aluminium Alloy. *Mater. Chem. Phys.* **2019**, *237*, 121866.
34. Baumgärtner, M.; Kaesche, H. Aluminum Pitting in Chloride Solutions: Morphology and Pit Growth Kinetics. *Corros. Sci.* **1990**, *31*, 231–236.
35. Srinivasan, H. S.; Mital, C. K. Studies on the Passivation Behaviour of Al-Zn-Mg Alloy in Chloride Solutions Containing Some Anions and Cations Using Electrochemical Impedance Spectroscopy. *Electrochim. Acta* **1994**, *39*, 2633–7.
36. Zor, S.; Özkazanc,, H.; Arslan, T.; Kandemirli, F. Inhibition Effects of 4-phenyl-3-thiosemicarbazide on the Corrosion of Aluminum in 0.1 M HCl: Theoretical and Experimental Studies. *Corrosion* **2010**, *66*, 045006–045006-7.
37. Yurt, A.; Ulutas, S.; Dal, H. Electrochemical and Theoretical Investigation on the Corrosion of Aluminium in Acidic Solution Containing Some Schiff Bases. *Appl. Surf. Sci.* **2006**, *253*, 919–925.
38. Li, X.; Deng, S.; Lin, T.; Xie, X.; Du, G. 2-Mercaptopyrimidine as an Effective Inhibitor for the Corrosion of Cold Rolled Steel in HNO_3 Solution. *Corros. Sci.* **2017**, *118*, 202–216.
39. Rahal, H. T.; Abdel-Gaber, A. M.; Younes, G. O. Inhibition of Steel Corrosion in Nitric Acid by Sulfur Containing Compounds. *Chem. Eng. Commun.* **2016**, *203*, 435–445.
40. Turnbull, J.; Szukalo, R.; Zagidulin, D.; Shoesmith, D. Nitrite Effects on Copper Corrosion in Nitric Acid Solutions. *Corros. Sci.* **2021**, *179*, 109147.
41. Turnbull, J.; Szukalo, R.; Behazin, M.; Hall, D.; Zagidulin, D.; Ramamurthy, S.; Wren, J.; Shoesmith, D. The Effects of Cathodic Reagent Concentration and Small Solution Volumes on the Corrosion of Copper in Dilute Nitric Acid Solutions. *Corrosion* **2018**, *74*, 326–336.
42. Li, X.; Deng, S.; Lin, T.; Xie, X. Cassava Starch Graft Copolymer as a Novel Inhibitor for the Corrosion of Aluminium in HNO_3 Solution. *J. Mol. Liq.* **2019**, *282*, 499–514.
43. Saha, S. K.; Murmu, M.; Murmu, N. C.; Obot, I. B.; Banerjee, P. Molecular Level Insights for the Corrosion Inhibition Effectiveness of Three Amine Derivatives on the Carbon Steel Surface

in the Adverse Medium: A Combined Density Functional Theory and Molecular Dynamics Simulation Study. *Surf. Interfaces* **2018**, *10*, 65–73.

44. Saha, S. K.; Murmu, M.; Murmu, N. C.; Banerjee, P. Evaluating Electronic Structure of Quinazolinone and Pyrimidinone Molecules for Its Corrosion Inhibition Effectiveness on Target Specific Mild Steel in the Acidic Medium: A Combined DFT and MD Simulation Study. *J. Mol. Liq.* **2016**, *224*, 629–638.
45. Chauhan, D. S.; Singh, P.; Quraishi, M. A. Quinoxaline Derivatives as Efficient Corrosion Inhibitors: Current Status, Challenges and Future Perspectives. *J. Mol. Liq.* **2020**, *320*, 114387.
46. Verma, C. K.; Rhee, Y.; Quraishi, M. A.; Ebenso, E. E. Pyridine Based N-heterocyclic Compounds as Aqueous Phase Corrosion Inhibitors: A Review. *J. Taiwan Inst. Chem. Eng.* **2020**, *117*, 265–277.
47. Verma, C.; Quraishi, M. A.; Ebenso, E. E. Quinoline and Its Derivatives as Corrosion Inhibitors: A Review. *Surf. Interfaces* **2020**, *21*, 100634.
48. Vernack, E.; Costa, D.; Tingaut, P.; Marcus, P. DFT Studies of 2-mercaptobenzothiazole and 2-mercaptobenzimidazole as Corrosion Inhibitors for Copper. *Corros. Sci.* **2020**, *174*, 108840.
49. Yadav, M.; Sinha, R. R.; Kumar, S.; Sarkar, T. K. Corrosion Inhibition Effect of Spiropyrimidinethiones On Mild Steel in 15% HCl Solution Insight from: Electrochemical And Quantum Studies. *RSC Adv.* **2015**, *5*, 70832–70848.
50. Tang, L.; Li, X.; Li, L. The Effect of 1-(2-pyridylazo)-2-naphthol on the Corrosion of Cold Rolled Steel in Acid Media: Part 2: Inhibitive Action in 0.5 M Sulfuric Acid. *Mater. Chem. Phys.* **2006**, *97*, 301–307.
51. Rehim, S. S. A.; Hazzazi, O. A.; Amin, M. A.; Khaled, K. F. On the Corrosion Inhibition of Low Carbon Steel in Concentrated Sulfuric Acid Solutions. Part I: Chemical and Electrochemical (AC and DC) Studies. *Corros. Sci.* **2008**, *50*, 2258–2271.
52. El-Lateef, H. M. A.; Abo-Riy, M. A.; Tantawy, A. H. Empirical and Quantum Chemical Studies on the Corrosion Inhibition Performance of Some Novel Synthesized Cationic Gemini Surfactants on Carbon Steel Pipelines in Acid Pickling Processes. *Corros. Sci.* **2016**, *108*, 94–110.
53. Abd-Elaal, A. A.; Elbasiony, N. M.; Shaban, S. M.; Zaki, E. G. Studying the Corrosion Inhibition of Some Prepared Nonionic Surfactants Based on 3-(4-hydroxyphenyl) Propanoic Acid and Estimating the Influence of Silver Nanoparticles on the Surface Parameters. *J. Mol. Liq.* **2018**, *249*, 304–317.
54. Zhou, T.; Yuan, J.; Zhang, Z.; Xin, X.; Xu, G. The Comparison of Imidazolium Gemini Surfactant [C14-4-C14im]Br_2 and Its Corresponding Monomer as Corrosion Inhibitors for A3 Carbon Steel in Hydrochloric Acid Solutions: Experimental and Quantum Chemical Studies. *Colloids Surf., A* **2019**, *575*, 57–65.
55. Tawfik, S. M. Ionic Liquids Based Gemini Cationic Surfactants as Corrosion Inhibitors for Carbon Steel in Hydrochloric Acid Solution. *J. Mol. Liq.* **2016**, *216*, 624–635.
56. Kaczerewska, O.; Leiva-Garcia, R.; Akid, R.; Brycki, B. Efficiency of Cationic Gemini Surfactants with 3-azamethylpentamethylene Spacer as Corrosion Inhibitors for Stainless Steel in Hydrochloric Acid. *J. Mol. Liq.* **2017**, *247*, 6–13.

57. Kaczerewska, O.; Leiva-Garcia, R.; Akid, R.; Brycki, B.; Kowalczyk, I.; Pospieszny, T. Effectiveness of O-Bridged Cationic Gemini Surfactants as Corrosion Inhibitors for Stainless Steel in 3 M HCl: Experimental and Theoretical Studies. *J. Mol. Liq.* **2018**, *249*, 1113–1124.
58. Pakiet, M.; Kowalczyk, I. H.; Garcia, R. L.; Akid, R.; Brycki, B. E. Influence of Different Counterions on Gemini Surfactants with Polyamine Platform as Corrosion Inhibitors for Stainless Steel AISI 304 in 3 M HCl. *J. Mol. Liq.* **2018**, *268*, 824–831.
59. Mobin, M.; Aslam, R.; Aslam, J. Nontoxic Biodegradable Cationic Gemini Surfactants as Novel Corrosion Inhibitor for Mild Steel in Hydrochloric Acid Medium and Synergistic Effect of Sodium Salicylate: Experimental and Theoretical Approach. *Mater. Chem. Phys.* **2017**, *191*, 151–167.
60. Aslam, R.; Mobin, M.; Aslam, J.; Lgaz, H.; Chung, I.-M. Inhibitory Effect of Sodium Carboxymethylcellulose and Synergistic Biodegradable Gemini Surfactants as Effective Inhibitors for MS Corrosion in 1 M HCl. *J. Mater. Res. Technol.* **2019**, *8*, 4521–4533.
61. El-Lateef, H. M. A.; Shalabi, K.; Tantawy, A. H. Corrosion Inhibition of Carbon Steel in Hydrochloric Acid Solution Using Newly Synthesized Urea-Based Cationic Fluorosurfactants: Experimental and Computational Investigations. *New J. Chem.* **2020**, *44*, 17791–17814.
62. Talari, M.; Nezhad, S. M.; Alavi, S. J.; Mohtashamipour, M.; Davoodi, A.; Hosseinpour, S. Experimental and Computational Chemistry Studies of Two Imidazolebased Compounds as Corrosion Inhibitors for Mild Steel in HCl Solution. *J. Mol. Liq.* **2019**, *286*, 110915.
63. Faydy, M. E.; Benhiba, F.; Lakhrissi, B.; Touhami, M. E.; Warad, I.; Bentiss, F.; Zarrouk, A. The Inhibitive Impact of Both Kinds of 5-isothiocyanatomethyl-8-hydroxyquinoline Derivatives on the Corrosion of Carbon Steel in Acidic Electrolyte. *J. Mol. Liq.* **2019**, *295*, 111629.
64. Ouakki, M.; Galai, M.; Rbaa, M.; Abousalemd, A. S.; Lakhrissi, B.; Touhami, M. E.; Cherkaoui, M. Electrochemical, Thermodynamic and Theoretical Studies of Some Imidazole Derivatives Compounds as Acid Corrosion Inhibitors for Mild Steel. *J. Mol. Liq.* **2020**, *319*, 114063.
65. Murthy, R.; Gupta, P.; Sundaresan, C. N. Theoretical and Electrochemical Evaluation of 2-thioureidobenzheteroazoles as Potent Corrosion Inhibitors for Mild Steel in 2 M HCl Solution. *J. Mol. Liq.* **2020**, *319*, 114081.
66. Rbaa, M.; Ouakki, M.; Galai, M.; Berishad, A.; Lakhrissi, B.; Jama, C.; Warad, I.; Zarrouk, A. Simple Preparation and Characterization of Novel 8-Hydroxyquinoline Derivatives as Effective Acid Corrosion Inhibitor for Mild Steel: Experimental and Theoretical Studies. *Colloids Surf., A* **2020**, *602*, 125094.
67. El aoufir, Y.; Zehra, S.; Lgaz, H.; Chaouiki, A.; Serrar, H.; Kaya, S.; Salghi, R.; AbdelRaheem, S. K.; Boukhris, S.; Guenbour, A.; Chung, I.-M. Evaluation of Inhibitive and Adsorption Behavior of Thiazole-4-carboxylates on Mild Steel Corrosion in HCl. *Colloids Surf., A* **2020**, *606*, 125351.
68. Rbaa, M.; Abousalem, A. S.; Rouifi, Z.; Lakhrissi, L.; Galai, M.; Zarrouk, A.; Lakhrissi, B.; Lakhrissi, Y. Selective Synthesis of New Sugars Based on 8-Hydroxyquinoline as Corrosion Inhibitors for Mild Steel in HCl Solution: Effect of the Saturated Hydrocarbon Chain: Theoretical and Experimental Studies. *Inorg. Chem. Commun.* **2020**, *118*, 108019.

69. Aslam, J.; Aslam, R.; Alrefaee, S. H.; Mobin, M.; Aslam, A.; Parveen, M.; Hussain, C. M. Gravimetric, Electrochemical, and Morphological Studies of an Isoxazole Derivative as Corrosion Inhibitor for Mild Steel in 1 M HCl. *Arabian J. Chem.* **2020**, *13*, 7744–7758.
70. Aslam, J.; Aslam, R.; Lone, I. H.; Radwan, N. R. E.; Mobin, M.; Aslam, A.; Parveen, M.; Al-Freedid, A. A.; Alzulaibani, A. A. Inhibitory Effect of 2-Nitroacridone on Corrosion of Low Carbon Steel in 1 M HCl Solution: An Experimental and Theoretical Approach. *J. Mater. Res. Technol.* **2020**, *9*, 4061–4075.
71. Singh, P.; Kumar, M.; Quraishi, M. A.; Haque, J.; Singh, G. Bispyranopyrazoles as Green Corrosion Inhibitors for Mild Steel in Hydrochloric Acid: Experimental and Theoretical Approach. *ACS Omega* **2018**, *3*, 11151–11162.
72. Erami, R. S.; Amirnasr, M.; Meghdadi, S.; Talebian, M.; Farrokhpour, H.; Raeissi, K. Carboxamide Derivatives as New Corrosion Inhibitors for Mild Steel Protection in Hydrochloric Acid Solution. *Corros. Sci.* **2019**, *151*, 190–197.
73. Corrales-Luna, M.; Manh, T. L.; Romero-Romo, M.; Palomar-Pardavé, M.; Arce-Estrada, E. M. 1-Ethyl 3-methylimidazolium Thiocyanate Ionic Liquid as Corrosion Inhibitor of API 5L X52 Steel in H_2SO_4 and HCl Media. *Corros. Sci.* **2019**, *153*, 85–99.
74. Tamany, E. S. H. E.; Elsaeed, S. M.; Ashour, H.; Zaki, E. G.; Nagy, H. A. E. Novel Acrylamide Ionic Liquids as Anti-Corrosion for X-65 Steel Dissolution in Acid Medium: Adsorption, Hydrogen Evolution and Mechanism. *J. Mol. Struct.* **2018**, *1168*, 106–114.
75. El-Hajjaji, F.; Messali, M.; Aljuhani, A.; Aouad, M. R.; Hammouti, B.; Belghiti, M. E.; Chauhan, D. S.; Quraishi, M. A. Pyridazinium-Based Ionic Liquids as Novel and Green Corrosion Inhibitors of Carbon Steel in Acidmedium: Electrochemical and Molecular Dynamics Simulation Studies. *J. Mol. Liq.* **2018**, *249*, 997–1008.
76. Azeez, F. A.; Al-Rashed, O. A.; Nazeer, A. A. Controlling of Mild-Steel Corrosion in Acidic Solution Using Environmentally Friendly Ionic Liquid Inhibitors: Effect of Alkyl Chain. *J. Mol. Liq.* **2018**, *265*, 654–663.
77. Kannan, P.; Varghese, A.; Palanisamy, K.; Abousalem, A. S. Evaluating Prolonged Corrosion Inhibition Performance of Benzyltributylammonium Tetrachloroaluminate Ionic Liquid Using Electrochemical Analysis and Monte Carlo Simulation. *J. Mol. Liq.* **2020**, *297*, 111855.
78. Nkuna, A. A.; Akpan, E. D.; Obot, I. B.; Verma, C.; Ebenso, E. E.; Murulana, L. C. Impact of Selected Ionic Liquids on Corrosion Protection of Mild Steel in Acidic Medium: Experimental and Computational Studies. *J. Mol. Liq.* **2020**, *314*, 113609.
79. Aslam, R.; Mobin, M.; Huda; Obot, I. B.; Alamri, A. H. Ionic Liquids Derived from α-Amino Acid Ester Salts as Potent Green Corrosion Inhibitors for Mild Steel in 1 M HCl. *J. Mol. Liq.* **2020**, *318*, 113982.
80. Ardakani, E. K.; Kowsari, E.; Ehsani, A. Imidazolium-Derived Polymeric Ionic Liquid as a Green Inhibitor for Corrosion Inhibition of Mild Steel in 1.0 M HCl: Experimental and Computational Study. *Colloids Surf., A* **2020**, *586*, 124195.
81. Aoun, S. B. On the Corrosion Inhibition of Carbon Steel in 1 M HCl with a Pyridinium-Ionic Liquid: Chemical, Thermodynamic, Kinetic and Electrochemical Studies. *RSC Adv.* **2017**, *7*, 36688–36696.
82. Kannan, P.; Varghese, A.; Mathew, A. T.; Palanisamy, K.; Abousalem, A. S.; Kalaiyarasan, M.; Rajendran, N. Exploring the Inhibition Performance of Tetrachloroferrate Ionic Liquid in Acid

Environment Using Scanning Electrochemical Microscope and Theoretical Approaches. *Surf. Interfaces* **2020**, *20*, 100594.
83. bot, I. B.; Macdonald, D. D.; Gasem, Z. M. Density Functional Theory (DFT) as a Powerful Tool for Designing New Organic Corrosion Inhibitors. Part 1: An Overview. *Corros. Sci.* **2015**, *99*, 1–30.
84. Verma, C.; Lgaz, H.; Verma, D. K.; Ebenso, E. E.; Bahadur, I.; Quraishi, M. A. Molecular Dynamics and Monte Carlo Simulations as Powerful Tools for Study of Interfacial Adsorption Behavior of Corrosion Inhibitors in Aqueous Phase: A Review. *J. Mol. Liq.* **2018**, *260*, 99–120.
85. Verma, C.; Quraishi, M. A.; Singh, A. 5-Substituted 1H-tetrazoles as Effective Corrosion Inhibitors for Mild Steel in 1 M Hydrochloric Acid. *Journal of Taibah University for Science* **2016**, *10*, 718–733.
86. Gutiérrez, E.; Rodríguez, J. A.; Cruz-Borbolla, J.; Alvarado-Rodríguez, J. G.; Thangarasu, P. Development of a Predictive Model for Corrosion Inhibition of Carbon Steel by Imidazole and Benzimidazole Derivatives. *Corros. Sci.* **2016**, *108*, 23–35.
87. Olasunkanmi, L. O.; Obot, I. B.; Ebenso, E. E. Adsorption and Corrosion Inhibition Properties of N-{n-[1-R-5-(quinoxalin-6-yl)-4,5-dihydropyrazol-3-yl]phenyl}methane Sulfon-Amides on Mild Steel in 1 M HCl: Experimental and Theoretical Studies. *RSC Adv.* **2016**, *6*, 86782–86797.
88. Yilmaz, N.; Fitoz, A.; Ergun, Ÿ.; Emregddotul, K. C. A Combined Electrochemical and Theoretical Study into the Effect of 2-((thiazole-2-ylimino) methyl) Phenol as a Corrosion Inhibitor for Mild Steel in a Highly Acidic Environment. *Corros. Sci.* **2016**, *111*, 110–120.
89. Yıldız, R. An Electrochemical and Theoretical Evaluation of 4,6-diamino-2-pyrimidinethiol as a Corrosion Inhibitor for Mild Steel in HCl Solutions. *Corros. Sci.* **2015**, *90*, 544–555.
90. Khadiri, G. A.; Saddik; Bekkouche, R. K.; Aouniti, A.; Hammouti, B.; Benchat, N.; Bouachrine, M.; Solmaz, R. Gravimetric, Electrochemical and Quantum Chemicalstudies of Some Pyridazine Derivatives as Corrosion Inhibitors for Mild Steel in 1 M HCl Solution. *J. Taiwan Inst. Chem. Eng.* **2016**, *58*, 552–564.
91. Sığırcık, G.; Tüken, T.; Erbil, M. Assessment of the Inhibition Efficiency of 3,4-diaminobenzonitrile against the Corrosion of Steel. *Corros. Sci.* **2016**, *102*, 437–445.
92. Antropov, L. I. Kinetics of Electrode Processes and Null Points of Metals. *Council of Scientific & Industrial Research, New Delhi* **1960**.
93. Popova, A.; Sokolova, E.; Raicheva, S.; Christov, M. AC and DC Study of the Temperature Effect on Mild Steel Corrosion in Acid Media in the Presence of Benzimidazole Derivatives. *Corros. Sci.* **2003**, *45*, 33–58.
94. Ramya, K.; Anupama, K. K.; Shainy, K. M.; Abraham, J. Synergistic and Hydrogen Bonded Interaction of Alkyl Benzimadazoles and Urea Pair on Mild Steel in Hydrochloric Acid: Adsorption, Electroanalytical and Theoretical Studies. *J. Taiwan Inst. Chem. Eng.* **2016**, *58*, 517–527.
95. Ramya, K.; Anupama, K. K.; Shainy, K. M.; Joseph, A. Corrosion Protection of Mild Steel in Hydrochloric Acid Solution through the Synergistic of Alkylbenzimidazoles and Semicarbazide Pair—Electroanalytical and Computational Studies. *Egypt. J. Pet.* **2017**, *26*, 421–437.
96. Hamani, H.; Douadi, T.; Daoud, D.; Al-Noaimi, M.; Rikkouh, R. A.; Chafaa, S. 1-(4-Nitrophenyloimino)-1-(phenylhydrazono)-propan-2-one as Corrosion Inhibitor for Mild

Steel in 1 M HCl Solution: Weight Loss, Electrochemical, Thermodynamic and Quantum Chemical Studies. *J. Electroanal. Chem.* **2017**, *801*, 425–438.

97. Snehalatha, M.; Ravikumar, C.; Joe, I. H.; Sekar, N.; Jayakumar, V. S. Spectroscopic Analysis and DFT Calculations of a Food Additive Carmoisine. *Spectrochim. Acta, Part A* **2009**, *72*, 654–662.
98. Daoud, D.; Douadi, T.; Hamani, H.; Al-Noaimi, M.; Chafaa, S. Corrosion Inhibition of Mild Steel by Two New S-Heterocyclic Compounds in 1 M HCl: Experimental and Computational Study. *Corros. Sci.* **2015**, *94*, 21–37.
99. Obot, I. B.; Madhankumar, A. Synergistic Effect of Iodide Ion Addition on the Inhibition of Mild Steel Corrosion in 1 M HCl by 3-amino-2-methylbenzylalcohol. *Mater. Chem. Phys.* **2016**, *177*, 266–275.
100. Abousalem, A. S.; Ismail, M. A.; Fouda, A. S. A Complementary Experimental and In Silico Studies on the Action of Fluoropheny-2,2′-bichalcophenes as Ecofriendly Corrosion Inhibitors and Biocide Agents. *J. Mol. Liq.* **2018**, *276*, 255–274.
101. Dutta, A.; Saha, S. K.; Banerjee, P.; Sukula, D. Correlating Electronic Structure with Corrosion Inhibition Potentiality of Some Bis-benzimidazole Derivatives for Mild Steel in Hydrochloric Acid: Combined Experimental and Theoretical Studies. *Corros. Sci.* **2015**, *98*, 541–550.
102. Obot, I. B.; Edouk, U. M. Benzimidazole: Small Planar Molecule with Diverse Anti-Corrosion Potentials. *J. Mol. Liq.* **2017**, *246*, 66–90.
103. Ech-chihbi, E.; Belghiti, M. E.; Salim, R.; Oudda, H.; Taleb, M.; Benchat, N.; Hammouti, B.; El-Hajjaji, F. Experimental and Computational Studies on the Inhibition Performance of the Organic Compound "2-phenylimidazo [1,2-a]pyrimidine-3-carbaldehyde" against the Corrosion of Carbon Steel in 1.0 M HCl Solution. *Surf. Interfaces* **2017**, *9*, 206–217.
104. Verma, C.; Quraishi, M. A.; Ebenso, E. E.; Obot, I. B.; Assyry, A. E. 3-Amino Alkylated Indoles as Corrosion Inhibitors for Mild Steel in 1M HCl: Experimental and Theoretical Studies. *J. Mol. Liq.* **2016**, *219*, 647–660.
105. Dohare, P.; Quraishi, M. A.; Verma, C.; Lgaz, H.; Salghi, R.; Ebenso, E. E. Ultrasound Induced Green Synthesis of Pyrazolo-Pyridines as Novel Corrosion Inhibitors Useful for Industrial Pickling Process: Experimental and Theoretical Approach. *Results Phys.* **2019**, *13*, 102344.
106. Fouda, A. S.; El-Ewady, Y.; Abo-El-Enien, O.; Agizah, F. Cinnamoylmalononitriles as Corrosion Inhibitors for Mild Steel in Hydrochloric Acid Solution. *Anti-Corros. Methods Mater.* **2008**, *55*, 317–323.
107. Saha, S. K.; Dutta, A.; Ghosh, P.; Sukul, D.; Banerjee, P. Novel Schiff-Base Molecules as Efficient Corrosion Inhibitors for Mild Steel Surface in 1 M HCl Medium: Experimental and Theoretical Approach. *Phys. Chem. Chem. Phys.* **2016**, *18*, 17898–17911.
108. Zhang, W.; Zhang, Z.; Li, W.; Huang, X.; Ruan, L.; Wu, L. Synergistic Effect of Phytic Acid and Benzyltrimethyl Ammonium Bromide on Corrosion Inhibition of Carbon Steel in 0.5 M HCl. *Int. J. Electrochem. Sci.* **2019**, *14*, 7348–7362.
109. Han, P.; Li, W.; Tian, H.; Gao, X.; Ding, R.; Xiong, C.; Chen, C. Designing and Fabricating of Time-Depend Self-Strengthening Inhibitor Film: Synergistic Inhibition of Sodium Dodecyl Sulfate and 4-mercaptopyridine for Mild Steel. *J. Mol. Liq.* **2018**, *268*, 425–437.

110. Fouda, A. S.; Elmorsi, M. A.; Elmekkawy, A. Eco-Friendly Chalcones Derivatives as Corrosion Inhibitors for Carbon Steel in Hydrochloric Acid Solution. *Afr. J. Pure Appl. Chem.* **2013**, *7*, 337–349.
111. Guo, L.; Qi, C.; Zheng, X.; Zhang, R.; Shen, X.; Kaya, S. Toward Understanding the Adsorption Mechanism of Large Size Organic Corrosion Inhibitors on an Fe (110) Surface Using the DFTB Method. *RSC Adv.* **2017**, 7, 29042–29050.
112. Hahn, J. R.; Kang, H. S. Role of Molecular Orientation in Vibration, Hopping, and Electronic Properties of Single Pyridine Molecules Adsorbed on Ag (110) Surface: A Combined STM and DFT Study. *Surf. Sci.* **2010**, *604*, 258–264.
113. Saha, S. K.; Banerjee, P. Introduction of Newly Synthesized Schiff Base Molecules as Efficient Corrosion Inhibitors for Mild Steel in 1 M HCl Medium: An Experimental, Density Functional Theory and Molecular Dynamics Simulation Study. *Mater. Chem. Front.* **2018**, *2*, 1674–1691.
114. Saha, S. K.; Dutta, A.; Ghosh, P.; Sukul, D.; Banerjee, P. Adsorption and Corrosion Inhibition Effect of Schiff Base Molecules on the Mild Steel Surface in 1 M HCl Medium: A Combined Experimental and Theoretical Approach. *Phys. Chem. Chem. Phys.* **2015**, *17*, 5679–5690.
115. Murmu, M.; Saha, S. K.; Murmu, N. C.; Banerjee, P. Effect of Stereochemical Conformation into the Corrosion Inhibitive Behaviour of Double Azomethine Based Schiff Bases on Mild Steel Surface in 1 mol L^{-1} HCl Medium: An Experimental, Density Functional Theory and Molecular Dynamics Simulation Study. *Corros. Sci.* **2019**, *146*, 134–151.
116. Murmu, M.; Saha, S. K.; Bhaumick, P.; Murmu, N. C.; Hirani, H.; Banerjee, P. Corrosion Inhibition Property of Azomethine Functionalized Triazole Derivatives in 1 mol L^{-1} HCl Medium for Mild Steel: Experimental and Theoretical Exploration. *J. Mol. Liq.* **2020**, *313*, 113508.
117. Barouni, K.; Bazzi, L.; Salghi, R.; Mihit, M.; Hammouti, B.; Albourine, A.; Issami, S. E. Some Amino Acids as Corrosion Inhibitors for Copper in Nitric Acid Solution. *Mater. Lett.* **2008**, *62*, 3325–3327.
118. Milošev, I.; Pavlinac, J.; Hodoščec, M.; Lesar, A. Amino Acids as Corrosion Inhibitors for Copper in Acidic Medium: Experimental and Theoretical Study. *J. Serb. Chem. Soc.* **2013**, *78*, 2069–2086.
119. Ibrahimi, B. E.; Jmiai, A.; Mouaden, K. E.; Oukhrib, R.; Soumoue, A.; Issami, S. E.; Bazzi, L. Theoretical Evaluation of Some α-Amino Acids for Corrosion Inhibition of Copper in Acidic Medium: DFT Calculations, Monte Carlo Simulations and QSPR Studies. *Journal of King Saud University-Science* **2018**, *32*, 163–171.
120. Zhang, D.; Xie, B.; Gao, L.; Cai, Q.; Joo, H.; Lee, K. Intramolecular Synergistic Effect of Glutamic Acid, Cysteine and Glycine against Copper Corrosion in Hydrochloric Acid Solution. *Thin Solid Films* **2011**, *520*, 356–361.
121. Kumar, D.; Jain, N.; Jain, V.; Rai, B. Amino Acids as Copper Corrosion Inhibitors: A Density Functional Theory Approach. *Appl. Surf. Sci.* **2020**, *514*, 145905.
122. Boutouil, A.; Laamari, M. R.; Elazhary, I.; Bahsis, L.; Anane, H.; Stiriba, S.-E. Towards a Deeper Understanding of the Inhibition Mechanism of a New 1,2,3-triazole Derivative for Mild Steel Corrosion in the Hydrochloric Acid Solution Using Coupled Experimental and Theoretical Methods. *Mater. Chem. Phys.* **2019**, *241*, 122420.

123. Pareek, S.; Jain, D.; Hussain, S.; Biswas, A.; Shrivastava, R.; Parida, S. K.; Behera, D. A New Insight into Corrosion Inhibition Mechanism of Copper in Aerated 3.5 wt.% NaCl Solution by Eco-Friendly Imidazopyrimidine Dye: Experimental and Theoretical Approach. *Chem. Eng. J.* **2019**, *358*, 725–742.
124. Blanco-Díaz, E. G.; Vázquez-Montelongo, E. A.; Cisneros, G. A.; González, E. O. C. Computational Investigation of Non-Covalent Interactions in 1-butyl 3-methylimidazolium/bis(trifluoromethylsulfonyl)imide [bmim][Tf2N] in EMD and NEMD. *The Journal of Chemical Physics* **2018**, *148*, 054303.
125. Aramaki, K.; Hackerman, N. Inhibition mechanism of medium-sized polymethyleneimine. *Journal of Electrochemical Society* **1969**, *116* (5), 568.
126. Tripathy, D. B.; Murmu, M.; Banerjee, P.; Quraishi, M. A. Palmitic Acid Based Environmentally Benign Corrosion Inhibiting Formulation Useful during Acid Cleansing Process in MSF Desalination Plants. *Desalination* **2019**, *472*, 114128.
127. Aquino-Torres, E.; Camacho-Mendoza, R. L.; Gutierrez, E.; Rodriguez, J. A.; Feria, L.; Thangarasu, P.; Cruz-Borbolla, J. The Influence of Iodide in Corrosion inhibition by Organic Compounds on Carbon Steel: Theoretical and Experimental Studies. *Appl. Surf. Sci.* **2020**, *514*, 145928.
128. Issa, R. M.; Awad, M. K.; Atlam, F. M. Quantum Chemical Studies on the Inhibition of Corrosion of Copper Surface by Substituted Uracils. *Appl. Surf. Sci.* **2008**, *255*, 2433–2441.
129. Wang, W. H.; Li, Z.; Sun, Q.; Du, A. J.; Li, Y. L.; Wang, J.; Bi, S. W.; Li, P. Insights into the Nature of the Coupling Interactions between Uracil Corrosion Inhibitors and Copper: A DFT and Molecular Dynamics Study. *Corros. Sci.* **2012**, *61*, 101–110.
130. Gadowa, H. S.; Farghaly, T. A.; Eldesoky, A. M. Experimental and Theoretical Investigations for Some Spiropyrazoles Derivatives as Corrosion Inhibitors for Copper in 2 M HNO_3 Solutions. *J. Mol. Liq.* **2019**, *294*, 111614.
131. Karthik, G.; Sundaravaduvelu, M. Investigations of the Inhibition of Copper Corrosion in Nitric Acid Solutions by Levetiracetam Drug. *Egypt. J. Pet.* **2016**, *25*, 481–493.
132. Fouda, A. S.; Fouad, R. R. New Azonitrile Derivatives as Corrosion Inhibitors for Copper in Nitric Acid Solution. *Cogent Chem.* **2016**, *2*, 1221174.

Chapter 8

Corrosion Inhibitors for Basic Environments

Mahmood Shariatmadar,[1] Nariman Alipanah,[1] Mohammad Mahdavian,[1,*] Bahram Ramezanzadeh,[1,*] and Eiman Alibakhshi[1,2]

[1]Surface Coatings and Corrosion Department, Institute for Color Science and Technology, Tehran, Iran

[2]Atlas Protecting Coating, Tehran, Iran

[*]Email: mahdavian.m@icrc.ac.ir.

[*]Email: ramezanzadeh-bh@icrc.ac.ir.

Alkaline corrosion of metals is usually considered less dangerous than acidic corrosion. However, alkaline corrosion may lead to severe corrosion under specific conditions or after long-term exposure. Aluminum and its alloys are vulnerable to alkaline corrosion due to hydroxide ions attacking the naturally formed inert oxide film at pHs greater than 9. Contrarily, mild steel is usually stable in mild alkaline media; however, it corrodes when the electrolyte is contaminated with chloride, which is very common for steel corrosion in a concrete pore solution. Copper and nickel are corroded in highly alkaline electrolytes and the mild alkaline condition when in the presence of aggressive ions. Zinc is usually stable at mild alkaline media. It undergoes corrosion at a pH greater than 10.5, which is very common in metal and air batteries. In this chapter, the sustainable corrosion inhibitors, which are usually based on plant extracts, expired drugs, amino acids, and fatty acids, introduced for each metal have been reviewed.

Aluminum

Aluminum alloys (AAs) are utilized in many industries such as production of household items, aerospace technology, and marine equipment. Al possesses properties such as low cost and high negative potential (-2.3 V compared to a standard hydrogen electrode), amperage delivery, and energy density (8.1 kW h/kg), that makes it a potential candidate for the galvanic anode in batteries with alkaline electrolytes. Al and air batteries have similar commercial applications to Al when in basic environments (*1–6*). In addition, alkaline etching is required for surface pretreatment of Al to develop the nanoscale surface texture and properties that improve the adhesion of organic coatings to the Al surface and enhance the corrosion protection (*5, 7, 8*).

Al is known as a highly corrosion resistant metal in many media thanks to its protective and passive film-forming via oxidation on the surface. A dense oxide film covers the Al metal in oxygen-containing environments such as air or water. This inert film is protective in the pH range of approximately 4 to 8.5. It even forms rapidly when a scratch or mechanical defect occurs. However, Al has less stability in alkaline media because of the high negative potential and energy density, especially in NaOH and KOH solutions. The stable film is not effective in alkaline media, and it is dissolved by attacking OH^- ions. Many types of sustainable inhibitors have been introduced for aluminum protection without significantly interfering with any of its functions. The organic compounds consisting of heteroatom (O, S, P, and N), conjugated double bonds, π bonds, and benzene rings in their structure can be potentially efficient corrosion inhibitors (*1*, *2*, *9*, *10*).

Xhanari et al. (*11*) reviewed the reports about the mechanism of Al corrosion in basic media. Through these reports, it is noted that the OH^- ions migrate through oxide film to the Al surface, and the Al hydroxide forms on the surface as seen in eq 1. This layer is dissolved chemically by aggressive OH^- as seen in eq 2. The overall reaction of aluminum in alkaline media (eq 3) can be obtained by the combination of eqs 1 and 2. It shows the formation of soluble aluminate on the anodic areas of aluminum.

$$Al + 3OH^- \rightarrow Al(OH)_3 + 3e^- \quad (1)$$

$$Al(OH)_3 + OH^- \rightarrow Al(OH)_4^- \quad (2)$$

$$Al + 4OH^- \rightarrow Al(OH)_4^- + 3e^- \quad (3)$$

Reduction of water or oxygen can occur as a cathodic reaction. The production of hydrogen gas in the process of Al corrosion in basic media shows reduction of water as the main cathodic reaction (eq 4). The combination of anodic and cathodic reactions leads to the overall corrosion reaction, according to eq 5.

$$3H_2O + 3e^- \rightarrow 3/2H_2 + 3OH^- \quad (4)$$

$$Al + 3H_2O + OH^- \rightarrow 3/2H_2 + Al(OH)_4^- \quad (5)$$

The oxygen reduction can also occur as a cathodic reaction (eq 6). The overall corrosion reaction can be expressed by eq 7 (2, *11*–*13*).

$$3/4O_2 + 3/2H_2O + 3e^- \rightarrow 3OH^- \quad (6)$$

$$Al + OH^- + 3/4O_2 + 3/2H_2O \rightarrow Al(OH)_4^- \quad (7)$$

Inhibitors Based on Plant Extracts

Abiola et al. (9) showed that *Gossypium hirsutum* leaves and seed extracts (GLE and GSE) are organic compounds with inhibition properties. These plant extracts have heteroatoms like O, N, and S in their structures, and they can be adsorbed on the Al surface in basic solutions. Adsorption of the extract inhibitive species restricts hydroxide anions access to the surface. The GLE was more effective than GSE for the protection of AA. It showed 82% and 97% inhibition efficiency in the 2 M of NaOH solution with 10% and 52% v/v concentration of leave extract, which was determined using the weight loss method.

Abiola et al. (*14*) also reported *Phyllanthus amarus* leaf extract can inhibit the 3SR AA by forming a barrier film consisting of inhibitor molecules according to the Langmuir adsorption isotherm. *P. amarus* contains numerous organic compounds such as triterpenoids, steroids, proteins, carbohydrates, cyanogenic glycosides, flavonoids, alkaloids, and sugar. There are also tannins and saponin with surface-active compounds with more than 500 g.mol^{-1} molecular weights. In addition, *P. amarus* leaf extract is rich in heterocyclic lignans (hypophyllathin and phyllathin), which contain many oxygen heteroatoms. All these organic compounds have π-electron centers and functional groups that can adsorb on the Al surface in alkaline media. A higher surface coverage and lower corrosion rate (CR) were also observed by an increase in the extract concentration. A 76% inhibition efficiency was obtained in a 2-M of NaOH solution at a 20% v/v concentration of extract. The weight loss measurements showed zero-order kinetics established in the corrosion reaction of Al in the basic media without and with the *P. amarus* leaf extract. As a result, they concluded that the existence of the extract reduced the CR but did not change its kinetics. The studies indicated that the Al CR in basic media is controlled by the anodic reaction as seen in eq 3.

Sirajunnisa et al. (*15*) assessed the effect of *Vitex negundo (nochi)* leaf extract on the corrosion behavior of pure Al in 1M of NaOH solution by AC-impedance measurements, potentiodynamic polarization, gasometric, and weight loss methods. Studies suggest that the active surface area was blocked, and both anodic and cathodic reactions were controlled by *V. negundo* leaf extract. The inhibition efficiency of more than 70% was achieved in the concentration of 900 ppm. A smoother surface in the inhibited sample was detected by field emission scanning electron microscopy due to the formation of the protective layer. Also, the impact of temperature was surveyed. The CR increased when the temperature rises, but the inhibition efficiency dropped from 72.6% to 64.6% when the temperature rises from 30°C to 60°C. This demonstrates that the inhibitor adsorption physically occurred on the Al surfaces. The CR on uninhibited panels was significantly more than the inhibited samples in each test temperature.

The various commercial brands of henna *(Lawsonia inermis)* leaf water extracts are utilized by Al-Sehaibani for inhibition of metals in the solutions with different pH levels (*16*). The henna is a cheap and eco-friendly product with a natural origin. It is a common dye staff for hair, and it is used in dermatological applications. Its maximum inhibition of around 99.8% was achieved at 20 g/L concentration in NaOH media. Henna leaf extract contains some active substances such as dextrose, 2-hydroxy-1,4-naphthoquinone, and gallic acid that form 18% of its dry mass. These active substances can adsorb chemically on the Al surface.

Singh et al. (*17*) studied the inhibition effect of *Piper longum* seed extracts in a 1 M of NaOH environment. This material is used to inhibit the Al anode in power sources like Al and air batteries. The highest concentration was 400 mg/L, and its inhibition potency was 94% by weight loss experiments and 89% by polarization method. Inhibitors adsorbed on the Al-1060 surface according to the Langmuir model of isotherm and acted mix-type inhibition. The effect of temperature was also studied by CR and activation energy (E_a) calculation. The CR increased by increases in temperature due to the corresponding decreases in the adsorption strength, as the equilibrium of adsorption-desorption shifted to desorption. The E_a value was 81.8 kJ/mg and 15.9 kJ/mg in the presence and absence of 400 mg/L inhibitors, respectively. The E_a was higher in the presence of extracts. As a result, the energy barrier was significantly higher for the establishing the corrosion reaction. It is suggested that the extract had a substantial inhibition at the higher temperature.

El-Deeb et al. (*5*) modified the lignin extract obtained from wheat straw that was an agricultural waste. Lignin is a natural polymer in the wall of plant cells that is produced in paper and pulping industries as a by-product. Its structure consists of various functional groups such as carboxyl, carbonyl, aliphatic hydroxyl, and phenolic hydroxyl. It was modified by hydrolysis and carboxylation. The corrosion current density decreased in the presence of lignin and modified lignin. Also, the corrosion potential shifted to a more noble area. The inhibition of carboxylated lignin was more than that of hydrolyzed lignin, and both had a more inhibitory effect than the unmodified lignin.

More inhibition originated from more active sites to adsorb on the Al surface. The carboxylated lignin showed a 81.1% inhibition efficiency in 1.0 M of NaOH media with a concentration of 300 ppm. The scanning electron microscopy (SEM) analysis proved that the roughness decreased and the homogeneity was enhanced by more adsorption and inhibition effects of inhibitors. Also, the energy-dispersive X-ray spectroscopy analysis showed the existence of the characteristic carbon element peak in the protective barrier layer formed on the Al surface. This peak was absent in the sample exposed to 1.0 M NaOH without any inhibitor.

Chaubey et al. (*6*) introduced *kalmegh (Andrographis paniculata)* leaf extract as a cheap and green inhibitor with a water-soluble active constituent that protects the AA in a 1 M of NaOH medium. This extract had various natural organic compounds such as paniculides, neoandrographolide, and andrographolide, which can be adsorbed on the AA surface due to their multiple bonds. For example, andrographolide exists predominantly in the extract, which is a lactone with bicyclic diterpenoid. The polarization analysis indicated 82.4% inhibition in a 1.0 g/L concentration, which was mainly cathodic type. The electrochemical impedance spectroscopy (EIS) results revealed that the charge transfer resistance was enhanced, and the double layer capacitance was reduced for the inhibited samples. This occurs due to a decrease in the local dielectric constant or an increase in the thickness of the electrical double layer by adsorption of inhibitors on the Al and solution interface. The adsorption followed Langmuir's isotherm. Also, atomic force microscopy (AFM) and SEM results showed a decrease in surface roughness and heterogeneity in the presence of the extract.

A novel combination of *Lupine* seed extract and a cationic surfactant was used by Abdel-Gaber et al. (*18*) to inhibit Al corrosion in a 2 M of NaOH environment. The *Lupine* seeds contain protein (approximately 50%), lipid (approximately 20%), and quinolizidine alkaloids (approximately 5%). Sparteine, albine, multiflorine, and lupanine exist as quinolizidine alkaloids in the *Lupine* seed extract. It has been shown that sparteine has medicinal properties. Inhibition efficiencies of 70.0% and 87.1% were achieved respectively by the inclusion of approximately 2.9 g/L and 4.4 g/L of *Lupine* seed extract according to polarization measurements. With more inhibition effect assigned to the plant extract, there was no appreciable synergism between the *Lupine* extract and cationic surfactant.

Fu et al. (*19*) used caffeine to inhibit AA2024, which is utilized frequently in the aerospace industry. Caffeine is an environmentally friendly, nontoxic, and natural inhibitor that can have an inhibitory effect in the alkaline environment due to various active sites in its structure. There are also many plant extracts employed for inhibition of Al and its alloys in various alkaline media. The extracts of *damsissa (Ambrosia maritima)* leaf (*20*), *Sinapis alba* (*21*), *brahmi (Bacopa monnieri)* stem (*22*), *Lupinus varius* leaf extract (*23*), *neolamarckia cadamba* bark (*24*), *Vigna unguiculata* (agricultural waste material) (*25*), gum Arabic (*26*), cantaloupe juice, and seed extracts (*27*) are some examples used in the reports. One of the best performances of these green extract inhibitors is collected in Table 1.

Inhibitors Based on Drugs

Edrah et al. (*10*) utilized thiourea derivatives to protect Al from corrosion in basic media. *Thiourea* derivatives as organic substances have been utilized in medicinal chemistry and drug design (*28*). They have functional groups with O, S, or N heteroatoms. The lone pair of electrons on these atoms facilitates the transfer of electrons from the inhibitor to the substrate, and as a result, the coordinate covalent bonds form between the organic inhibitor and Al substrate. The adsorption strength is influenced by the donor atoms' electron density and the functional groups' polarizability. The carboxy phenyl thiourea revealed 28.3% inhibition efficiency in 0.3 M NaOH solution at 0.2%

w/v concentration at room temperature. A similar pattern was seen in the weight loss graphs, but the values significantly decreased in the existence of inhibitors.

The expired drug theophylline (DT), also named 1,3 dimethylxanthine, was utilized by Su et al. (2) for corrosion inhibition of Al. Many drugs such as isoxazoles, thiophenes, imidazoles (IMZs), furans, and pyridines are organics that include heteroatoms. They are nontoxic and have no environmental threat. Expired medications are pharmaceutical substances that must be discarded, and their use as a corrosion inhibitor is very cost-effective and environmentally friendly. The DT is a drug for breathing diseases that originated from tea and cocoa. The inhibition of 7075 AAs in a 1M NaOH solution in the presence of DT was investigated by electrochemical and gravimetric tests and quantum calculations. The capacitance of the double layer decreased, and charge transfer resistance increased with the addition of DT to the solution according to impedance studies. The cathodic shift and inhibition efficiency of 90% was achieved in a concentration of 2.5% v/v from polarization studies. The smooth texture was also detected in the inhibited panel by SEM and energy-dispersive X-ray spectroscopy.

Other important heterocyclic organic compounds with tremendous biological application are pyridine derivatives. Pyridine is a benzene molecule with one of its C-H atoms replaced by a nitrogen atom. Pyridine derivatives have good activities against biological targets (a variety of microbial, viral, and cancerous diseased) (*29*). Patil et al. (*30*) investigated 3-nitropyridine and 3-methylpyridine influences on Al corrosion in a 1 M KOH solution. Adsorption of inhibitor molecules on the metal surface was the reason for inhibition, and it obeyed the Frumkin isotherm. The 3-methylpyridine had more protection and showed 83.59% inhibition efficiency in the presence of a 5.0×10^{-4} mol/L inhibitor (*31*). Padash et al. expressed that 2,6-dimethylpyridine had an approximate 59.9% inhibition efficiency for Al in a 1.0 M NaOH with 0.4 M concentration of inhibitor. They proved that the preferred group for adsorption on the Al surface is the electron pair on the N atom of pyridine according to interaction energy and natural bond orbital results. The nicotinamide was chosen by Asan et al. (*32*) to inhibit the Al in alkaline environments due to unpaired electrons of N and O atoms in its structure. It is a water-soluble form of vitamin B3 with medicinal applications and is found in numerous foods. A 46.9% inhibition efficiency was obtained in a 0.1 M NaOH using 500 ppm nicotinamide.

Inhibitors Based on Biosubstances

The gelatin was used by Abdallah et al. (*1*) to inhibit the Al substrate in a 0.1 M of NaOH solution. There were toxic materials for inhibition, but their use has been limited due to environmental issues. Natural and nontoxic products with an environmentally friendly nature from renewable sources are good alternatives to synthetic compounds. Gelatin has some compounds such as glutamic acid, alanine, and glycine with NH_2 and COOH functional groups. It can form a barrier of charge transfer and mass on the surface of Al and Al-Si to protect them from aggressive agents. The adsorption followed the Freundlich isotherm and had negative ΔG°_{ads} showing the spontaneity of the adsorption process.

Camel's urine (CU), as a nontoxic and antimicrobial natural waste, was applied by Noor et al. (*33*) to inhibit the Al–Cu alloy in a 0.05M NaOH solution. A more than 99% inhibition efficiency was recorded by EIS and polarization measurements at a concentration of 10% v/v. The inhibition behavior of CU was a mixed type in low concentrations, but anodic inhibition was affected more by adding more than 3% CU. Many metabolism products like amino acid derivatives, fatty acid derivatives, canavanine, and prostaglandins are present in CU. Also, many inorganic components, including Mg, Ca, Sr, Zn, P, Al, Mn, Na, Fe, Cr, Cu, and K, accompany them. CU is a mixture of inhibitors with a synergistic effect that can be more effective than a special inhibitor. The adsorption could be physical or chemical. The charge of the Al surface in the alkaline condition is negative. As a result, the positive constituents of the inhibitor physically adsorb on the metal surface. Moreover, the components with a negative charge, such as amino acids, can form complexes with metallic cations that can be positive, negative, or uncharged. These various charged structures produce multilayer compounds consisting of CU components on the surface. The chemisorption can also occur by electron transferring between the surface and the inhibitors, especially in species with heteroatoms.

Umoren et al. (*34*) also used humic acids extracted from poultry, goat, and cow dung to inhibit 3SR Al in a 0.1 M of NaOH solution. The humic acid was a mixed-type inhibitor and showed 77.9% inhibition in the concentration of 0.10 g/L. The adsorption isotherm obeyed the Langmuir isotherm. It indicated that a monolayer adsorbed on the Al surface, and no interaction existed between the adsorbate molecules.

Wang et al. (*4*) selected L-cysteine to inhibit AA5052 in a 4 M of NaOH solution. This alloy and electrolyte are utilized in Al and air batteries. L-cysteine is an α-amino acid with mercapto, carboxylic, and amino groups. It is industrially produced by the hydrolysis of animal origin materials and used in food applications and pharmaceutical industries (*35*). L-cysteine has a thiol functional group that can be involved in redox reactions (*36*). The EIS and polarization analysis showed maximum inhibition in the concentration of 30 mmol/L. The isotherm of L-cysteine adsorption on the AA surface was Langmuir. Also, the measurements indicated that the cathodic reaction was mainly inhibited by this amino acid.

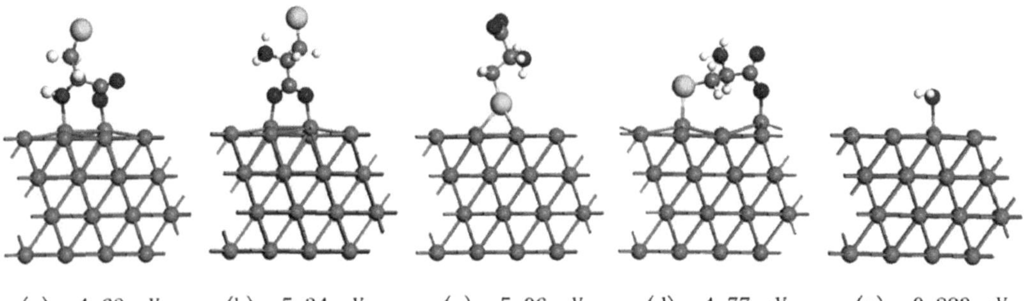

Figure 1. Adsorption energy and optimized structure of different adsorption modes of L-cysteine and water molecules on Al surface 4. (a) The optimized structure of a L-cysteine molecule adsorbed on an Al surface via N and O atoms. (b) The optimized structure of a L-cysteine molecule adsorbed on an Al surface via O atoms. (c) The optimized structure of a L-cysteine molecule adsorbed on Al via N and S atoms. (d) The optimized structure of a L-cysteine molecule adsorbed on Al via S and O atoms. (e) The optimized structure of a water molecule adsorbed on Al. Reprinted with permission from Ref. (4). Copyright 2016 Elsevier.

The adsorption of all three active sites (S in the mercapto group, N in the amino group, and O in the carboxylic group) on the Al surface was confirmed by quantum chemical calculations. Figure 1 shows some stable adsorption modes of *L-cysteine* molecules on Al. There are optimized structures of adsorbed L-cysteine molecules (via different inhibitive atoms) and water molecules on the Al surface. The energies of the clean Al surface and the isolated L-cysteine molecules were subtracted from the total energies of the adsorption system, and the adsorption energies were calculated. The adsorption energy results are also presented in Figure 1. The more negative values suggested a more stable state of adsorption. The carboxyl group had more adsorption ability. The stronger adsorption ability of

L-cysteine inhibitive molecules, when compared to water molecules, confirms that inhibitors were adsorbed on the Al surface instead of water molecules.

The charge density difference was also calculated for further investigation. As seen in Figure 2, the red and blue areas are assigned to the accumulation and deficit of charge regions, respectively. The reddish area around the heteroatoms ascribes to lone pair electrons, and the blue regions around the Al atoms associate to empty orbitals. The differences in charge density visually suggest the L-cysteine inhibitive molecules interact with the Al surface. It confirms that the coordination bonds form between the L-cysteine heteroatoms and Al.

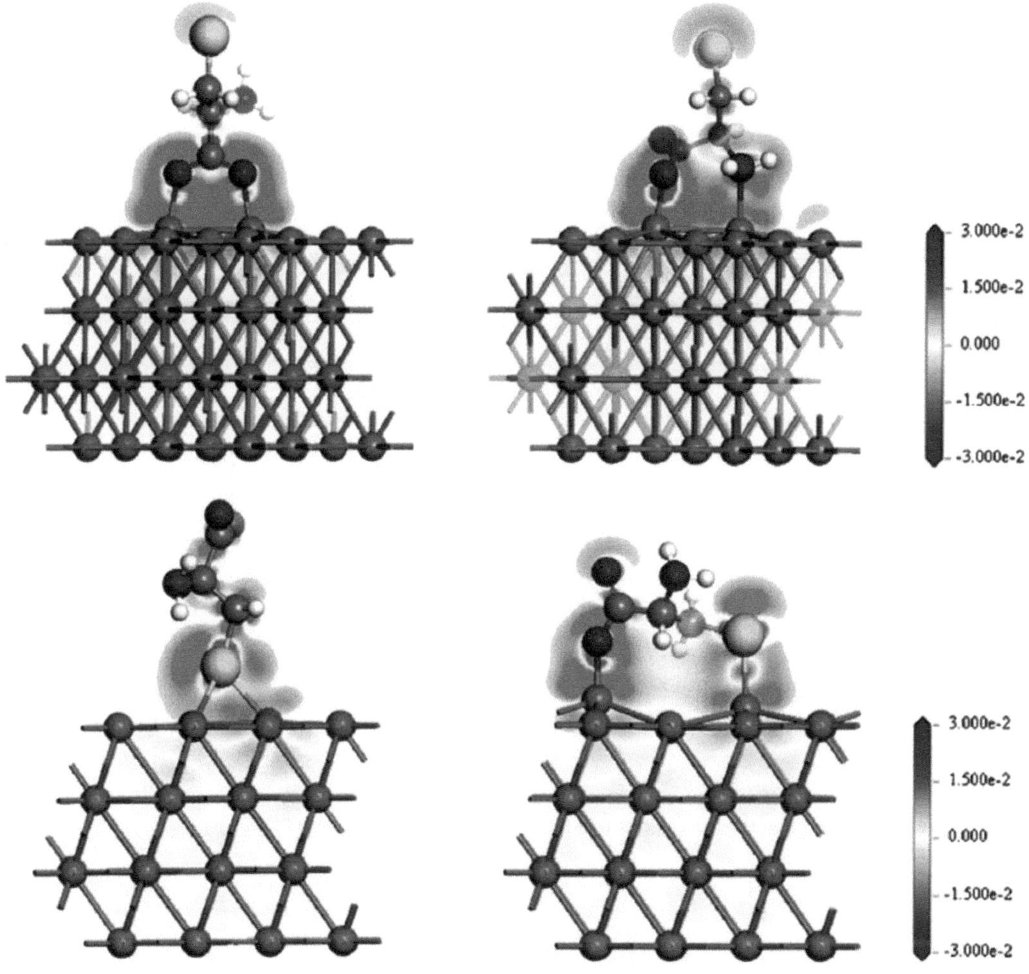

Figure 2. Charge density difference of various adsorption modes of L-cysteine inhibitive molecules on an Al surface.

The alkyl polyglucoside, known as a safe and biodegradable material, was used by Wu et al. (37) to inhibit the Al anode. They made a hybrid inhibitor system with alkyl polyglucoside and potassium stannate. This inhibitor showed a synergistic effect with 94.14% inhibition efficiency. The alkyl polyglucoside individually had an 80.56% inhibition efficiency in 4 M KOH solution at a concentration of 2 mM.

Inhibitors Based on Other Organic Compounds

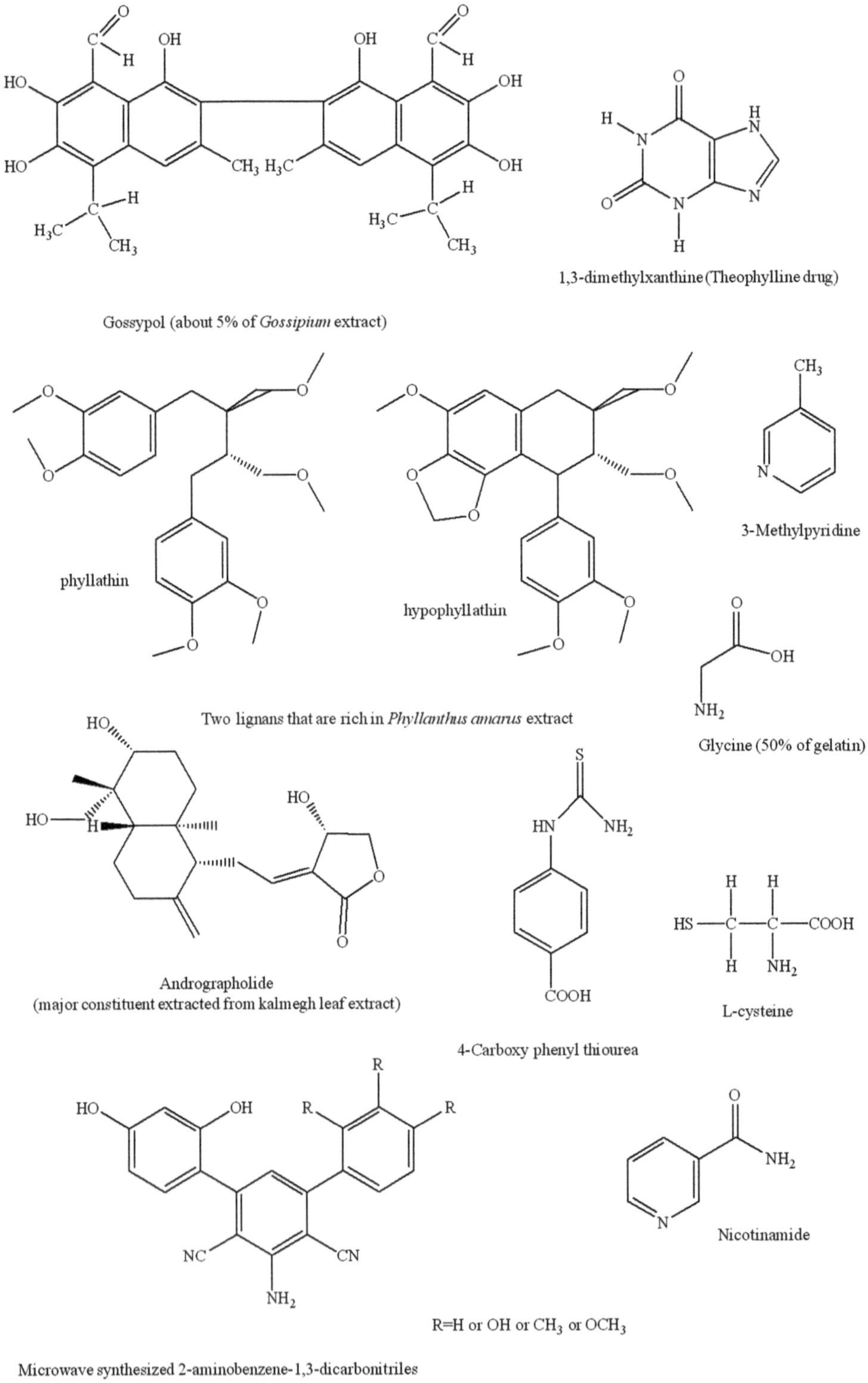

Figure 3. The structures of some inhibitors for aluminum used in alkaline media.

Table 1. The Inhibitors of Al and AAs in the Alkaline Environment and Their Inhibition Efficiency

Inhibitor	*Concentration*	*NaOH Concentration*	*Inhibitive Efficiency (%)*
G. hirsutum **leaf and seed extracts** (*9*)	52% v/v	2 M	97
P. amarus **leaves extract** (*14*)	20% v/v	2 M	76
V. negundo (nochi) **leaf extract** (*15*)	900 ppm	1 N	72.6
Henna (L. inermis) **leaf water extracts** (*16*)	20 g/L	0.1 M	99.8
Piper longum **seed extract** (*17*)	400 mg/L	1 M	94
Carboxylated lignin extracts obtained from wheat straw (*5*)	300 ppm	1.0 M	81.1
Kalmegh (A. paniculata) **leaf extract** (*6*)	1.0 g/L	1 M	82.4
Lupine **seed extract** (*18*)	2.976 g/L	2 M	70.0
Damsissa(A. maritima) **leaf extract** (*20*)	1.650 g/L	2 M	71.8
S. Alba **extract** (*21*)	1.00 v/v	1.0 M	98.41
Brahmi (B. monnieri) **stem extract** (*22*)	400 mg/L	0.5 M	96
L. varius **leaf extract** (*23*)	100% v/v	1 M	93.73
N. cadamba **bark extract** (*24*)	0.6 g/L	1 M	87
V. unguiculata **extract (agricultural waste)** (*25*)	0.5 g/L	0.5 M	77.78
Gum arabic (*26*)	0.5 g/L	1 M	50.24
Cantaloupe juice extracts (*27*)	50% v/v	1.0 M	69.95
Carboxyphenyl thiourea (*10*)	0.2% w/v	0.3 M	28.3
Expired DT (*2*)	2.5% v/v	1.0 M	90
2,6-dimethylpyridine (*30*)	0.4 M	1.0 M	59.93
Nicotinamide (*32*)	500 ppm	0.1 M	46.9
Gelatin (*1*)	2000 ppm	0.1 M	65.71
Camel's urine (*33*)	10% v/v	0.05 M	99.49
Humic acids extracted from poultry, goat, and cow dung (*34*)	0.1 g/L	0.1 M	77.9
L-cysteine (*4*)	30 mM	4 M	Not calculated
Microwave synthesized 2-aminobenzene-1,3-dicarbonitriles derivatives (*38*)	300 ppm	0.5 M	94.50
6-dehydroabietic acylamino sodium (*39*)	1×10^{-3} M	0.05 M	92.95

The 2-aminobenzene-1,3-dicarbonitriles derivatives were synthesized by Verma et al. (*38*) with microwave irradiation. This rapid and powerful method is a good way to transform functional groups. Low operating costs and environmental heat loss are also advantages of microwave irradiation synthesis. These advantages make microwave-assisted synthesis a green procedure. The synthesized inhibitor with the name of 5′-amino-2,3″,4-trihydroxy-1,3-methoxy, 1′:3′,1″-terphenyl-4′,6′ dicarbonitrile at a concentration of 300 ppm recorded 94.50% inhibition efficiency when protecting Al-1060 exposed in 0.5 M of NaOH solution. The Langmuir isotherm of adsorption and mixed-type inhibition was also concluded from these results.

Zheng et al. (39) utilized 6-dehydroabietic acylamino sodium as an eco-friendly rosin-based inhibitor of AA2024-T3 in basic media. This anionic surfactant, which contains conjugated π bonds and carboxylic groups, was obtained from low cost and renewable rosin. The measurements showed 92.95% inhibition in a 0.01 M NaOH with a 1×10^{-3} M inhibitor. The AFM, SEM, and X-ray photoelectron spectroscopy (XPS) analyses suggested the covered surfaces by inhibitors were smooth, and coordination bonds could be formed by electron-donating the benzene ring and O atom from the inhibitor.

Table 1 indicates some performances of inhibitors used for inhibition of Al and its alloys in basic environments. All reported values in Table 1 are assigned to room temperature analysis (from 25 °C to 30 °C). Also, the structures of some inhibitors are collected in Figure 3. Many heteroatoms, conjugated systems, and benzene rings exist in the structure of these sustainable inhibitors.

Steel

Steel has been used in the construction, water cooling system, gas transportation, and oil industries. In all these industries, the steel is exposed to alkaline and neutral media (*40*).

One of the main alkaline media for mild steel in construction is concrete pore solution (with a pH greater than 12). A protective oxide layer forms on the steel surface in this medium. However, diffusion of water with corrosive ions can initiate rebar corrosion at the steel-concrete interface and cause cracking and reduction of the load-carrying capacity. Using inhibitors in concrete or mild steel coatings is an economical way to protect the rebars from corrosion.

The passive film forms on the carbon steel's surface in alkaline media consisting of two layers. The interlayer's main composition was reported as magnetite (Fe_3O_4), which was the mixture of FeO and Fe_2O_3 oxides. The possible reactions for creating the passive layer are shown in eqs 8 to 13 (*41–43*).

$$Fe \rightarrow Fe^{2+} + 2e^- \quad (8)$$

$$Fe^{2+} + 2OH^- \rightarrow Fe(OH)_2 \quad (9)$$

$$4Fe(OH)_2 + 2H_2O + O_2 \rightarrow 4Fe(OH)_3 \quad (10)$$

$$2Fe(OH)_3 \rightarrow Fe_2O_3.H_2O + 2H_2O \quad (11)$$

$$3Fe(OH)_2 + 2OH^- \rightarrow Fe_3O_4 + 4H_2O + 2e^- \quad (12)$$

$$3FeO + 2OH^- \rightarrow Fe_3O_4 + H_2O + 2e^- \quad (13)$$

5,8-Dihydroxy-2-(2-phenylethyl) chromone		hyperosid	
Astragalin		Licochalcone A	
Buddlenoid A		Pinocembrin	
cyclo-(Phe-Phe)		Procyanidin B1	
cyclo-(Tyr-Leu)		DMEA	
Docosanoic acid			

Figure 4. The configuration of inhibiting molecules of PAL extract and DMEA on the steel surface. Reprinted with permission from Ref. (45). Copyright 2020 Elsevier.

Ben Harb et al. (44) used olive leaf extract in an alkaline solution containing 0.1 M of NaOH and 0.5 M of NaCl (pH of 13) to simulate sea water-polluted concrete pore solution. The olive leaf extract was prepared using four different solvents including methanol, ethyl acetate, dichloromethane, and hexane. The composition of the extract was investigated by gas chromatography. The results showed that this green inhibitor contained N atoms, O atoms, hydroxyl, and amine functional groups. The inhibition efficiency increased by the polarity of solvents used in the extracting process of the inhibitor, which achieved 91.9% in methanol.

Liu et al. (45) studied *Platanus acerifolia* leaf extract (PAL) in a simulated chloride-induced concrete pore solution. The inhibitor was extracted by four different methods using NaOH and ethanol. The inhibitive properties of PAL were compared to sodium nitrite (NIT) (1 wt %) and N, N-dimethylethanolamine (DMEA) (3 wt %). In this study, the mild steel was immersed in an alkaline solution containing inhibitors for seven days. The NaCl was then added to the solution at a rate of 0.01 mol/L^{-1}/day^{-1} for 10 days. After that, potential dynamic polarization, linear polarization (LP), and EIS were used to calculate the inhibition efficiency. Potential dynamic polarization test results showed a mixed-type inhibition effect for PAL and DMEA and an anodic-type inhibition effect for the NIT inhibitor. The inhibition efficiencies of NIT, DMEA, and PAL at different concentrations and extraction methods were calculated using corrosion current density. The PAL at 5 wt % extracted by ethanol in an alkaline solution using an ultrasonic method showed high efficiency around 98.2% and 99.56% obtained by LP and EIS measurements, respectively, which was higher than DMEA and the same as NIT. XPS was used to analyze the rebar surface. The results showed that PAL formed a nitrogen-carbon film on the metal surface. Ultra-performance liquid chromatography–mass spectrometry results indicated more than 10 types of flavonoids and oligopeptides in the composition of PAL. Quantum mechanics and molecular dynamics simulations showed that molecules in PAL had higher ΔE_{ads} and stronger donor-acceptor interactions with metal compared to DMEA. The molecular dynamic simulations were implemented on 10 kinds of molecules in PAL and DMEA. The results showed that the PAL molecules horizontally aligned on the surface, which increases the surface coverage ability and enhances the protection of the steel surface. The orientation of the molecules on the steel surface is depicted in Figure 4.

Torres-Acosta used *Opuntia ficus indica* (nopal) mucilage in a contaminated saturated $Ca(OH)_2$ solution as a reinforced steel inhibitor (46). The rebars were placed in a $Ca(OH)_2$ solution containing nopal (0.5, 1, and 2 wt %) until potential stabilization, which took 25 days. Then, 5 g/L of NaCl was added to the solution at 30 days. The electrode potentials increased to higher than -10 mV before the addition of Cl^- ions. This indicated the formation of the passive layer on the steel surface and oxygen depletion in the solution (before 25 days of immersion). After the addition of Cl^- in each step, the potentials dropped to active corrosion, showing initiation of pitting corrosion. In each step, the nopal-containing solution had more positive potentials than the control solution. The inhibition efficiency was obtained using a LP test. The inhibition efficiency was 85% and 83% in 0.5 wt % and 2 wt % of nopal, respectively, but it decreased to 50% in 1 wt % in the presence of 5 g/L of NaCl.

El-Etre et al. studied the inhibition efficiency of henna extract in a 0.1 M of NaOH solution of carbon steel (47). The inhibition efficiency showed an increase after an increase in henna extract concentration and reached 69.56% at 800 ppm. The inhibitor showed mixed-type inhibition and a Langmuir adsorption isotherm. The henna extract consisted of tannin and lawsone (hydroxyl aromatic molecules), which made complexes on the metal surface.

Kumar et al. (48) studied inhibition behavior of leaves, latexes, and fruit extract of *Calotropis procera* and *Calotropis gigantea* in 0.5 M and 1 M NaOH solutions. The plants were extracted in boiling ethanol. The inhibition efficiency was obtained by mass loss and thermometric techniques after 0.5-h and 24-h immersions. The efficiency of all extracts was increased by rising inhibitor concentration. *C. procera's* latex and *C. gigantea's* latex extract showed maximum inhibition efficiency of approximately 79.2% and 79.7%, respectively, at 0.6 wt % in 1 M NaOH solution after 24 h of immersion.

Raghavendra et al. used mature arecanut husk (MAH) extract in a 0.5 M NaOH solution as the mild steel inhibitor (*49*). The adsorption isotherm was Langmuir, and the Gibbs energy adsorption (ΔG°_{ads} = -40 kJ/mol) showed chemisorption of the inhibitor onto the metal surface. The polarization results showed mixed-type inhibition and 83.02% efficiency at 12 g/L of concentration.

Inhibitors Based on Fatty Acids

Lin et al. (*50*) studied the inhibition properties of five different carboxylates (with different alkylene chain lengths) for steel in a simulated carbonation concrete pore solution with a pH of 11.5 (0.0021 M of NaOH, 0.0042 M of KOH, and 0.02 M of NaCl). 6-Heptenoic acid (C7), undecylenic acid (C11), and oleic acid (C18) showed sufficient inhibition behavior in mild steel. The inhibition efficiency, pitting corrosion, and polarization resistance (R_p) were evaluated using a polarization test. The inhibition efficiency increased by increasing the concentration. The efficiencies of C11, C18, and C7 at 1000 ppm after seven days of immersion were 90.2%, 83.8%, and 55.6%, respectively. The pitting potential (E_{pit}) shifted to a positive value by increasing the concentration of additives, thus illustrating the inhibitor adsorbed onto the metal surface at the early stage of immersion, which prevents the Cl^- anions attacks. The interaction of carboxylate and steel was evaluated by a quantum chemical calculation. The carboxylic group was an electron donor, and C=C was an electron acceptor. The undecylenic acid showed the minimum ΔE (E_{LUMO}-E_{HOMO}), which related to adsorption inversely. The adsorption isotherm of inhibitors was Langmuir. ΔG°_{ads} of 6-heptenoic acid, undecylenic acid, and oleic acid was less than -20 kJ/mol showing a chemisorption mechanism created by the interaction of a free electron pair of oxygen and d orbital of Fe. XPS and Raman spectrometry results of carbon steel immersed in the solution for 24 h showed that the inhibitor made a protective film containing Fe-COO-C_x compounds adsorbed by electrostatic and donor-acceptor interactions. The inhibition efficiency increased by increasing the alkylene chain. The quantum calculation results showed that the E_{HOMO} increased slightly as the fatty acid's carbon chain length increased. In contrast, E_{LUMO} decreased by increasing the carbon chain length and the distance between the C=C bonds and –COO^- functional groups revealing incremental electron-accepting ability of the molecular structure. It was shown that the C=C bonds in the fatty acid chain facilitate the adsorption by accepting the electron from the d orbital of Fe.

Inhibitors Based on Other Compounds

Monticelli et al. investigated large groups of additives in saturated calcium hydroxide containing 0.1 M of chloride solution that simulated a polluted reinforced concrete pore solution (*51*). Corrosion potential and pitting corrosion of rebars were measured. The difference of pitting potential and corrosion potential (E_{pit}-E_{cor}), known as the passive range of mild steel, was increased in the presence of inhibitors. The inhibition efficiency of additives in a highly alkaline solution was measured by the gravimetric method. This study showed that 0.005 M of 5-hexyl-benzotriazole, 0.05 M of sodium b-glycerophosphate (GPH), and saturated dicyclohexylammonium nitrite

(DCHAMN) had an inhibitory effect on steel rebar eliminated by pitting corrosion after 30 days of immersion in a corrosive solution. In the next phase of the study, the mentioned inhibitors were used in concrete test blocks (adding to the mortar mixture or in coating rebar). The EIS was carried out in chloride-contaminated concrete. The equivalent circuit reported in this study consisted of a three-time constant, in which R_p was reported as the charge transfer resistance. GPH- and DCHAMN-containing concrete showed good inhibiting results with 90% and 88% efficiency, respectively. After one year, concretes impregnated with tungstosilicic acid hydrate achieved an inhibitory efficiency of 62% (*51*).

Table 2. List of Green Corrosion Inhibitors for Mild Steel in Alkaline Media

Inhibitor	*Concentration*	*Alkaline Media*	*Inhibitive Efficiency (%)*
5-Hexyl-benzotriazole (*51*)	0.005 M	saturated $Ca(OH)_2$ + 0.1 M Cl^-	67%
Sodium b-glycerophosphate (*51*)	0.05 M 0.05[a]	saturated $Ca(OH)_2$ + 0.1 M Cl^- pH > 12[b]	92% 90%
Dicyclohexylammonium nitrite (*51*)	Saturated 0.05[a]	saturated $Ca(OH)_2$ + 0.1 M Cl^- pH > 12[b]	97% 88%
Tungstosilicic acid hydrate (*51*)	0.0035[a]	pH > 12[b] Cl^-/cement = 0.02[c]	62%
6-Heptenoic acid (*50*) Undecylenic acid (*50*) **Oleic acid** (*50*)	1000 ppm 1000 ppm 1000 ppm	pH = 11.5 (0.0021 M NaOH, 0.0042 M KOH, and 0.02 M NaCl)	55.6% 90.2% 83.8%
PVP (*40*)	1000 ppm 1000 ppm 1000 ppm 1000 ppm	pH = 9 + 0.1 M NaCl pH = 9 + 0.1 M NaCl + 0.06 M KI pH = 9 + 0.1 M NaCl pH = 9 + 0.1 M NaCl + 0.06 M KI	51.68% 67.25% 63.7% 88.76%
C. procera (*48*) ***C. gigantea*** (*48*)	0.6 wt % 0.6 wt %	1 M NaOH 1 M NaOH	79.92% 79.68%
Olive leaf extract (*44*)	-	0.1 M NaOH + 0.5 M NaCl (pH=13)	91.9%
***P. acerifolia* leaves** (*45*)	5 wt %	Saturated limewater + 0.1 M NaCl	98.2%[d] 99.56%[e]
Henna extract (Lawsonia) (*47*)	800 ppm	0.1 M NaOH	69.56%
***O. ficus indica* (Nopal) mucilage** (*46*)	0.5 wt %	saturated $Ca(OH)_2$ + 5 g/L NaCl	85%
MAH (*49*)	12 g/L	0.5 M NaOH	83.02%

[a] [ln(inhibitor)/Cl^-] molar ratio. [b] Concrete impregnated with alkaline containing inhibitor solution. [c] Weight ratio. [d] Using LP measurement. [e] Using EIS measurement.

Polyvinyl pyrrolidone (PVP) is a widely used polymer in the food, medicine, cosmetics, and paint industries (52). Juhaiman et al. used PVP as a green inhibitor of carbon steel in concentrated $Ca(OH)_2$ containing 0.1 M of NaCl solution (pHs of 9 and 10) (*40*). The adsorption isotherm of PVP on a carbon steel surface fitted well by a Langmuir isotherm. The results showed that the addition of 1000 ppm of PVP in an alkaline solution covered 52–62% of the carbon steel surface. The surface coverage of the steel increased by increasing the concentration of the inhibitor up to 1500 ppm due to steric hindrance. The $\Delta G°_{ads}$ of the inhibitor was -36.35 kJ/mol showing the chemisorption of PVP through the free electron pair of oxygen on the carbon steel surface. Also, the synergic effect of PVP and potassium iodide (KI) was investigated. It was shown that the inhibition efficiency of PVP increased from 51.68% to 67.25% at a pH of 9 and from 63.7% to 88.76% at a pH of 10 in the presence of 0.06 M of KI. The efficiency of PVP in the presence of KI was stable after 45 h of immersion, which was about 80% at a pH of 10.

The green inhibitors used for steel corrosion inhibition in alkaline media are listed in Table 2.

Copper and Brass Alloys

Copper and its alloys are widely used in industries such as cooling water systems, electronics, and marine applications due to good electrical conductivity, good thermal conductivity, and strong antifouling properties.

The results from several studies showed that Cu_2O was formed immediately on the metal surface (eq 14) through the adsorption of OH^- and formation of $CuOH^-_{ads}$ intermediates (53).

$$2Cu + 2OH^- \rightarrow Cu_2O + H_2O + 2e^- \quad (14)$$

Other thermodynamically possible reactions are shown in eqs 15 and 16.

$$Cu_2O + 2OH^- \rightarrow 2CuO + H_2O + 2e^- \quad (15)$$

$$Cu_2O + 2OH^- + H_2O \rightarrow 2Cu(OH)_2 + 2e^- \quad (16)$$

The $\Delta G°$ads of interaction between dissolved oxygen in alkaline solution and copper was reported as -72.52 kJ/mol. So the reaction shown in eq 17 is thermodynamically favored (54).

$$2Cu + \tfrac{1}{2} O_2 \rightarrow Cu_2O \quad (17)$$

Inhibitors Based on Azoles

Azoles are one of the main inhibitors used for copper in alkaline media (55). Azoles are organic compounds that contain heteroatoms (N, S, and O) that adsorb to the metal substrate and make strong interactions due to the free electron pair of heteroatoms (56). Altaf et al. (57) studied the effect of five azoles (benzotriazole, mercaptobenzothiazole, benzimidazole, mercaptobenzimidazole, and thiadiazole) on corrosion inhibition of pure copper and brass (CuZnAl, CuZnSn, CuZnPb, and CuZnMn) at a pH of 10.4 by potentiodynamic polarization and EIS. It was revealed that the adsorption followed the Langmuir isotherm. The azoles create a stable inhibitor film through the interaction of electron pairs of heteroatoms (N and S) with d-orbitals of Cu and Zn. All azoles showed an excellent inhibition effect on Cu and its alloys in potentiodynamic polarization, decreased the cathodic and anodic currents, and increased E_{corr} to higher potentials. The inhibitory efficiency of azoles increased as the concentration rose. In the absence of inhibitors, the other elements in copper alloys shifted the E_{cor} to more positive values as the anodic and cathodic currents decreased. In this case, the CuZnAl alloy showed higher corrosive resistance than the other alloys (*57*). The mercaptobenzothiazole showed the highest inhibitory efficiency for a CuZnAl alloy at 0.0003 M (96%).

Inhibitors Based on Plant Extracts

The inhibitory effect of MAH extract on copper in an alkaline medium (0.5 M of NaOH) was investigated by Raghavendra et al. (*49*). This inhibitor consists of lignins, cellulose, and hemicelluloses as the main components. The adsorption of MAH was fitted well by the Langmuir isotherm. The calculated $\Delta G°_{ads}$ of MAH was -40 kJ/mol, revealing chemical interaction and physical adsorption between the inhibitor and the copper substrate (*49, 58*). Potentiodynamic results showed that MAH acted as an anodic inhibitor. The corrosion current density (i_{corr}) of Cu in 0.5 M of NaOH (3.417 $mA.cm^2$) decreased significantly (0.037 $mA.cm^2$) in the presence of 12 g/L of MAH showing 98.8% inhibitory efficiency.

Inhibitors Based on Amino Acids

Methionine is an essential amino acid that plays a vital role in the protein structures of the human body and metabolism (*59*). Wu et al. studied the inhibitive effect of the combination of methionine and copper (I) oxide film on Cu in an alkaline solution (1 M of NaOH) (*60*). This system, known as the core shell, consisted of Cu as the core and the organic inhibitor (methionine) as the outer layer. The passive layer mainly consisted of Cu_2O. The polarization study showed the i_{corr} of Cu decreased over 10 orders of magnitude in a 10-mM concentration of methionine. In the presence of methionine, the shift of E_{corr} to negative potentials showed cathodic inhibition by adsorbing on the passive film of copper (I) oxide and slowing the oxygen diffusion to the cathodic sites. The inhibitory efficiency increased by increasing the concentration, eventually reaching 90.8% at 10 mM of concentration.

Another amino acid, cysteine, was studied for brass in a weak alkaline solution (pH of 9) in the presence of sulfate (SO_4^{2-}) by Radovanović et al. (*61*). Polarization studies showed that cysteine acted as a mixed inhibitor in 0.5 mol/dm^3 of Na_2SO_4 and 1×10^{-4} mol/dm^3 of cysteine. The corrosion potential of brass became more negative by raising the concentration of the inhibitor at the concentration of 1×10^{-4} mol/dm^3 of adsorption and forming a stable Cu(I)–cysteine complex, which is confirmed by chronoamperometric measurements. Langmuir was reported as the adsorption isotherm of this inhibitor. The Gibbs energy adsorption value of cysteine (-42.9 kJ/mol) indicated chemisorption of inhibitor on the brass surface in a weakly alkaline solution.

The green inhibitors used for copper and brass corrosion inhibition in alkaline media are listed in Table 3.

Table 3. List of Green Corrosion Inhibitors for Copper and Brass in Alkaline Media

Inhibitor	*Concentration*	*Alkaline Media*	*Inhibitive Efficiency (%)*
MAH (*49*)	12 g/L	0.5 M NaOH	98.8%
Methionine (*60*)	10 mM	1 M NaOH	90.8%
Cysteine (*61*)	1×10^{-4} mol/dm^3	pH = 9	83.6%
Mercaptobenzothiazole (*57*) **(for CuZnAl alloy)**	0.3 mol/dm^3	pH = 10.4	96%

Nickel

Nickel is used in corrosion resistance alloys mainly in stainless steel, electroplating, electroforming, and metal coatings (*62–65*). Nickel has been used in anticorrosive steels, electrocatalytic materials (*66*), fuel cells (*67*), and batteries (*68*) where it can be exposed to alkaline media.

In highly alkaline media, nickel corrodes according to eq 18 (*69*).

$$Ni + 2H_2O \rightarrow HNiO_2^- + 3H^+ + 2e^- \quad (18)$$

At a pH range of 9 to 12, a passive layer based on nickel oxides and hydroxides ($Ni(OH)_2$, NiO_2, and Ni_3O_4) forms on the surface.

Inhibitors Based on Plant Extract

Abdallah et al. investigated the inhibition effect of some natural oils, including sesame oil, watercress oil, wheat germ oil, and almond oil for Ni at 0.01 M of NaOH solution (*70*). The main components of the studied natural oils are reported in Table 4.

Table 4. The Main Components of Natural Oils Used as Green Inhibitors (*70*)

Natural Oil	*Main Components*
Sesame oil	Vitamin A
Wheat cress oil	Oleic acid (31%) – Palmitic acid (12%) – Linolinic acid (1%)
Watercress oil	γ linolinic acid
Almond oil	Oleic acid (68%) – Linolinic acid (25%) – Palmitic acid (4%)

The inhibition efficiency and the surface coverage were calculated by potentiostatic polarization measurements. The inhibition efficiency of natural oils increased by increasing the inhibitor's concentration and reached 72%, 66%, 62%, and 56% for sesame, watercress, wheat germ, and almond oils, respectively, at 150 ppm for individual inhibitors. The open-circuit potential results of nickel immersed in different NaOH concentrations showed that the potential shifted to positive values by increasing the pH of the solution. This indicated the formation of a Ni-O film. The adsorption isotherm of all natural oils was fitted by Freundlich isotherm. Also, the ΔG°_{ads} of the inhibitors was in the same range (-2.43, -2.39, -2.36, -2.29 kJ/mol for sesame oil, watercress oil, wheat germ oil, and almond oil, respectively), which showed spontaneous adsorption on the nickel surface. In the next phase of this study, the inhibitory effect of natural oils was investigated in the presence of Cl^- ions. The potential of pitting corrosion against inhibitor concentration was recorded. In the presence of inhibitors, the E_{pit} shifted to more positive amounts indicating the protection of inhibitors in corrosive media. The positive shifting of E_{pit} decreased in the following order: sesame oil > watercress oil > wheat germ oil > almond oil (*70*).

El-Etre et al. used henna extract in 0.1 M of NaOH solution for nickel (*47*). The efficiency was measured by potentiostatic polarization and reached 73.91% at 800 ppm. The inhibitor became complex on a nickel surface and was fitted well by the Langmuir adsorption isotherm. The green inhibitors used for nickel corrosion inhibition in alkaline media are listed in Table 5.

Table 5. List of Reviewed Inhibitors for Nickel in Alkaline Media

Inhibitor	*Concentration*	*Alkaline Media*	*Inhibitive Efficiency (%)*
Henna extract (*Lawsonia*) (*47*)	800 ppm	0.1 M NaOH	73.91%
Sesame oil (*70*)	150 ppm	0.01 M NaOH	72%
Watercress oil (*70*)	150 ppm	0.01 M NaOH	66%
Wheat germ oil (*70*)	150 ppm	0.01 M NaOH	62%
Almond oil (*70*)	150 ppm	0.01 M NaOH	56%

Zinc

Zinc is widely used as an anode in alkaline batteries (including Zn and MnO_2, Zn and NiO, Zn and Ag_2O, and Zn and air batteries) due to high specific energy, low cost, and nontoxicity. These alkaline batteries contain concentrated alkaline solutions as electrolytes and powder zinc as an anode. The corrosion of a zinc electrode results in the reduction of active zinc, which increases the impedance for battery discharge and decreases battery capacity (*71*, *72*).

The formation of a passive layer on the zinc surface in alkaline media occurs through three different models (dissolution-precipitation, adsorption, and nucleation and growth models) (*73*).

In the dissolution-precipitation model, ZnO was precipitated on the metal surface by supersaturating the $Zn(OH)_4^{2-}$ near the zinc surface (eqs 19 and 20).

$$Zn + 4OH^- \rightarrow Zn(OH)_4^{2-} + 2e^- \quad (19)$$

$$Zn(OH)_4^{2-} \rightarrow ZnO + H_2O + 2OH^- \quad (20)$$

In the adsorption model, the $Zn(OH)_{ads}$ is deprotonated at specific potentials and converted to the passive layer.

$$ZnOH(ads) \rightarrow ZnO + H^+ + 2e^- \quad (21)$$

In the nucleation and growth model, the metal directly reacts with OH^- and forms the oxide layer.

$$Zn + 2OH^- \rightarrow ZnO + H_2O + 2e^- \quad (22)$$

Zhou et al. investigated the synergism inhibition effect of imidazole (IMZ) and polyethylene glycol 600 (PEG) for zinc in the 3 M of KOH alkaline solution, which simulates the alkaline battery solution (72). The corrosion of Zn immersed in 3 M of KOH containing different additive concentrations was examined using the weight loss method and polarization after 10 days. The optimum efficiency for individual additive-containing solutions was obtained at 0.1 wt %. The inhibition efficiency (conducted by weight loss) of 0.1 wt % of IMZ, 0.1 wt % of PEG, and 0.05% IMZ and 0.05% PEG was 27%, 50%, and 83%, respectively, indicating the synergistic effect of IMZ and PEG. The polarization results showed IMZ and PEG were anodic and cathodic types of inhibitors, and their composition acted as a mixed-type inhibitor. SEM images of the Zn surface immersed in an alkaline solution for 30 days were obtained to study the surface morphology. The metal surface immersed in the alkaline solution containing 0.05% IMZ and 0.05% PEG solution had minimal roughness and showed the best corrosion inhibition. PEG exists as $k^+[PEG^-]$ in KOH solutions. The alkoxide anion adsorbed to the metal surface via an oxygen atom. Also, IMZ turned into IMZ potassium salt in a KOH solution and adsorbed to the anodic sites of the surface due to the stronger interaction of the nitrogen atoms to Zn than oxygen atoms. The PEG adsorbed to the sites that IMZ could not adsorb due to steric obstacles.

Deyab et al. studied controlling corrosion of zinc in alkaline battery solutions (7 M of KOH) by nonionic surfactant polyoxyethylene (40) nonylphenyl ether (PNE) as an inhibitor (*74*). The critical micelle concentration (CMC) of PNE was measured by contact angle measurements and reported as 0.25 mM. The inhibition efficiency of PNE was calculated using potentiodynamic polarization. The efficiency of the inhibitor was achieved at a maximum (98%) near the CMC. Also, the polarization results showed that the current density of anodic and cathodic branches decreased, and E_{corr} shifted less than 85 mV as the concentration of PNE increased, showing mixed-type inhibition behavior. The b_a and b_c were not changed significantly, which indicates adsorption of PNE at the zinc surface by overall blocking the zinc surface. From EIS test results, the R_{ct} and C_{dl} reached the maximum and minimum values at the CMC, respectively. This illustrated that the PNE adsorbed at the metal and electrolyte interface reduced electrical capacity and increased the thickness of the electrical double layer on zinc by replacing H_2O and ions on the metal surface. The adsorption isotherm was fitted by the Freundlich equation. The ΔG°_{ads} of PNE was -20 kJ/mol, indicating the adsorption of the inhibitor was physical. The SEM images of zinc were captured after 24 h of immersion in 7 M of KOH with and without the PNE (0.25 mM) solution. The surface morphology of Zn was rough, thus showing that corrosion takes place in an alkaline solution. In contrast, the morphology of the metal in the presence of PNE was smooth. The effect of PNE on batteries was investigated by the addition of 0.25 mM of PNE to the battery's anode gel. The battery containing 0.25 mM of PNE showed better capacity after 30 days of storage at 100 mA (*74*).

El-Etre et al. investigated the inhibitory effect of the henna extract in an alkaline solution (0.1 M of NaOH) (47). The efficiency increased with a rising concentration and became 73.91% at 800 ppm. The adsorption isotherm was fitted by Langmuir. The green inhibitors used for zinc corrosion inhibition in alkaline media are listed in Table 6.

The structures of some inhibitors used for steel, copper, nickel, and zinc are illustrated in Figure 5.

Table 6. List of Reviewed Inhibitors for Zinc in Alkaline Media

Inhibitor	*Concentration*	*Alkaline Media*	*Inhibitive Efficiency (%)*
Henna extract (*Lawsonia*) (47)	800 ppm	0.1 M NaOH	76.92%
PNE	0.25 mM	7 M KOH	98%
IMZ + PEG	0.05 wt % of each inhibitor	3 M KOH	83%

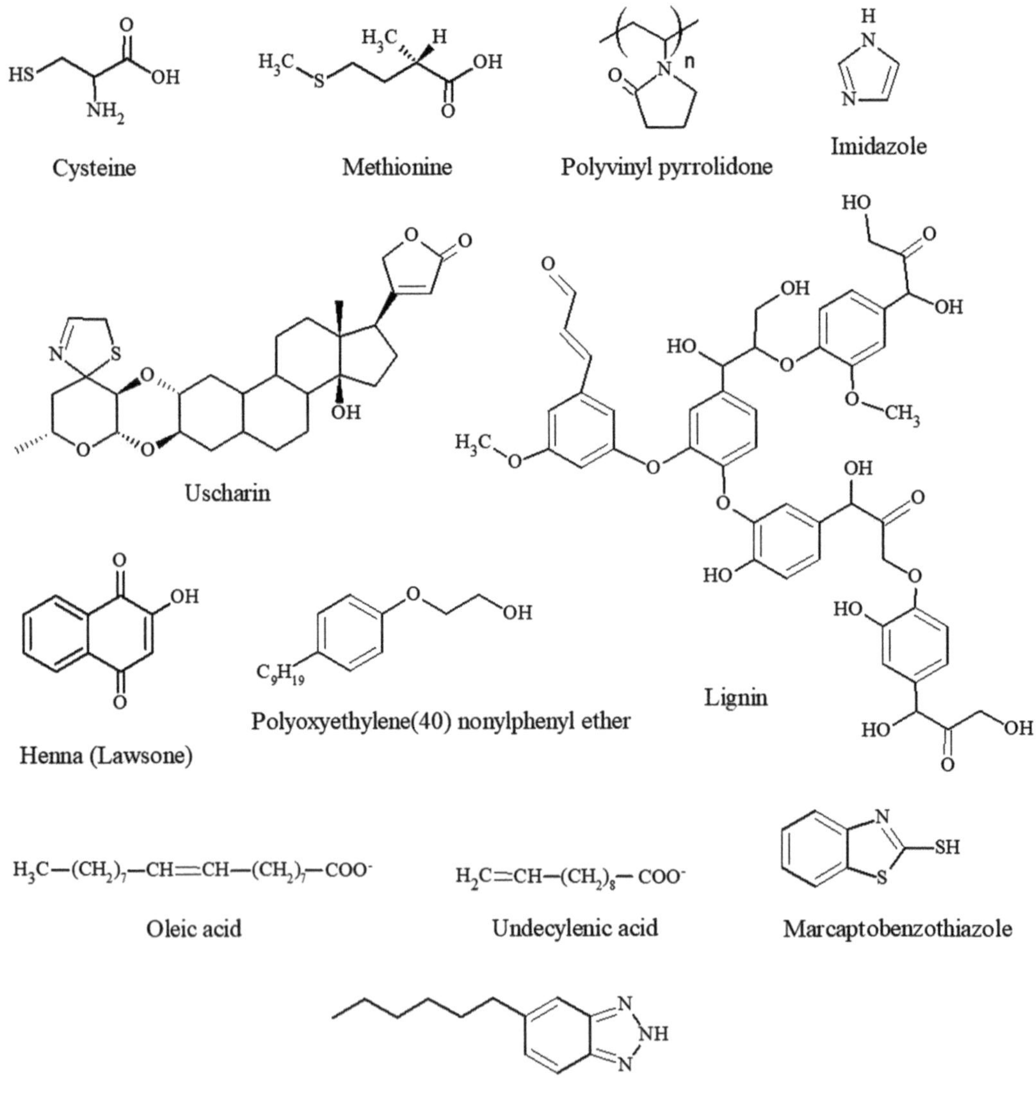

Figure 5. The structures of some inhibitors for metals (except aluminum) used in alkaline media.

Conclusion

Different sustainable substances reported in the literature that are effective corrosion inhibitors of various metals have been reviewed. These corrosion inhibitors are mainly based on plant extracts, expired drugs, amino acids, and fatty acids. Adsorption of the active inhibitive species on the metallic substrate and blocking the active anodic or cathodic areas has been introduced as the corrosion inhibition mechanism. It has been shown that adsorption generally followed the Langmuir isotherm, while in some cases, Freundlich and Frumkin were introduced as the best fit. Interaction of the inhibitive species with the metal surface was reported to be physical and chemical, respectively, where the Gibbs energy adsorption was more positive and more negative than the threshold energy (-20 kJ/mole). Despite the fact that heteroatoms (e.g., O, N, and S) are essential for inhibitory properties, it has been shown that being an appropriate inhibitor for a metallic substrate in an alkaline condition does not necessarily mean that it can be an effective inhibitor for other metals. Therefore, a specific substance (or combination of substances) is required to serve as a corrosion inhibitor for certain metals in alkaline media.

References

1. Abdallah, M.; Kamar, E. M.; El-Etre, A. Y.; Eid, S. Gelatin as Corrosion Inhibitor for Aluminum and Aluminum Silicon Alloys in Sodium Hydroxide Solutions. *Prot. Met. Phys. Chem. Surfaces* **2016**, *52* (1), 140–148. https://doi.org/10.1134/S2070205116010020.
2. Su, P.; Li, L.; Li, W.; Huang, C.; Wang, X.; Liu, Y.; Singh, A. Expired Drug Theophylline as Potential Corrosion Inhibitor for 7075 Aluminum Alloy in 1 M NaOH Solution. *Int. J. Electrochem. Sci.* **2020**, *15*, 1412–1425. https://doi.org/10.20964/2020.02.25.
3. Elango, A.; Periasamy, V. M.; Paramasivam, M. Study on Polyaniline-ZnO Used as Corrosion Inhibitors of 57S Aluminium in 2 M NaOH Solution. *Anti-Corrosion Methods Mater.* **2009**, *56* (5), 266–270. https://doi.org/10.1108/00035590910989561.
4. Wang, D.; Gao, L.; Zhang, D.; Yang, D.; Wang, H.; Lin, T. Experimental and Theoretical Investigation on Corrosion Inhibition of AA5052 Aluminium Alloy by L-Cysteine in Alkaline Solution. *Mater. Chem. Phys.* **2016**, *169*, 142–151. https://doi.org/10.1016/j.matchemphys.2015.11.041.
5. El-Deeb, M. M.; Ads, E. N.; Humaidi, J. R. Evaluation of the Modified Extracted Lignin from Wheat Straw as Corrosion Inhibitors for Aluminum in Alkaline Solution. *Int J Electrochem Sci* **2018**, *13*, 4123–4138. https://doi.org/10.20964/2018.05.49.
6. Chaubey, N.; Singh, V. K.; Quraishi, M. A. Electrochemical Approach of Kalmegh Leaf Extract on the Corrosion Behavior of Aluminium Alloy in Alkaline Solution. *Int. J. Ind. Chem.* **2017**, *8* (1), 75–82. https://doi.org/10.1007/s40090-016-0103-y.
7. Ma, Y.; Zhou, X.; Thompson, G. E.; Skeldon, P. Surface Texture Formed on AA2099 Al-Li-Cu Alloy during Alkaline Etching. *Corros. Sci.* **2013**, *66*, 292–299. https://doi.org/10.1016/j.corsci.2012.09.032.
8. Sharifi Golru, S.; Attar, M. M.; Ramezanzadeh, B. Effects of Different Surface Cleaning Procedures on the Superficial Morphology and the Adhesive Strength of Epoxy Coating on Aluminium Alloy 1050. *Prog. Org. Coatings* **2015**, *87*, 52–60. https://doi.org/10.1016/j.porgcoat.2015.05.005.

9. Abiola, O. K.; Otaigbe, J. O. E.; Kio, O. J. Gossipium Hirsutum L.Extracts as Green Corrosion Inhibitor for Aluminum in NaOH Solution. *Corros. Sci.* **2009**, *51* (8), 1879–1881. https://doi.org/10.1016/j.corsci.2009.04.016.
10. Edrah, S.; Hasan, S. K. Studies on Thiourea Derivatives as Corrosion Inhibitor for Aluminum in Sodium Hydroxide Solution. *J Appl Sci Res* **2010**, *6* (8), 1045–1049.
11. Xhanari, K.; Finšgar, M. Organic Corrosion Inhibitors for Aluminum and Its Alloys in Chloride and Alkaline Solutions: A Review. *Arab. J. Chem.* **2019**, *12* (8), 4646–4663. https://doi.org/10.1016/j.arabjc.2016.08.009.
12. Moon, S.-M.; Pyun, S.-I. The Formation and Dissolution of Anodic Oxide Films on Pure Aluminium in Alkaline Solution. *Electrochim. Acta* **1999**, *44* (14), 2445–2454. https://doi.org/10.1016/S0013-4686(98)00368-5.
13. Pyun, S.-I.; Moon, S.-M. Corrosion Mechanism of Pure Aluminium in Aqueous Alkaline Solution. *J. Solid State Electrochem.* **2000**, *4* (5), 267–272. https://doi.org/10.1007/s100080050203.
14. Abiola, O. K.; Otaigbe, J. O. E. The Effects of Phyllanthus Amarus Extract on Corrosion and Kinetics of Corrosion Process of Aluminum in Alkaline Solution. *Corros. Sci.* **2009**, *51* (11), 2790–2793. https://doi.org/10.1016/j.corsci.2009.07.006.
15. Sirajunnisa, A.; Fazal Mohamed, M. I.; Subramania, A. Vitex Negundo Leaves Extract as Green Inhibitor for the Corrosion of Aluminium 1N NaOH Solution. *J. Chem. Pharm. Res.* **2014**, *6* (1), 580–588.
16. Al-Sehaibani, H. Evaluation of Extracts of Henna Leaves as Environmentally Friendly Corrosion Inhibitors for Metals. *Mater. und Werkstofftechnik Mater. Sci. Eng. Technol.* **2000**, *31* (12), 1060–1063.
17. Singh, A.; Ahamad, I.; Quraishi, M. A. Piper Longum Extract as Green Corrosion Inhibitor for Aluminium in NaOH Solution. *Arab. J. Chem.* **2016**, *9*, S1584–S1589. https://doi.org/10.1016/j.arabjc.2012.04.029.
18. Abdel-Gaber, A. M.; Khamis, E.; Abo-Eldahab, H.; Adeel, S. Novel Package for Inhibition of Aluminium Corrosion in Alkaline Solutions. *Mater. Chem. Phys.* **2010**, *124* (1), 773–779. https://doi.org/10.1016/j.matchemphys.2010.07.059.
19. Fu, J.; Chen, T.; Wang, M.; Yang, N.; Li, S.; Wang, Y.; Liu, X. Acid and Alkaline Dual Stimuli-Responsive Mechanized Hollow Mesoporous Silica Nanoparticles as Smart Nanocontainers for Intelligent Anticorrosion Coatings. *ACS Nano* **2013**, *7* (12), 11397–11408. https://doi.org/10.1021/nn4053233.
20. Abdel-Gaber, A. M.; Khamis, E.; Abo-ElDahab, H.; Adeel, S. Inhibition of Aluminium Corrosion in Alkaline Solutions Using Natural Compound. *Mater. Chem. Phys.* **2008**, *109* (2), 297–305. https://doi.org/https://doi.org/10.1016/j.matchemphys.2007.11.038.
21. Bataineh, T.; Al-Qudah, M.; Nawafleh, E.; Al-Rawashdeh, N. Sinapis Alba Extract as Green Corrosion Inhibitor for Aluminum in Alkaline Media. *Int. J. Electrochem. Sci.* **2014**, *9*, 3543–3557.
22. Singh, A.; Ebenso, E. E.; Quraishi, M. A. Stem Extract of Brahmi (Bacopa Monnieri) as Green Corrosion Inhibitor for Aluminum in NaOH Solution. *Int. J. Electrochem. Sci* **2012**, *7*, 3409–3419.

23. Irshedat, M. K.; Nawafleh, E. M.; Bataineh, T. T.; Muhaidat, R.; Al-Qudah, M. A.; Alomary, A. A. Investigations of the Inhibition of Aluminum Corrosion in 1 M NaOH Solution by Lupinus Varius l. Extract. *Port. Electrochim. Acta* **2013**, *31* (1), 1–10. https://doi.org/10.4152/pea.201301001.
24. Chaubey, N.; Quraishi, M. A.; Ebenso, E. E. Corrosion Inhibition of Aluminium Alloy in Alkaline Media by Neolamarkia Cadamba Bark Extract as a Green Inhibitor. *Int. J. Electrochem. Sci* **2015**, *10*, 504–518.
25. Umoren, S. A.; Obot, I. B.; Akpabio, L. E.; Etuk, S. E. Adsorption and Corrosive Inhibitive Properties of Vigna Unguiculata in Alkaline and Acidic Media. *Pigment Resin Technol.* **2008**, *37*, 98–105. https://doi.org/10.1108/03699420810860455.
26. Umoren, S. A.; Obot, I. B.; Ebenso, E. E.; Okafor, P. C.; Ogbobe, O.; Oguzie, E. E. Gum Arabic as a Potential Corrosion Inhibitor for Aluminium in Alkaline Medium and Its Adsorption Characteristics. *Anti-Corrosion Methods Mater.* **2006**, 53 (5), 277–282. https://doi.org/10.1108/00035590610692554.
27. Emran, K. M.; Ahmed, N. M.; Torjoman, B. A.; Al-Ahmadi, A. A.; Sheekh, S. N. Cantaloupe Extracts as Eco Friendly Corrosion Inhibitors for Aluminum in Acidic and Alkaline Solutions. *J. Mater. Environ. Sci* **2014**, *5* (6), 1940–1950.
28. Shakeel, A.; Altaf, A. A.; Qureshi, A. M.; Badshah, A. Thiourea Derivatives in Drug Design and Medicinal Chemistry: A Short Review. *J. Drug Des. Med. Chem* **2016**, *2* (1), 10. https://doi.org/10.11648/j.jddmc.20160201.12.
29. Altaf, A. A.; Shahzad, A.; Gul, Z.; Rasool, N.; Badshah, A.; Lal, B.; Khan, E. A Review on the Medicinal Importance of Pyridine Derivatives. *J. Drug Des. Med. Chem.* **2015**, *1* (1), 1–11. https://doi.org/10.11648/j.jddmc.20160201.12.
30. Padash, R.; Sajadi, G. S.; Jafari, A. H.; Jamalizadeh, E.; Rad, A. S. Corrosion Control of Aluminum in the Solutions of NaCl, HCl and NaOH Using 2,6-Dimethylpyridine Inhibitor: Experimental and DFT Insights. *Mater. Chem. Phys.* **2020**, *244*, 122681. https://doi.org/10.1016/j.matchemphys.2020.122681.
31. Patil, D. B.; Sharma, A. R. Inhibition of Corrosion of Aluminium in Potassium Hydroxide Solution by Pyridine Derivatives. *Int. Sch. Res. Not.* **2014**, *2014*. https://doi.org/10.1155/2014/154285.
32. Asan, G.; Asan, A. Inhibitor Effect of Nicotinamide on Corrosion of Aluminum. *J. Mol. Struct.* **2020**, *1201*, 127184. https://doi.org/10.1016/j.molstruc.2019.127184.
33. Noor, E. A.; Al-Moubaraki, A. H.; Al-Ghamdi, A. A. Continuous Studies on Using Camel's Urine as Nontoxic Corrosion Inhibitor–Corrosion Inhibition of Al–Cu Alloy in Alkaline Solutions. *Arab. J. Sci. Eng.* **2019**, *44* (1), 237–250. https://doi.org/10.1007/s13369-018-3489-3.
34. Umoren, S. A.; Inam, E. I.; Udoidiong, A. A.; Obot, I. B.; Eduok, U. M.; Kim, K.-W. Humic Acid from Livestock Dung: Ecofriendly Corrosion Inhibitor for 3SR Aluminum Alloy in Alkaline Medium. *Chem. Eng. Commun.* **2015**, *202* (2), 206–216. https://doi.org/10.1080/00986445.2013.836635.
35. Yin, J.; Ren, W.; Yang, G.; Duan, J.; Huang, X.; Fang, R.; Li, C.; Li, T.; Yin, Y.; Hou, Y. L-Cysteine Metabolism and Its Nutritional Implications. *Mol. Nutr. Food Res.* **2016**, *60* (1), 134–146. https://doi.org/10.1002/mnfr.201500031.

36. Wang, K.; Li, R.; Ma, J. H.; Jian, Y. K.; Che, J. N. Extracting Keratin from Wool by Using L-Cysteine. *Green Chem.* **2016,** *18* (2), 476–481. https://doi.org/10.1039/c5gc01254f.
37. Wu, S.; Zhang, Q.; Sun, D.; Luan, J.; Shi, H.; Hu, S.; Tang, Y.; Wang, H. Understanding the Synergistic Effect of Alkyl Polyglucoside and Potassium Stannate as Advanced Hybrid Corrosion Inhibitor for Alkaline Aluminum-Air Battery. *Chem. Eng. J.* **2020**, *383*, 123162. https://doi.org/10.1016/j.cej.2019.123162.
38. Verma, C. B.; Singh, P.; Bahadur, I.; Ebenso, E. E.; Quraishi, M. A. Electrochemical, Thermodynamic, Surface and Theoretical Investigation of 2-Aminobenzene-1,3-Dicarbonitriles as Green Corrosion Inhibitor for Aluminum in 0.5 M NaOH. *J. Mol. Liq.* **2015**, *209*, 767–778. https://doi.org/10.1016/j.molliq.2015.06.039.
39. Zheng, T.; Liu, J.; Wang, L.; Jia, G.; Wang, J. Eco-Friendly Rosin-Based 6-Dehydroabietic Acylamino Sodium as Corrosion Inhibitor for AA2024-T3 in Alkaline Solution by Experimental and Theoretical Studies. *J. Mol. Liq.* **2021**, *328*, 115506. https://doi.org/10.1016/j.molliq.2021.115506.
40. Al Juhaiman, L. A.; Abu Mustafa, A.; Mekhamer, W. K. Polyvinyl Pyrrolidone as a Green Corrosion Inhibitor for Carbon Steel in Alkaline Solutions Containing NaCl. *Anti-Corrosion Methods Mater.* **2013,** *60* (1), 28–36. https://doi.org/10.1108/00035591311287429.
41. Broomfield, J. P. *Corrosion of Steel in Concrete: Understanding, Investigation and Repair 2nd Edition*; 2557; Vol. 7; CRC Press: Baton Rouge, FL.
42. Freire, L.; Catarino, M. A.; Godinho, M. I.; Ferreira, M. J.; Ferreira, M. G. S.; Simões, A. M. P.; Montemor, M. F. Electrochemical and Analytical Investigation of Passive Films Formed on Stainless Steels in Alkaline Media. *Cem. Concr. Compos.* **2012,** *34* (9), 1075–1081. https://doi.org/10.1016/j.cemconcomp.2012.06.002.
43. Sánchez-Moreno, M.; Takenouti, H.; García-Jareño, J. J.; Vicente, F.; Alonso, C. A Theoretical Approach of Impedance Spectroscopy during the Passivation of Steel in Alkaline Media. *Electrochim. Acta* **2009**, *54* (28), 7222–7226. https://doi.org/10.1016/j.electacta.2009.07.013.
44. Ben Harb, M.; Abubshait, S.; Etteyeb, N.; Kamoun, M.; Dhouib, A. Olive Leaf Extract as a Green Corrosion Inhibitor of Reinforced Concrete Contaminated with Seawater. *Arab. J. Chem.* **2020,** *13* (3), 4846–4856. https://doi.org/10.1016/j.arabjc.2020.01.016.
45. Liu, Q.; Song, Z.; Han, H.; Donkor, S.; Jiang, L.; Wang, W.; Chu, H. A Novel Green Reinforcement Corrosion Inhibitor Extracted from Waste *Platanus acerifolia* Leaves. *Constr. Build. Mater.* **2020,** *260*, 119695. https://doi.org/10.1016/j.conbuildmat.2020.119695.
46. Torres-Acosta, A. A. Opuntia-Ficus-Indica (Nopal) Mucilage as a Steel Corrosion Inhibitor in Alkaline Media. *J. Appl. Electrochem.* **2007**, *37* (7), 835–841. https://doi.org/10.1007/s10800-007-9319-z.
47. El-Etre, A. Y.; Abdallah, M.; El-Tantawy, Z. E. Corrosion Inhibition of Some Metals Using *Lawsonia* Extract. *Corros. Sci.* **2005,** *47* (2), 385–395. https://doi.org/10.1016/j.corsci.2004.06.006.
48. Kumar, S.; Arora, S.; Sharma, M.; Arora, P.; Prakash Mathur, S. Synergistic Effect of *Calotropis* Plant in Controlling Corrosion Of Mild Steel In Basic Solution. *J. Chil. Chem. Soc.* **2009,** *54* (1), 83–88. https://doi.org/10.4067/S0717-97072009000100020.

49. Raghavendra, N.; Bhat, J. I. Chemical Components of Mature Areca Nut Husk Extract as a Potential Corrosion Inhibitor for Mild Steel and Copper in Both Acid and Alkali Media. *Chem. Eng. Commun.* **2018**, *205* (2), 145–160. https://doi.org/10.1080/00986445.2017.1370709.
50. Lin, B.; Zuo, Y. Corrosion Inhibition of Carboxylate Inhibitors with Different Alkylene Chain Lengths on Carbon Steel in an Alkaline Solution. *RSC Adv.* **2019**, *9* (13), 7065–7077. https://doi.org/10.1039/C8RA10083G.
51. Monticelli, C.; Frignani, A.; Trabanelli, G. A Study on Corrosion Inhibitors for Concrete Application. *Cem. Concr. Res.* **2000**, *30* (4), 635–642. https://doi.org/10.1016/S0008-8846(00)00221-0.
52. Robinson, B. V.; Sullivan, F. M.; Borzelleca, J. F.; Schwartz, S. L. *PVP: A Critical Review of the Kinetics and Toxicology of Polyvinylpyrrolidone (Povidone)*; Routledge: Boca Raton, FL 2018. https://doi.org/10.1201/9780203741672.
53. Burke, L. D.; Ahern, M. J. G.; Ryan, T. G. An Investigation of the Anodic Behavior of Copper and Its Anodically Produced Oxides in Aqueous Solutions of High PH. *J. Electrochem. Soc.* **1990**, *137* (2), 553–561. https://doi.org/10.1149/1.2086496.
54. Ganzha, S. V; Maksimova, S. N.; Grushevskaya, S. N.; Vvedenskii, A. V. Formation of Oxides on Copper in Alkaline Solution and Their Photoelectrochemical Properties. *Prot. Met. Phys. Chem. Surfaces* **2011**, *47* (2), 191. https://doi.org/10.1134/S2070205111020080.
55. Allen, D.; Wilson, D.; Drew, R.; Perfect, J. Azole Antifungals: 35 Years of Invasive Fungal Infection Management. *Expert Rev. Anti. Infect. Ther.* **2015**, *13* (6), 787–798. https://doi.org/10.1586/14787210.2015.1032939.
56. Mihit, M.; Laarej, K.; Abou El Makarim, H.; Bazzi, L.; Salghi, R.; Hammouti, B. Study of the Inhibition of the Corrosion of Copper and Zinc in HNO3 Solution by Electrochemical Technique and Quantum Chemical Calculations. *Arab. J. Chem.* **2010**, *3* (1), 55–60. https://doi.org/10.1016/j.arabjc.2009.12.009.
57. Altaf, F.; Qureshi, R.; Yaqub, A.; Ahmed, S. Electrochemistry of Corrosion Mitigation of Brasses by Azoles in Basic Medium. *Chem. Pap.* **2019**, 73 (5), 1221–1235. https://doi.org/10.1007/s11696-018-00673-x.
58. Zarrouk, A.; Hammouti, B.; Dafali, A.; Bentiss, F. Inhibitive Properties and Adsorption of Purpald as a Corrosion Inhibitor for Copper in Nitric Acid Medium. *Ind. Eng. Chem. Res.* **2013**, *52* (7), 2560–2568. https://doi.org/10.1021/ie301465k.
59. Brosnan, J. T.; Brosnan, M. E.; Bertolo, R. F. P.; Brunton, J. A. Methionine: A Metabolically Unique Amino Acid. *Livest. Sci.* **2007**, *112* (1–2), 2–7. https://doi.org/10.1016/j.livsci.2007.07.005.
60. Wu, J.; Zheng, X.; Li, W.; Yin, L.; Zhang, S. Copper Corrosion Inhibition by Combined Effect of Inhibitor and Passive Film in Alkaline Solution. *Res. Chem. Intermed.* **2015**, *41* (11), 8557–8570. https://doi.org/10.1007/s11164-014-1910-4.
61. Radovanović, M. B.; Petrović, M. B.; Simonović, A. T.; Milić, S. M.; Antonijević, M. M. Cysteine as a Green Corrosion Inhibitor for Cu37Zn Brass in Neutral and Weakly Alkaline Sulphate Solutions. *Environ. Sci. Pollut. Res.* **2013**, *20* (7), 4370–4381. https://doi.org/10.1007/s11356-012-1088-5.
62. Parshutin, V. V.; Bogdashkina, N. L.; Chernova, G. P. The Effect of Medium on the Corrosion and Electrochemical Behavior of Nickel. *Prot. Met.* **2007**, *43* (1), 59–65. https://doi.org/10.1134/S0033173207010080.

63. Ahn, K.-H.; Song, K.-G.; Cha, H.-Y.; Yeom, I.-T. Removal of Ions in Nickel Electroplating Rinse Water Using Low-Pressure Nanofiltration. *Desalination* **1999**, *122* (1), 77–84. https://doi.org/10.1016/S0011-9164(99)00029-6.
64. McGeough, J. A.; Leu, M. C.; Rajurkar, K. P.; De Silva, A. K. M.; Liu, Q. Electroforming Process and Application to Micro/Macro Manufacturing. *CIRP Ann.* **2001**, *50* (2), 499–514. https://doi.org/10.1016/S0007-8506(07)62990-4.
65. Rossi, S.; Deflorian, F.; Venturini, F. Improvement of Surface Finishing and Corrosion Resistance of Prototypes Produced by Direct Metal Laser Sintering. *J. Mater. Process. Technol.* **2004**, *148* (3), 301–309. https://doi.org/10.1016/j.jmatprotec.2003.02.001.
66. Kuleshov, V. N.; Kuleshov, N. V.; Grigoriev, S. A.; Udris, E. Y.; Millet, P.; Grigoriev, A. S. Development and Characterization of New Nickel Coatings for Application in Alkaline Water Electrolysis. *Int. J. Hydrogen Energy* **2016**, *41* (1), 36–45. https://doi.org/10.1016/j.ijhydene.2015.10.141.
67. Bidault, F.; Brett, D. J. L.; Middleton, P. H.; Abson, N.; Brandon, N. P. A New Application for Nickel Foam in Alkaline Fuel Cells. *Int. J. Hydrogen Energy* **2009**, *34* (16), 6799–6808. https://doi.org/10.1016/j.ijhydene.2009.06.035.
68. Fauvarque, J. F.; Guinot, S.; Bouzir, N.; Salmon, E.; Penneau, J. F. Alkaline Poly(Ethylene Oxide) Solid Polymer Electrolytes. Application to Nickel Secondary Batteries. *Electrochim. Acta* **1995**, *40* (13–14), 2449–2453. https://doi.org/10.1016/0013-4686(95)00212-W.
69. Cattaneo, E.; Riegel, B. Chemistry, Electrochemistry, and Electrochemical Applications | Nickel. In *Encyclopedia of Electrochemical Power Sources*; Elsevier, 2009; pp 796-809. https://doi.org/10.1016/B978-044452745-5.00062-9.
70. Abdallah, M.; Zaafarany, I. A.; Abd El Wanees, S.; Assi, R. Corrosion Behavior of Nickel Electrode in NaOH Solution and Its Inhibition by Some Natural Oils. *Int. J. Electrochem. Sci.* **2014**, *9* (3), 1071–1086.
71. Dell, R.; Rand, D. A. J. *Understanding Batteries*; Royal Society of Chemistry: London, UK; 2001.
72. Zhou, H.; Huang, Q.; Liang, M.; Lv, D.; Xu, M.; Li, H.; Li, W. Investigation on Synergism of Composite Additives for Zinc Corrosion Inhibition in Alkaline Solution. *Mater. Chem. Phys.* **2011**, *128* (1–2), 214–219. https://doi.org/10.1016/j.matchemphys.2011.02.061.
73. Thomas, S.; Birbilis, N.; Venkatraman, M. S.; Cole, I. S. Corrosion of Zinc as a Function of PH. *CORROSION* **2012**, *68* (1), 015009-1-015009–9. https://doi.org/10.5006/1.3676630.
74. Deyab, M. A. Application of Nonionic Surfactant as a Corrosion Inhibitor for Zinc in Alkaline Battery Solution. *J. Power Sources* **2015**, *292*, 66–71. https://doi.org/10.1016/j.jpowsour.2015.05.040.

Chapter 9

Corrosion Inhibitors for Sweet (CO_2 Corrosion) and Sour (H_2S Corrosion) Environments

Anita Kumari,[1,2] Bhawna Chugh,[3] Sheetal,[3] Gurmeet Singh,[2,4] and Ashish Kumar Singh[5,*]

[1]Department of Chemistry, Miranda House, University of Delhi, New Delhi-110007, India

[2]Department of Chemistry, University of Delhi, New Delhi-110007, India

[3]Department of Chemistry, Netaji Subhas University of Technology, Sector 3, Dwarka, New Delhi-110078, India

[4]Pondicherry University, Puducherry-605014, India

[5]Department of Applied Sciences, Bharati Vidyapeeth's College of Engineering, Paschim Vihar, New Delhi-110063, India

[*]Email: ashish.singh.rs.apc@itbhu.ac.in.

Corrosion is a severe problem, and it is stated as the gradual destruction of materials by chemical or electrochemical processes. Mild steel (MS) is extensively employed for construction purposes in several industries due to its mechanical strength and other advantages. Still, corrosion degrades the useful properties MS. The aggressive environment, due to CO_2 and H_2S in manufactured fluids, cause the corrosion of the metallic surface. Corrosion inhibitors play a crucial role in protecting metals from corrosion. This chapter elaborates on the analysis of corrosion control by using inhibitors on metal surfaces in CO_2 and H_2S corrosive environments. Several factors that influence sweet (CO_2) and sour (H_2S) corrosion are illustrated, such as temperature, pH, humidity, and more. Iron carbonates and iron sulfides act as a barrier on the metal surface that prevents corrosion. The chapter broadly focuses on the various categories of inhibitors used for mitigation of such corrosion, for example, imidazoline-based inhibitors, amine-based inhibitors, polymer-based inhibitors, and green inhibitors.

Introduction

Today, corrosion is considered a well-known phenomenon touching all the aspects of life. It is a spontaneous process where disintegration or decay of material or its properties by chemical, electrochemical, or biochemical reactions occurs naturally. Most common electrolytic media include water, soil, air, etc. Corrosion is considered an influential force that is very harmful and destroys human as well as natural resources and resulting economic harm. The direct economic expense of corrosion has been determined to be in between 2–4% of a country's gross national product (*1*, *2*). Though the indirect cost of corrosion cannot be estimated, it can be visualized in the costs associated with plant shutdown, contamination, overdesign, and lower equipment efficiency.

Corrosion is an unwanted process that annihilates the luster of materials and reduces their lifetime. It affects the quality of daily lives and their technical progress (3). Corrosion occurs due to association between the metal and the environment, which leads to its spontaneous degradation. U. R. Evans—"the father of corrosion science"—stated that "*Corrosion is largely an electrochemical phenomenon, which may be defined as the destruction of metals by electrochemical or chemical agencies*" (*4*). In order to return to lower energy states, metals (generally iron or alloy steel) commonly react with oxygen and water present in the atmospheric or natural environments to form rust (i.e., hydrated iron oxide), which is equivalent to the original iron ore in terms of chemical composition.

Corrosion leads to equipment failure, service interruption, and other economic disruptions. According to National Association of Corrosion Engineers (NACE)International, annual corrosion costs for each industrial country are about 24% gross national product. Corrosion occurs more profoundly in the presence of impurities/pollutants in oil and gas industrial areas due to gases (e.g., CO_2, H_2S, SO_2, etc.). Sulfur dioxide, the gaseous product obtained from combustion of fuels comprising sulfur (i.e., coal, diesel fuel, gasoline, and natural gas) has proven among the most prominent air pollutants contributing to the metallic corrosion (*5*, *6*).

Metal is corroded because of metal's chemical reaction with surrounding gases, which cause the corrosion process. Methods to reduce sweet (CO_2corrosion) and sour (H_2S corrosion), sulfur, chloride, or oxygen content in the surrounding conditions could decrease the extent of metallic corrosion. The rate of corrosion can be controlled through adjustment of the environment by the following factors (7):

- Change of environmental variables—temperature and pH.
- Removal of corrosive components—CO_2 , H_2S, sulfur, chloride, or oxygen and moisture.
- Use of inhibitors—adding chemical substances that slow down a chemical reaction.

Inhibitors, (e.g., sodium carbonates or regular amines structures) have been used to decrease the damaging outcome of CO_2, H_2S, and common acids, allowing the use of dynamically more affordable substances and less impenetrable to corrosion in wells removing the crude oil. Sulphonated oils and sodium nitrite are regularly used for channels of gas and kerosene oil. Oil wells utilize oily amines, unsaturated fats, and quaternary ammonium salts. Internal pipe disintegration occurs in wet gas transportation because of the condensed form of water containing dissolved gases, such as carbon dioxide and hydrogen sulfide, similar to the development of destructive vapors (*8*, 9). Organic substances containing heteroatoms (e.g., O, N, S) have been widely used as corrosion inhibitors (*10*–*13*).

Corrosion Mechanism on Metal and Inhibition

A corrosion procedure can be treated as a short-circuited electrochemical cell in which the anodic oxidation process can be written as:

$$M(s) \rightarrow M^{2+}(aq) + 2e^{-} \quad (1)$$

The cathodic reactions can be any of the following, depending on depolarization (*14*).

$$\text{Oxygen: } O_2 + 2\ H_2O + 4\ e^{-} \rightarrow 4\ OH^{-} \quad (2)$$

$$\text{Acid: } H^{+} + e^{-} \rightarrow \tfrac{1}{2}\ H_2\ (g) \quad (3)$$

$$\text{Cation of more noble metal: } M^{2+} + 2e^{-} \rightarrow M(s) \quad (4)$$

Where, M is a more noble metal or cathodic center of the same metal.

In this reaction, M is the cathodic center of the metal. Electrochemical corrosion of iron occurs at the anode region where the metal is experiencing tension. The metal particles separate in the clamminess film, and the electrons move to the cathodic site, where a depolarizer takes them up. Oxygen is a common depolarizer present in the atmosphere that results in rust ($Fe_2O_3.xH_2O$) formation as a combination of hydrous iron-oxides (*15*). Because of the electrochemical nature of the corrosion processes, the electrochemical methods are more beneficial for studying corrosion, determining the kinetics of corrosion processes in the desired environment, and to determine or control oxidizing power or potential of the environment.

The metals and their alloys can be prevented from corrosion by employing corrosion-resistant materials and regulating the surrounding environment. Corrosion inhibitors are defined as chemical substances or materials whose addition to the environment will decrease the extent of corrosion.

The Economic Impact of Corrosion

According to the World Corrosion Organization (WCO), Corrosion awareness day, 2019 highlights the estimated annual cost of corrosion globally is 2.5 trillion-dollar which is 3 to 4% of the gross national product of industrialized countries. In India, the loss of metal due to corrosion is 2 to 2.5 billion dollars yearly (*16*).

Corrosion represents a tremendous economic loss, and much can be done to reduce it. The prominent economic figures are not unforeseen as the extent of corrosion is so severe at the areas where such metals have been used (*17*, *18*). To decrease annual losses due to corrosion, civil/industrial and government sectors should apply effective analytical strategies, develop reliable science and technology, and enforce suitable corrosion mitigation technologies (*19*). Some of these reports divide the cost of corrosion as "avoidable costs" and "unavoidable costs". Costs that can be minimized by applying available practices of corrosion control comprise "avoidable costs", whereas costs required for new corrosion technologies and new materials comprise "unavoidable costs" (*20*, *21*).

Presence of CO_2 and H_2S in Oil and Gas Industry

Corrosion in oilfields occurs at all stages, from downhole to surface equipment and processing facilities, appearing as leaks in tanks, tubing, pipelines, and other machineries (*22*, *23*). Garcia-Arriaga et al. reported the economic loss as 10–30% of maintenance budget as a result of natural gas sweetening and oil refining plants. Thus, corrosion problems associated with operation and equipment maintenance of oil and gas industries costs are severe. Hence, the method for the control of corrosion is worth inspecting in oilfield areas. Internal corrosion is another significant problem in operating pipe flow lines which occurs because of stress corrosion cracking (*24*).

Appropriate corrosion control by suitable inhibitors is needed to evade any possible damage that can result in severe problems (e.g., loss of life, negative social impacts, water resources, and environmental pollution). The oil and gas field inhibition are more complicated due to aggressive gases (i.e., H_2S, CO_2, organic acids, etc.). Thus, particular inhibitors for refinery application (i.e., wells, recovery units, pipelines, etc.). are needed to mitigate corrosion. Naphthenic acid constituents and sulfur compounds in crude oils are a reason for corrosion problems in petroleum refining operations (*25*).

Corrosion Mechanism in CO_2 Environment

The increase of CO_2 from oil and gas industries results in corrosion, which is termed "sweet corrosion". Carbonic acid (H_2CO_3) is produced by the chemical reaction of CO_2 with groundwater due to the presence of CO_2 in-ground as represented in equations (5) and (6) (*26, 27*):

$$CO_2\,(g) \rightarrow CO_2\,(aq) \quad (5)$$

$$CO_2 + H_2O \rightarrow H_2CO_3 \quad (6)$$

Further, carbonic acid dissociates into two components; bicarbonate and carbonate, in equations (7) and (8):

$$H_2CO_3 \rightarrow H^+ + HCO_3^- \quad (7)$$

$$HCO^{3-} \rightarrow H^+ + CO_3^{2-} \quad (8)$$

The overall electrochemical reaction of CO_2 corrosion is given in equation (9):

$$Fe + CO_2 + H_2O \rightarrow FeCO_3 + H_2 \quad (9)$$

Anodic reaction involves the oxidation of Fe to Fe^{2+} ion as presented in equation (10):

$$Fe \rightarrow Fe^{2+} + 2e^- \quad (10)$$

Cathodic reaction contains the reduction of hydrogen ions that are present in the ground water, as given in equations (11) and (12):

$$2H^+ + 2e^- \rightarrow H_2 \quad (11)$$

$$2H_2CO_3 + 2e^- \rightarrow H_2 + 2HCO_3^- \quad (12)$$

The thin layer is formed by organic inhibitors or a proactive corrosion layer, (i.e., iron carbonate) on the metallic surface that decreases the sweet CO_2 corrosion.

Corrosion Mechanism in H_2S Environment

Hydrogen sulfide is a very harmful, toxic, and explosive gas. It causes corrosion when metals contact with moisture, known as "sour corrosion" and H_2S corrosion. It is highly more soluble in water than oxygen and carbon dioxide. Iron sulfides and hydrogen are corrosion products. The occurrence of sour corrosion in oil and gas industries is exceptionally high and damage its pipes. At low temperatures, iron sulfides behave as a barrier on the metal surface to decrease the degree of corrosion. The sour corrosion can be expressed as the following relations (*28*, *29*):

$$H_2S \rightarrow H^+ + HS^- \quad (13)$$

$$HS^- \rightarrow H^+ + S^{2-} \quad (14)$$

$$H_2S + Fe + H_2O \rightarrow FeHS_{adsorption} + H_3O^+ \quad (15)$$

$$FeHS_{adsorption} \rightarrow FeHS^+_{adsorption} + e^- \quad (16)$$

$$FeHS^+_{adsorption} + H_2O \rightarrow Fe^{2+} + H_2S + H_2O \quad (17)$$

$$Fe^{2+} + HS^- \rightarrow FeS + H^+ \quad (18)$$

Factors Affecting CO_2 /H_2S Corrosion

The factors affecting corrosion are considered for the corrosion of unprotected metals, but when the same metal has a surface coating, these factors may turn out to be better. This depends on the environmental conditions in which that particular metal is placed, for example, temperature, pH, relative humidity, and presence of impurities in the atmosphere) (*30*, *31*).

Temperature

Temperature portrays a vital role in CO_2/H_2S corrosion, such as the transport of substances and chemical reactions in the bulk of the solutions at the metallic interface. Initially there is the regular rise in corrosion activity which could theoretically double for each 10 degree rise in temperature. Secondly, due to the heat capacity of metallic objects, a small lag in temperature is seen. As the ambient temperature decreases at night, metallic surfaces tend to remain warmer than the humid air surrounding them. They do not begin to collect condensation until sometime after the dew point has been reached. With the rise in the temperature of the atmosphere, the diffusion rate increases and because of this increased diffusion rate, the corrosion rate is ordinarily improved.

Relative Humidity

Relative humidity is described as the ratio of the quantity of water vapor existing in the atmosphere to the saturation quantity at a particular temperature (shown as percent). A necessity for atmospheric corrosion phenomenon is the presence of a thin film electrolyte that develops on metal interface when exposed to a critical level of humidity. Moreover, the oxide film on a metal surface can absorb moisture. In the presence of this moisture, all requirements for electrochemical or CO_2/H_2S type corrosion exist, and hence corrosion rate is enhanced.

pH Value

Highly acidic and basic conditions causes corrosion of the metal considerably The corrosion rate of iron in the oxygen-free water is not increased till pH <5. The relative corrosion rate of oxygen is much greater. At lower pH < 2, iron can be dissolved, and precipitations not occurred by iron sulfide on the metal surface, due to higher solubility of iron sulfide, thus increased corrosion rate.

Presence of Atmospheric Impurities

In industrial locations, the atmosphere comprises corrosive gases (e.g., CO_2, H_2S, SO_2, fumes of HCl and H_2SO_4, etc.). In the corrosive medium, the acidity of the liquid in vicinity to the metal surface increases and therefore increases corrosion (i.e., pollutants increase the rate of corrosion).

Corrosion Inhibitors in Sweet CO_2 Environment

Imidazoline-Based Inhibitors

Imidazoline-based inhibitors possess surfactant properties, which is the main reason they are used to prevent pipelines and gas wells from an inevitable threat of carbon dioxide corrosion. Imidazoline-based inhibitors also have high adsorption tendencies and they easily adhere to the metal surfaces and alloys, thereby protecting them via barrier formation.

Several imidazoles and amides were tested for their anti-corrosive potential for mild steel (MS) by employing different techniques like linear polarization resistance and rotating cylinder electrode. Basically, the inhibitor is added to the brine solution containing the carbon dioxide environment using constant concentration and variable slug concentration. Most of the corrosion inhibition data were produced in the case of continual concentration treatment. Further, to deepen insight into how the chain lengths and head groups affected the corrosion inhibiting performance, a series of imidazolines and amides were synthesized using pure fatty acids having (C8, C12, C16, C18, and C20) in their hydrocarbon chains and diethylenetriamine. Essentially, C8, C12, C16, C18, and C20 stand for the hydrocarbon chains having the specified number of carbon atoms. A difference in the activity was observed and the lauric precursor of imidazole was found to be the least effective in this regard. In contrast, the stearic demonstrated maximum efficiency. The hydrophobic chain lengths with carbon atoms less than 12 were observed to have no significant corrosion inhibitive capacity (32).

In one study, imidazoles were prepared by utilizing stearic acid, diethylenetriamine, and the corrosion inhibitive performance of the respective compound were explored and tested via electrochemical studies. The results concluded that the self-assembled monolayers of the imidazole on the metallic surface (iron) had a higher corrosion inhibition tendency than the stearic acid. Also it was illustrated that the imidazoles blocked the surface of iron by lowering the double-layer capacitance, proving to be an excellent anti-corrosive substance for metal like iron in an acidic environment (33).

Further, many more researchers have proven imidazoles as potent inhibitors for CO_2 corrosion for different metals and alloys. 2-undecyl-1-aminoethyl imidazoline (AEI-11) and 2-undecyl-1-aminoethyl1-hydroxyethyl quaternary imidazoline (AQI-11) prove to have anticorrosive properties specifically for CO_2 corrosion in the case of MS. The results described that the anticorrosion tendency of studied imidazoline inhibitors is specifically affected by the flow conditions and the particulate sand particles. It has been noticed that the flow conditions decreased the corrosion inhibition efficiency of the inhibitor, and thus the rate of corrosion increased due to the mechanical abrasion. Further, it has been evaluated that the inhibition efficiency is primarily based on the capability of the adsorbent to get adhered to the metal surface and on the availability of the adsorption sites, which are constituted mainly by the heteroatoms like oxygen (34).

Wang et al. (35) synthesized Thioureido imidazoline inhibitor and investigated its anticorrosive effect against carbon dioxide corrosion for MS alloy via weight loss and electrochemical studies. It was concluded that this inhibitor is worth preventing the metal from corrosion. The possible mechanism involves the inhibitor being adsorbed on the metal surface via the coordinate bonds between the heteroatom of the imidazoline compound and iron on the metallic surface (35). It has been evaluated from detailed research in this field that the corrosion protection capability of the respective imidazoline compounds are enhanced due to the introduction of electron-donating substituent, specifically one with the conjugated system (36).

Amine-Based Inhibitors

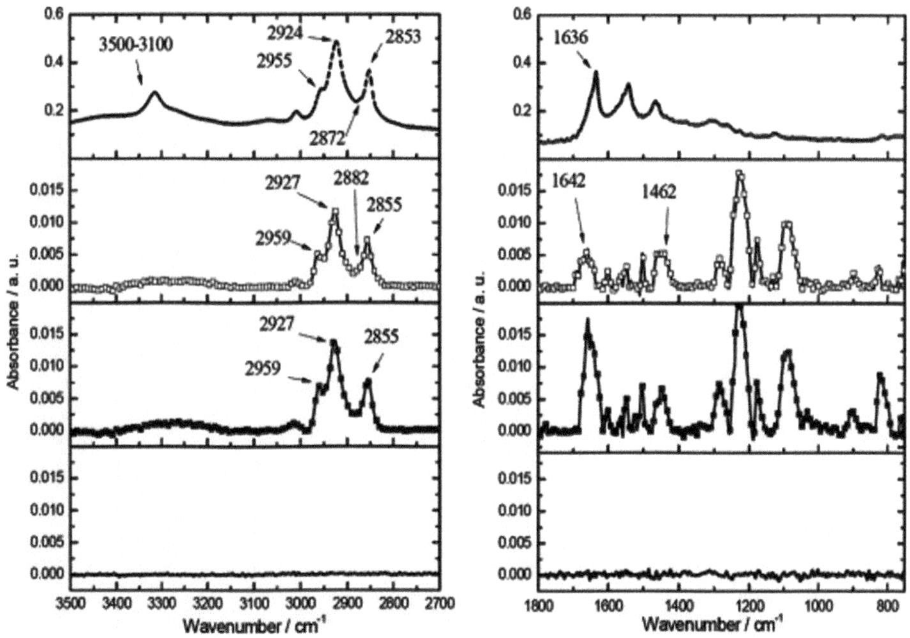

Figure 1. PM-IRRAS spectra of the corrosion inhibitor on MS substrate. Reproduced with permission from ref (37). Copyright 2011 Elsevier.

Ample evidence in the literature has affirmed the potency of amine-containing inhibitors for various metals and their alloys in the carbon dioxide environment. N-[2-[(2-aminoethyl) amino] ethyl]-9-octadecenamide (AAOA) has been tested for its anticorrosive properties against MS metallic alloy in NaCl-CO_2 environment via electrochemical and polarization techniques at a specific temperature of 25^{0}C. The experimental results elucidated that this sort of inhibitor helped in decreasing the rate of corrosion by reducing the active sites for adsorption over the metallic surface so that further reactions can be blocked. This inhibitor was found to have high inhibition efficiency and behaved as a mixed type, predominantly an anodic inhibitor. Electrochemical studies gave the best results to reflect inhibition. Polarization Modulation Infrared Reflection Absorption Spectroscopy (PM-IRRAS) results further provided details like the mechanism of inhibition involving chemical interactions between the metal empty d-orbitals and the heteroatoms nitrogen, oxygen, and others (37). Figure 1 demonstrates PM-IRRAS spectra of AAOA adsorbed on a MS substrate and the data reveal that AAOA monolayer has a highly disordered structure.

The anticorrosive effect of 12-aminododecanoic acid was seen on carbon steel in carbon dioxide saturated hydrochloric acid solution. Results revealed this as a mixed-type corrosion inhibitor. Furthermore, corrosion inhibition efficiency enhanced with the rise in pH of the solution (less than 3) and temperature below 318K. Thus, this inhibitor was highly feasible, and the nature of interactions formed is termed as physical (38).

Another example in this class includes Quaternary alkynoxymethyl amine (IMC-80-Q) and imidazoline in a 3% NaCl environment which showed anticorrosive properties against CO_2 corrosion. The stagnant and dynamic conditions are involved in testing and conducting experiments. The adsorption of this IMC-80-Q was found to obey Langmuir adsorption isotherm under the static environment when the concentration is taken less than 150 mg/L. But on increasing the value beyond 150mg/L, Freundlich adsorption isotherm followed. Flowrate along with the inhibitor concentration can be altered for maximum output of inhibition efficiency (39).

Polymer-Based Inhibitors

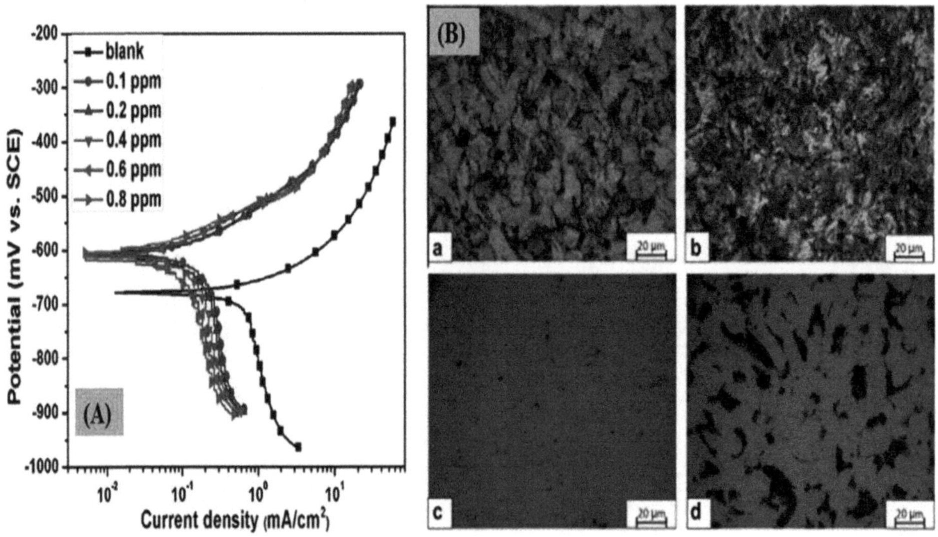

Figure 2. (A) Polarization plots for corrosion of SAE1018 with and without inhibitor in stagnant conditions. (B) Optical microscopy images in the presence of (a) blank solution in stagnant conditions; (b) blank solution in hydrodynamic conditions; (c) inhibited solution in stagnant conditions; (d) inhibited solution in hydrodynamic conditions. Reproduced with permission from ref (40). Copyright 2011 Elsevier.

Water-based acrylic terpolymers (ATP) were investigated for their corrosion preventing activity with the help of electrochemical tests. The experimental conditions provided here constituted two variable conditions: stagnant and hydrodynamic. The experimental results in both the cases resulted in different effects (e.g., increase in rotation speed).An increase in corrosion rate and corrosion inhibition tendency has also been observed. The possible mechanism involved in the prevention was the adsorption of these polymeric materials over the metallic surface. Results with ATP were obtained as a mixed-type inhibitor, and also, the high tendency of this material was linked with the speed of the working electrode. Also, the Nyquist plots for ATP signified an enhancement in the double layer charge transfer resistance on the metal surface and thus developed a protective coating on the metallic surface. Optical microscopy of the metallic surface in both dynamic and static cases showed efficient corrosion inhibition (*40*). Figure 2 demonstrates polarization curves and optical microscopy images showing corrosion of SAE1018 (carbon steel) with and without inhibitor in a stagnant environment with hydrodynamic conditions. In these dynamic conditions, the grain boundary of ferrite is not visible, while in static solution it seems to be unaltered. Inhibited solutions with static as well as dynamic conditions show remarkable inhibition efficiency compared to blank solution.

Also, polymeric materials are multifunctional because they are efficient and improve corrosion inhibitors. Further deep research in this field came up with certain conclusions that polymeric substances with less than 10 repeating units are observed to be adequate corrosion inhibitors. Additionally, it can be concluded that the polymers are large, have several attaching sites which can adhere to the metallic surface, and thus could be much more efficient corrosion inhibitors (*41*). Commonly the corrosion inhibitors, which are small in size, degrade at higher temperatures, and thus contaminate the main product (*42*).

Natural Inhibitors in Sweet CO_2 Environments

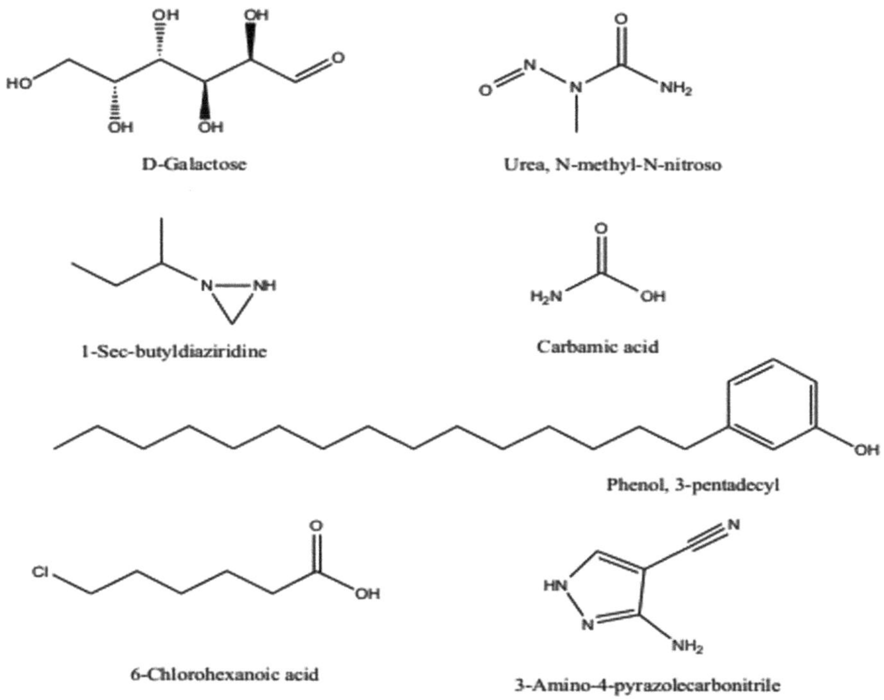

Figure 3. Structure of the active constituents of GBFE. Reproduced with permission from ref (43). Copyright 2014 Taiwan Institute of Chemical Engineers, Elsevier Inc.

Natural inhibitors are also referred to as green inhibitors; they are called so due to their affordable prices and renewability. Plant extracts obtained from leaves and fruits, have been observed to show such anticorrosive properties. For example, *Gingko Biloba* (GBFE) extract was investigated for its corrosion inhibiting capacity on a MS surface in NaCl solution. GBFE is composed of constituents comprising heteroatoms (O, N, S) in their chemical moiety, as illustrated in Figure 3. Few conventional studies like electrochemical and weight loss were employed to check its potential. As a result, it has been confirmed that the corrosion prevention capacity enhanced with an incline in the concentration of inhibitor. Electrochemical studies illustrated it as a mixed-type inhibitor, and the adsorption plots followed, in this case, obeyed the Temkin adsorption curves. Further, studies like Scanning Electron Microscopy (SEM) concluded that the development of protective film took place over the surface of MS. As a result, a much smoother surface was confirmed (*43*).

Calotropis procera leaf extract is also a green and natural corrosion inhibitor against CO_2 corrosion for MS in NaCl solution. The extract of this leaf was taken and investigated with the help of several conventional methods. Results were so notable that, even at elevated temperatures, the incorporation of inhibitors enhanced the corrosion inhibition efficiency. In addition, *Calotropis procera* extract was proven to obey the Langmuir adsorption isotherm and behaved like a mixed-type inhibitor. Furthermore, this plant extract was observed to form physical interactions on the MS surface and facilitated the inhibition phenomenon (*44*).

Ilex kudingcha C.J. Tseng leaves have been analyzed as a carbon dioxide corrosion inhibitor for J55 Steel. The polarization studies with this experiment signified that this extract behaved as a mixed-type inhibitor, and further, its addition in the media increased the inhibition efficiency up to 96.53%. Surface studies revealed the formation of protective layers on the metallic surface, and this way, the smoother texture in the presence of inhibitor was observed (*45*). *Momordica charantia* (Karela) Seeds (MCS) have been validated as a green CO_2 corrosion inhibitor for one of the steel alloys named P110SS. The weight loss and polarization studies followed the experimental procedure with this inhibitor. MCS was proven to be a mixed-type inhibitor and its adsorption over the metallic surface followed the Langmuir isotherms. Further, after the application of this inhibitor, surface studies were carried out, and it was observed that this inhibitor served to obstruct the active sites for reaction on the metallic surface and, in this way, provided a solid inhibition for CO_2 corrosion (*46*).

Additionally, sulfated fatty acids diethanolamine complexes were synthesized based on various vegetable oils. Their corrosion preventing potential was tested at a specific temperature of 50 °C for MS. The individual results signified that these surfactants could also be used as efficient corrosion inhibitors, even at their low concentrations. Also, just like many other instances, the inhibition effect of these increased with the rise in concentrations, and also, the maximum peak was observed at critical micelle concentration. Furthermore, these surfactants were seen to form chemical bonds with the metal surface, and Langmuir plots were followed (*47*).

Corrosion Inhibitors for H_2S Corrosion

Imidazoline-Based Corrosion Inhibitors

Imidazoles, just like amines, are made up of nitrogen atoms, and along with that, they have two nitrogen atoms which differentiate them from the amines. Imidazolines and their compounds, according to researchers, have been proven as efficient sour corrosion inhibitors (*48*). Also, imidazolines have other characteristics as these are regarded as negatively charged cationic surfactants, and these don't alter their properties on changing pH (*49*). They have been observed to behave as suitable sour corrosion inhibitors. Some of the potential examples of this category have been discussed. Imidazoline (derivatives of cyclopentylic and cyclohexylonaphthenic acids) has been found to have anti-corrosive properties at a specific pH range of 5.5 and an optimum concentration of 25 ppm. Excellent inhibition efficiency at varying temperature ranges of 20–60 °C was obtained (*50*). Imidazoline inhibitors obtained from vegetable oil, sunflower, canola, soybean, safflower, corn, and mixtures thereof reacted with polyalkylated polyamines have also been categorized into this class of inhibitors. The temperature decided for work conditions of this inhibitor is 60 °C, and the concentration best recommended for use is 5 to 50 mg/L (*51*). Normal and acid salts of substituted imidazolines and saturated aliphatic mono and dicarboxylic acids also can prevent sour corrosion with a concentration of 25 ppm. The corrosion rate was observed to be 0.6–7.6 mpy in the presence of this inhibitor (*52*).

The product obtained from the reaction of polymeric acid with 1-aminoalkyl-2-alkyl-2-imidazoline was proven to show efficient anti-corrosive properties. The inhibitor concentration was seen to be 25 ppm, and the corrosion rate was observed to be 5.7 mpy (*53*). Naphthylethyl thioacylmorpholin was synthesized using the reactants methyl-α-naphthyl ketone, morpholine, and sulfur by employing Willgerodt–Kindler technique and the corrosion protection potential of this inhibitor has been tested. The results concluded that the inhibitor concentration to be utilized should be 100 mg/L, and inhibition efficiency was ascertained as 97.2% (*54*). Hydroxyethyl imidazoline with an optimum concentration of 5 and 10 ppm showed anticorrosive properties and gave an inhibition efficiency of 95% (*55*). Another experiment tested Carboxyethyl imidazoline as a sour corrosion inhibitor, and the experimental results elucidated the inhibitive performance of 98% (*56*). Hence it can be categorized as a suitable corrosion inhibitor. On the other hand, Hydroxyethyl imidazoline depicted an inhibition efficiency of 84.6%. This case's temperature conditions and optimum concentration were considered 50 °C and 25 ppm (*48, 57*).

It was observed that the corrosion inhibition mechanism by these kinds of imidazolines followed the formation of bonds between the imidazoline and the metal surface, and along with that, the hydrophobic part played an important role. The role of the pedant group is stated as minimal here (*58, 59*). But along with that, another thought came into existence that these imidazolines served as corrosion inhibitors by the combined effects of the processes like blocking, activation and energy-related factors. Also, some other approaches came into the light like these prevent the corrosion by the development of self-assembled monolayers on the metallic surface (*60*).

Amine-Based Corrosion Inhibitors

Amines are the derivatives of ammonia, and there is a further classification of amines, namely primary, secondary, and tertiary, based on the nature of the substituents linked to the nitrogen. Amines are further categorized into cyclic and aliphatic ones. Also, these serve many potential applications like in the pharma field, production of rubber, dyes, etc. Amines have been designated as much cheaper and more effective materials in this process of corrosion inhibition. Also, they have been observed to make a more profound effect by making the film over the metal surface (*61*). A large variety of inhibitors are found in the literature based on amines used to tackle sour corrosion.

Dibutylamine, Cyclohexylamine, Diphenylamine, Phenylenediamine, Tributylamine, etc. have been observed to have an efficient anticorrosive effect for H_2S corrosion. All these examples were found to have a good inhibition efficiency at varying temperatures. Furthermore, reaction mixtures of formaldehyde, acetophenone, cyclohexylamine or aliphatic carboxylic acid component have also been observed to have some corrosion inhibiting properties in the temperature range of 150–250 °F. Furthermore, α, β-ethylenically unsaturated aldehydes (with organic polyamines and monofunctional amines) and their reactions with organic halides and epoxides have also been observed to have a better inhibition effect on corrosion (*57*).

These amine-based inhibitors are ascertained to reflect anticorrosive performance due to the presence of a heteroatom like nitrogen which is capable of donating its electrons and therefore acts as an adsorption site. This H_2S corrosion inhibition by these amines takes place via the formation of chemical interactions. Further, this nitrogen donates its electron pair to the empty metal orbital, diminishing the urge of metal to react and corrode. The degree of chemisorption entirely depends on the strength of the chemical bond between the nitrogen of amine and metal. However, these amines are found to possess good anti-corrosive properties, but they have some drawbacks. If the aliphatic amine contains more than 14 carbon atoms, it isn't easy to dissolve them due to their solid nature at room temperature. Two processes, namely methylation and ethoxylation, can lower their melting point (*62*).

Polymer-Based Corrosion Inhibitors

Polymers are defined as long macromolecules formed by small subunits that generally have a repeating pattern. They have two origins: synthesized or found naturally like cellulose, wool, etc. Both types of polymers have a broad range of applications involving the packaging of food materials. In addition, naturally occurring polymers have some specific environmentally positive characteristics (e.g., biodegradable) and pose significantly less threat to the aquatic sources due to their low toxicity (*63*).

Many polymers have been commercially available with some specific product names and are used as sour corrosion inhibitors. VpCI-639 has been explicitly designed to prevent sour and sweet corrosion and can be used temperatures up to 400 °F (*57*). Another product from the same corporation, VpCI-637 TOL, has been produced with the same desirable qualities—it can tackle corrosion by hydrogen sulfide. It has been specifically designed for the internal protection of pipelines. Periodic injection should be done to have profound effects. CorrSorb 3600, synthesized by Baker Hughes, are solid inhibitors used for the protection of oil and gas wells [57]. CORTRON AR-505, developed by Champion Technologies, has been examined as a sour corrosion inhibitor for 1018 carbon steel in seawater. In this case, Optimum concentration was observed to be 10 ppm and the corrosion rate of 1mm/year can be achieved (*64*).

CORRTREAT by Clariant has been observed to prevent pitting corrosion. In addition to that, it also can provide high-temperature corrosion, high-shear protection, and high- sweet and sour corrosion protection. Ercabac 50 and Ercabac 80 have also been beneficial for this kind of sour corrosion protection. This acts by preventing the formation of sulfurous gases and also can be utilized to prevent the pipeline corrosion. It also has an interesting application of working as a demulsified in enhanced oil extraction. In recent years, these polymer-based sour corrosion inhibitors have gained the least attention (*48*), perhaps explained by their poor solubility in aqueous solutions. This property may lead to less corrosion inhibition under static conditions specifically (*65*). Also, these polymers are unstable at a very high temperature and cannot be used at a varying range of temperatures. However, several processes, (i.e., copolymerization, synergistic effects, etc.) have enhanced the corrosion resistance tendencies of these polymer-based corrosion inhibitors (*66*).

Natural Inhibitors for H_2S Corrosion

Synthetic corrosion inhibitors have been extensively used; however, recently the cost of manufacturing and high toxicity have eliminated their usage by researchers and scientists. Natural corrosion inhibitors appear in the wild, and their implications for practical purposes do not pose any threat to the environment. Natural corrosion inhibitors have been observed to produce good results as compared to their synthetic counterparts. They have specific qualities that encourage their usage, like being environment friendly, not accumulating in the media, and the least toxicity. Various plant extracts have been categorized to prevent corrosion in the hydrogen sulfide environment (*67–69*).

Conclusion and Outlook

This chapter summarized the effective corrosion inhibitors best suited for sweet and sour environments. The fluid reservoirs in the oil and gas industry involve gases like carbon dioxide, hydrogen sulfide, etc. These gases tend to cause severe issues to the different metals and their respective alloys. The main element utilized in almost every industrial sector is steel (low grade) due to its low cost and ease of availability. Incorporating gases (i.e., CO_2 and H_2S) into the metallic alloy can lead to severe corrosion attacks, which can be best combated by the utilization and application of corrosion inhibitors. Sweet and sour corrosion inhibitors have been categorized into different classes: imidazole based, amine-based, polymers, and natural-based inhibitors. Synthesis of sweet and sour inhibitors, which are environment friendly, are designed to tackle industrial corrosion. Though there are natural inhibitors, the experimental results of such green corrosion inhibitors are not notably efficient. Thus, there is significant benefit to properly understanding corrosion inhibition and working with inhibitors in different environments. Future research in this field should develop less toxic, environmentally friendly, and efficient corrosion inhibitors. Ionic liquids, plants extract, and more have been evidenced to have corrosion inhibition properties, and these also have the compatibility to match with the environment. As such, the research should focus on sustainable or environmentally friendly corrosion inhibitors.

References

1. Shreir, L. L.; Jarman, R. A.; Buratein, G. T.; *Corrosion*. Volume 1, Buttlerworth Heinemann, 3rd Ed., 1994, 1.3−1.15.
2. Scientific Publications Division, *Corrosion Reviews*, Volumes 10–11, Freund Publishing House, 1992.
3. Uhlig, H.; Winston, R. *Corrosion and Corrosion Control*, 3rd Ed., New York, 1985, 1−20.
4. Milella, P. P. *Fatigue and Corrosion in Metals*, Springer Science & Business Media, 2012.
5. Hackerman, N. *Review on corrosion inhibitor science and technology*. National Association of Corrosion Engineers, NACE, 1989.
6. Schweitzer, P. E., *Corrosion Engineering Handbook*, Volume 3, CRC Press, 2nd Ed., 1996, 23−33.
7. Ahmad, Z. *Principals of Corrosion Engineering and Corrosion Control*, ISBN: 0750659246, Elsevier Science & Technology Books, 2006, 121−270.
8. Pleith, W. *Electrochemistry for Material Science*, Elsevier B. V., 2008, ISBN: 978-0-444 52792-9, 291−323.
9. Shidong, Z.; Hui, L.; Zhenquan, B. Dynamic corrosion behavior of P110 steel in stimulated oil field CO2/H2S environment. *Chemical Engineering of Oil & Gas* **2009**, *38*, 65–68.
10. Saha, S. K.; Dutta, A.; Ghosh, P.; Sukul, D.; Banerjee, P. Adsorption and corrosion inhibition effect of Schiff base molecules on the mild steel surface in 1 M HCl medium: a combined experimental and theoretical approach. *Phys. Chem. Chem. Phys.* **2015**, *17* (8), 5679–5690.
11. Saha, S. K.; Dutta, A.; Ghosh, P.; Sukul, D.; Banerjee, P. Novel Schiff-base molecules as efficient corrosion inhibitors for mild steel surface in 1 M HCl medium: experimental and theoretical approach. *Phys. Chem. Chem. Phys.* **2016,** *18* (27), 17898–17911.
12. Murmu, M.; Saha, S. K.; Murmu, N. C.; Banerjee, P. Effect of stereochemical conformation into the corrosion inhibitive behaviour of double azomethine based Schiff bases on mild steel

surface in 1 mol L-1 HCl medium: An experimental, density functional theory and molecular dynamics simulation study. *Corros. Sci.* **2018**, *146*, 134–151.

13. Chugh, B.; Thakur, S.; Pani, B.; Murmu, M.; Banerjee, P.; Al-Mohaimeed, A. M.; Ebenso, E. E.; Singh, M.; Singh, J.; Singh, A. K. Investigation of phenol-formaldehyde resins as corrosion impeding agent in acid solution. *J. Mol. Liq.* **2021**, *330*, 115649.
14. Alagta, A.; Felhösi, I.; Bertoti, I.; Kálmán, E. Corrosion protection properties of hydroxamic acid self-assembled monolayer on carbon steel. *Corros. Sci.* **2008**, *50*, 1644–1649.
15. Nanan, K. *The basics of cathodic protection* http://www.corrosionpedia.com/2/1368/prevention/cathodic-protection/cathodic-protection-101. (accessed Jul 10, 2021)
16. Koch, G.; Varney, J.; Thompson, N.; Moghissi, O.; Gould, M.; Payer, J. *International Measures of Prevention, Application and Economics of Corrosion Technologies Study*, NACE International, 2016, 216.
17. Bhaskaran, R.; Palaniswamy, N.; Rengaswamy, S.; Jayachandran, M. A review of differing approaches used to estimate the cost of corrosion (and their relevance in the development of modern corrosion prevention and control strategies). *Anti-Corros. Method. M.* **2005**, *52*, 29–41.
18. Koch, G. H.; Brongers, M. P. H.; Thompson, N. G.; Virmani, Y. P.; Payer, J. H. *Corrosion Cost and Preventive Strategies in the United States*, Report No. FHWA-RD-01-156, Federal Highway Administration, U.S. Department of Transportation, Washington D. C., 2001.
19. Roberge, P. R. *Corrosion Engineering Principles and Practice*, McGraw Hill Companies, Inc. 2008, 1–47.
20. Uhlig, H. H.; Revie, W. R. *Corrosion and Corrosion Control*, John Wiley, New York, 1985, 2–4.
21. McCafferty, E. *Introduction to Corrosion Science*, Springer, 2010, ISBN 978-1-4419-0454-6, 1–11.
22. Quiong-xiao, C.; Changqing, C. Study on corrosion behavior of tubular steel N80 in CO2/H2S environment. Natural Gas Chemical Industry. **2009**, *34*, 51–54.
23. Yan, H.; Daode, L.; Guanfa, L. Corrosion resistance of 13Cr stainless steel in the Cl , CO_2 environment with microamount H_2S. *Physical Testing and Chemical Analysis Part A: Physical Testing* **2010**, *46* (3), 145–150.
24. Singh, A.; Lin, Y.; Liu, W.; Deng, K.; Pan, J.; Huang, B.; Ren, C.; Zeng, D. A study on the inhibition of N80 steel in 3.5% NaCl solution saturated with CO2 by fruit extract of Gingko biloba. *J. Taiwan. Inst. Chem. Eng.* **2014**, *45*, 1918–1926.
25. Jun, A. CO2/H2S corrosion behavior of N80 steel in heterogeneous medium. *Journal of Oil and Gas Technology* **2009**, *31* (4), 157–160.
26. Sun, W. *Kinetics of Iron Carbonate and Iron Sulfide Scale Formation in CO2/H2S Corrosion*. Ph.D dissertation, Ohio University , Athens, 2006, 193–200.
27. Han, J.; Yang, Y.; Nesic, S.; Brown, B. N. Roles of Passivation and Galvanic Effects in Localized CO2 Corrosion of Mild Steel Corrosion. *NACE International* **2008**, 08332.
28. Zhang, J. Corrosion Behavior of L360 Pipeline Steel and Its Welded Joints in H_2S/CO_2 Medium. *Beijing: Beijing University of Chemical Technology* **2011**.
29. Sun, W.; Nesic, S. A Mechanistic Model of H_2S Corrosion of Mild Steel. *NACE Corrosion.* **2007**.

30. Yin, Z. F.; Zhao, W. Z.; Bai, Z. Q.; Feng, Y. R.; Zhou, W. J. Corrosion behavior of SM 80SS tube steel in stimulant solution containing H_2S and CO_2. *Electrochim. Acta.* **2008**, *53* (10), 3690–3700.
31. Nesic, S.; Lee, K. L. J. A Mechanistic Model of Iron Carbonate Film Growth and the Effect on CO_2 Corrosion of Mild Steel. *Corrosion.* **2002**.
32. Jovancicevic, V.; Ramachandran, S.; Prince, P. Inhibition of Carbon Dioxide Corrosion of Mild Steel by Imidazolines and Their Precursors. *Corrosion* **1999**, *55* (5), 449–455.
33. Liu, X.; Chen, S.; Ma, H.; Liu, G.; Shen, L. Protection of Iron Corrosion by Stearic Acid and Stearic Imidazoline Self-Assembled Monolayers. *Appl. Surf. Sci.* **2006**, *253* (2), 814–820.
34. Liu, X.; Okafor, P. C.; Zheng, Y. G. The Inhibition of CO2 Corrosion of N80 Mild Steel in Single Liquid Phase and Liquid/Particle Two-Phase Flow by Aminoethyl Imidazoline Derivatives. *Corros. Sci.* **2009**, *51* (4), 744–751.
35. Wang, B.; Du, M.; Zhang, J.; Gao, C. J. Electrochemical and Surface Analysis Studies on Corrosion Inhibition of Q235 Steel by Imidazoline Derivative against CO2 Corrosion. *Corros. Sci.* **2011**, *53* (1), 353–361.
36. Wang, D.; Li, S.; Ying, Y.; Wang, M.; Xiao, H.; Chen, Z. Theoretical and Experimental Studies of Structure and Inhibition Efficiency of Imidazoline Derivatives. *Corros. Sci.* **1999**, *41* (10), 1911–1919.
37. Desimone, M. P.; Grundmeier, G.; Gordillo, G.; Simison, S. N. Amphiphilic Amido-Amine as an Effective Corrosion Inhibitor for Mild Steel Exposed to CO2 Saturated Solution: Polarization, EIS and PM-IRRAS Studies. *Electrochim. Acta.* **2011**, *56* (8), 2990–2998.
38. Ghareba, S.; Omanovic, S. Interaction of 12-Aminododecanoic Acid with a Carbon Steel Surface: Towards the Development of "Green" Corrosion Inhibitors. *Corros. Sci.* **2010**, *52* (6), 2104–2113.
39. Jiang, X.; Zheng, Y. G.; Ke, W. Effect of Flow Velocity and Entrained Sand on Inhibition Performances of Two Inhibitors for CO2 Corrosion of N80 Steel in 3% NaCl Solution. *Corros. Sci.* **2005**, *47* (11), 2636–2658.
40. Vakili Azghandi, M.; Davoodi, A.; Farzi, G. A.; Kosari, A. Water-Base Acrylic Terpolymer as a Corrosion Inhibitor for SAE1018 in Simulated Sour Petroleum Solution in Stagnant and Hydrodynamic Conditions. *Corros. Sci.* **2012**, *64*, 44–54.
41. Jeyaprabha, C.; Sathiyanarayanan, S.; Venkatachari, G. Polyaniline as Corrosion Inhibitor for Iron in Acid Solutions. *J. Appl. Polym. Sci.* **2006**, *101* (4), 2144–2153.
42. Straetmans, U.; Soltau, M.; Straetmans, F. *Polymeric Corrosion Inhibiter for Metal Surfaces and the Production Thereof.* U.S. Patent Application No. 14/000,772, Dec 12, 2013.
43. Singh, A.; Lin, Y.; Liu, W.; Kuanhai, D.; Pan, J.; Huang, B.; Ren, C.; Zeng, D. A Study on the Inhibition of N80 Steel in 3.5% NaCl Solution Saturated with CO2 by Fruit Extract of Gingko Biloba. *J. Taiwan Inst. Chem. Eng.* **2014**, *45* (4), 1918–1926.
44. Ibrahim, T.; Gomes, E.; Obot, I. B.; Khamis, M.; Abou Zour, M. Corrosion Inhibition of Mild Steel By Calotropisproceraleaves Extract in a CO2 saturated Sodium Chloride Solution. *J. Adhes. Sci. Technol.* **2016**, *30* (23), 2523–2543.
45. Chen, S. Inhibition Effect of Ilex Kudingcha C.J. Tseng (Kudingcha) Extract on J55 Steel in 3.5wt% NaCl Solution Saturated with CO2. *Int. J. Electrochem. Sci.* **2017**, 782–796.

46. Singh, A.; Lin, Y. H.; Liu, W. Y.; Zhou, L. J.; Deng, K. H.; Huang, B. A Study on the Inhibition of J55 and P110SS Steel in 3.5% NaCl Solution Saturated with CO_2 by Seed Extract of Momordica Charantia (Karela). *Solid State Phenom.* **2015**, *227*, 55–58.
47. El-Lateef, H. M. A. Green Surfactants from the Type of Fatty Acids as Effective Corrosion Inhibitors for Mild Steel in CO2- Saturated NaCl Solution. *Am. J. Phys. Chem.* **2013**, *2*, 16.
48. Obot, I. B.; Solomon, M. M.; Umoren, S. A.; Suleiman, R.; Elanany, M.; Alanazi, N. M.; Sorour, A. A. Progress in the Development of Sour Corrosion Inhibitors: Past, Present, and Future Perspectives. *J. Ind. Eng. Chem.* **2019**, *79*, 1–18.
49. Askari, M.; Aliofkhazraei, M.; Ghaffari, S.; Hajizadeh, A. Film Former Corrosion Inhibitors for Oil and Gas Pipelines - a Technical Review. *J. Nat. Gas Sci. Eng.* **2018**, *58*, 92–114.
50. Szyprowski, A. J. Relationship between Chemical Structure of Imidazoline Inhibitors and Their Effectiveness against Hydrogen Sulphide Corrosion of Steels. *British Corrosion Journal* **2000**, *35*, 155–160.
51. Aguilar, J. L. R. B. *Corrosion Inhibitors Derived from Vegetable Oils and It's Process of Obtaining.* 2017, 1–12.
52. Oppenlaender, K.; Stork, K.; Barthold, K.; S. *Imidazoline Based Corrosion Inhibitors Which Inhibit Corrosion Caused by CO2 and H2. U. S. Patent* 1983, No. 4,388,214.
53. Martin, R. L.; Mcmahon, J. A.; Alink, B. A. O. *Biodegradable Corrosion Inhibitors of Low Toxicity. U. S. Patent* 1998, No. 5,785,895.
54. Li, Z. J. High-Temperature Corrosion Inhibitor for H2S-HCl-H2O System and Preparation Method of High temperature Corrosion Inhibitor. *Materials Science* **2016**.
55. Lucio-Garcia, M. A.; Gonzalez-Rodriguez, J. G.; Martinez-Villafañe, A.; Dominguez-Patiño, G.; Neri-Flores, M. A.; Chacon-Nava, J. G. A Study of Hydroxyethyl Imidazoline as H2S Corrosion Inhibitor Using Electrochemical Noise and Electrochemical Impedance Spectroscopy. **2009**, *40* (2), 393–399.
56. Diaz, E. F.; Gonzalez-Rodriguez, J. G.; Martinez-Villafañe, A.; Gaona-Tiburcio, C. H2S Corrosion Inhibition of an Ultra High Strength Pipeline by Carboxyethyl-Imidazoline. *J. Appl. Electrochem.* **2010**, *40* (9), 1633–1640.
57. Umoren, S. A.; Solomon, M. M.; Saji, V. S. *Corrosion Inhibitors for Sour Oilfield Environment*; pp 229–254.
58. Edwards, A.; Osborne, C.; Webster, S.; Klenerman, D.; Joseph, M.; Ostovar, P.; Doyle, M. Mechanistic Studies of the Corrosion Inhibitor Oleic Imidazoline. *Corros. Sci.* **1994**, *36* (2), 315–325.
59. Gusmano, G.; Labella, P.; Montesperelli, G.; Privitera, A.; Tassinari, S. Study of the Inhibition Mechanism of Imidazolines by Electrochemical Impedance Spectroscopy. *Corrosion* **2006**, *62* (7), 576–583.
60. Dai, Y.-X.; Lv, F.-N.; Wang, B.; Chen, Y. Thermoresponsive Phenolic Formaldehyde Amines with Strong Intrinsic Photoluminescence: Preparation, Characterization and Application as Hardeners in Waterborne Epoxy Resin Formulations. *Polymer* **2018**, *145*, 454–462.
61. Askari, M. Film Former Corrosion Inhibitors for Oil and Gas Pipelines-A Technical Review. J. Nat. Gas Sci. Eng. **2018**, *58*, 92–114.
62. Blair, C. M.; Groves, W.; Gross, W. F. *Processes for Preventing Corrosion and Corrosion Inhibitors.* 1949, 1–7.

63. Umoren, S. A.; Solomon, M. M. Effect of Halide Ions on the Corrosion Inhibition Efficiency of Different Organic Species–A Review. *J. Ind. Eng. Chem.* **2015**, *21*, 81–100.
64. Al-Baker, N.; Shawabkeh, R.; Rihan, R. Kinetic Study of Effect of Amine Based Corrosion Inhibitor in Reducing Corrosion Rate of 1018 Carbon Steel in Seawater Solution. *Corros. Eng. Sci. Techn.* **2011**, *46*, 767–776.
65. Umoren, S. A.; Solomon, M. M.; Udosoro, I. I.; Udoh, A. P. Synergistic and Antagonistic Effects between Halide Ions and Carboxymethyl Cellulose for the Corrosion Inhibition of Mild Steel in Sulphuric Acid Solution. *Cellulose* **2010**, *17* (3), 635–648.
66. Fares, M. M.; Maayta, A. K.; Al-Mustafa, J. A. Corrosion Inhibition of Iota-Carrageenan Natural Polymer on Aluminum in Presence of Zwitterion Mediator in HCl Media. *Corros. Sci.* **2012**, *65*, 223–230.
67. El-Etre, A. Y.; Abdallah, M. Natural Honey as Corrosion Inhibitor for Metals and Alloys. II. C-Steel in High Saline Water. *Corros. Sci.* **2000**, *42*, 731–738.
68. Raja, P. B.; Sethuraman, M. G. Natural Products as Corrosion Inhibitor for Metals in Corrosive Media — a Review. *Mater. Lett.* **2008**, *62*, 113–116.
69. Banerjee, S.; Srivastava, V.; Singh, M. M. Chemically Modified Natural Polysaccharide as Green Corrosion Inhibitor for Mild Steel in Acidic Medium. *Corros. Sci.* **2012**, *59*, 35–41.

Chapter 10

Corrosion Inhibitors for Crude Oil Refinery Units

Jeenat Aslam,[1,*] Ruby Aslam,[2] Megha Basik,[2] and Chandrabhan Verma[3]

[1]Department of Chemistry, College of Science, Taibah University, Yanbu-30799, Al-Madina, Saudi Arabia

[2]Corrosion Research Laboratory, Department of Applied Chemistry, Faculty of Engineering and Technology, Aligarh Muslim University, Aligarh-202002, India

[3]Interdisciplinary Research Center for Advanced Materials, King Fahd University of Petroleum and Minerals, Dhahran 31261, Saudi Arabia

[*]Email:drjeenataslam@outlook.com.

Corrosion has constantly been a major hindrance in crude oil refinery units. It has become more abundant and complicated in the past few decades because of the increasing costs of materials and labor. This chapter discusses the failures encountered by corrosive moieties in crude oil refining processes and how these result in the deterioration of tools. It will also discuss protective measures for reducing corrosion influences, as well as corrosion prevention and examining techniques. Corrosion management plays a very important function in resolving these issues.

Introduction

Refining of crude oil involves partitioning it in different portions and their subsequent treatments using a series of unit processes in which every unit process performs a separate function. This process is important for yielding profits (*1–4*). A refinery is a collection of integrated industrialized plants that differ in number and produces a wide range of products (Figure 1) (*5*). Corrosion can cause tool failure and can dictate how refineries should take action against it, causing divisions, or even the whole refinery, to be shut down (*6, 7*). The problem with corrosion in refineries is that there is not a particular corrosion source, but a lot of factors together are involved (*8–12*). Several corrosive agents might be interactive and the corrosivity might become severe. Furthermore, the conditions of physical processes play an important role: flow and temperature have to be taken into account, as well as, and of no less significance, the refinery's infrastructure itself, as vessels, pipes, welding joints, and tools also contribute to the menace of corrosion.

Various corrosion materials arise in refinery units and they can occur in different forms. For example, these forms include hydrogen embrittlement, stress corrosion cracking from sulfide attacks, and pitting corrosion from acidic water. Carbon steel is utilized for the majority (>80%) of refinery constituents because of its resistive property to the most general forms of corrosion, mainly at temperatures below 400°F (205 °C) from hydrocarbon impurities. Other aggressive atmospheres and chemicals protect carbon steel use in all refineries. Other alternative materials are low-alloy steel, including molybdenum and chromium, with high-chromium stainless steel being most often applied in corrosion atmospheres. Copper, titanium, and nickel alloys are applied in the most challenging regions in which extremely corrosive chemicals and very high temperatures are present.

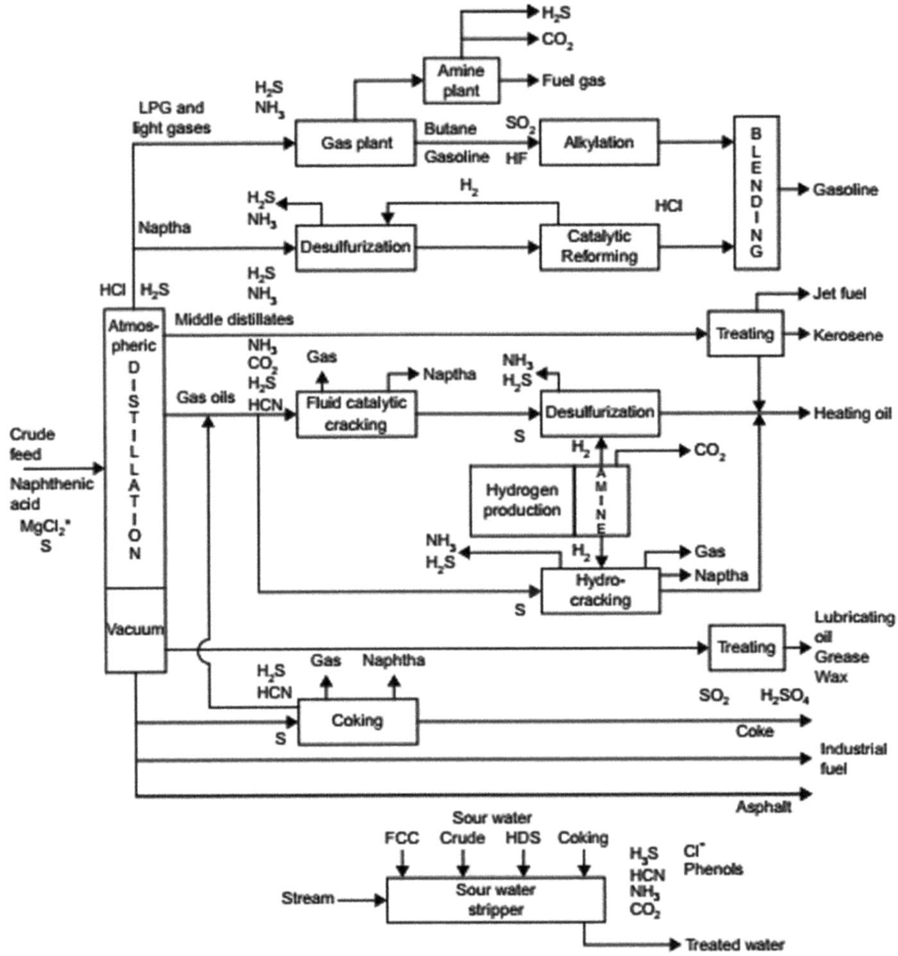

Figure 1. A representative refinery layout. Reproduced with permission from reference (5). Copyright 2014© Elsevier.

Corrosion is a common problem in crude oil refining units, mainly in the overhead system of crude oil distillation, where many of the metallic structures utilized for production and tools can be affected. Because of the variation in the chemical composition of the processed fluids and parameters such as pressure, temperature, moisture, and pH levels, the percentage of corrosive components is different in refining systems (*13*). Consequently, different corrosion issues will occur and ultimately management has to adopt diverse preventative methods for reducing corrosion influences. This chapter will suggest measures such as relevant dosing (water, inhibitor, and chemical), appropriate desalination by suitable materials, and sustaining correct monitoring operational parameters to efficiently reduce the corrosion influences.

Corrosion in Refining Processes

The problem of corrosion in crude oil refining processes can be divided in three parts; 1) environmental corrosion, 2) corrosion caused by crude oil composition, and 3) chemicals utilized in refinery processes (*13*).

Sources of Corrosion

The problem with corrosion is mainly ascribed to the presence of corrosive constituents such as mercaptans, organic sulfur compounds, inorganic and organic chlorides, hydrogen sulfide, hydrogen chloride, carbon dioxide, dissolved water, oxygen, nitrogen compounds ,and organic acids in the crude oil (*13*).

Types of Corrosion

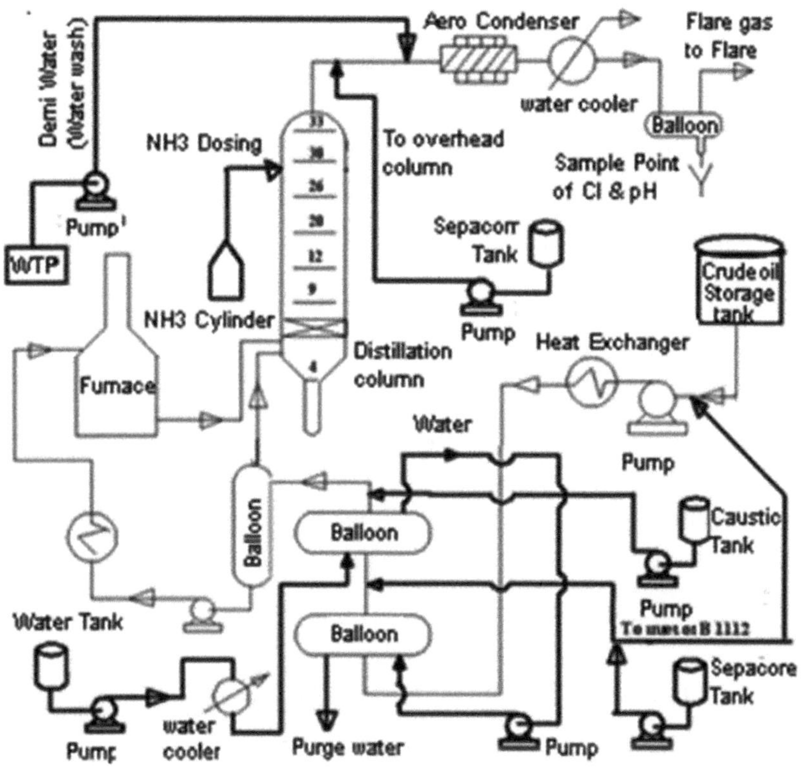

Figure 2. Chemical dosing in crude oil refining process. Reproduced with permission from reference (15). Copyright 2019 © IOP Publishing Ltd. Published by IOP Science under a Creative Commons Attribution License 3.0 (CC BY).

The list of corrosive species responsible in damaging different areas and tools of refining process of crude oil are shown in Table 1.

Table 1 illustrates that refining processes of crude oil mainly suffer from corrosion solely caused by H_2S, HCl, inorganic and organic chlorides, and sulfur compounds alone or in combinations of these compounds. This is because the various types of corrosive compounds have different corrosive mechanisms in refining units (*14*). To prevent this corrosion, appropriate chemical dosing needs to be properly applied to the corrosion process system, as shown in Figure 2 (*15*).

Table 1. Corrosive Compounds Generally Found in Tools in Refining Processes

Area/tools	*Corrosive compounds*
Furnace	S-compounds and H_2S
Tank	H_2S, HCl, dissolved O_2, and water
Preheat exchanger	H_2S and HCl
Top zone	HCl and water
Middle zone	H_2S
Flash zone	H_2S, S-compounds, and organic acids
Bottom exchanger	H_2S and S-compounds
Tower overhead system	H2S, HCl and water

Hydrogen Sulfide (H_2S) and Sulfur (S) Compounds Corrosion

Hydrogen sulfide and sulfur components are mainly present in crude compositions in nature, or they can be produced via the decomposition of sulfurous compounds during the production process. Here, the extent of corrosion mainly depends on the sulfur components available and the temperature of the process (*6, 16*).

Sulfur Corrosion at Low and Medium Temperatures

Corrosion caused by hydrogen sulfide can occur even at low temperatures. The initiation of corrosion arises from the corrosive species present in the crude composition or by decomposition of sulfur compounds while they are being processed.H_2S starts this reaction through iron and produces FeS scales in the overhead unit of the distillation plant, where water is mainly available.

$$Fe + H_2S \rightarrow FeS + H_2 \quad (1)$$

Presence of HCl dissolves the scales formed and produces more H_2S, which eventually reveals the clean metal surface.

$$FeS + 2HCl \rightarrow FeCl_2 + H_2S \quad (2)$$

Corrosion produced via H_2S includes hydrogen embrittlement and hydrogen blistering.

Sulfur Corrosion at High Temperatures

Temperature plays a significant role in corrosion via H_2S. Crude oil refineries that process the sour crude oil (high sulfur content) risk severe corrosion that may happen in distillation units of the plants where the temperatures exceed 450°F,and the decomposition of sulfur mainly occurs at or above this temperature. Corrosion rates as high as 0.5 in per year have been observed in carbon steel furnace tubes. At temperatures ranging from 450–900°F, metal generally suffers from severe corrosion. Above 900°F, the extent of corrosion is reduced because of the formation of a coke protective coating on the metal's surface (*17*).

Sulfur Corrosion at High Temperatures in the Presence of H_2

The presence of H_2—for instance, in hydro cracking and hydro treating operations—raises the harshness of sulfide corrosion at higher temperatures. Hydrogen converts approximately all detectable sulfur into H_2S. This results in corrosion initiated either by partial pressure or concentration of H_2S.

Hydrochloric Acid Organic and Inorganic Chloride Corrosion

Brine originates from crude, so separation of crude from brine is difficult or not fully effective, as it contains salts NaCl and $MgCl_2$,which dissolve in water and form HCl via hydrolysis at temperatures of approximately 300–400 °F. Dry HCl is generally not corrosive, but when it combines with water, it becomes extremely corrosive. After forming HCl, it passes through the preheat exchanger, then into the furnace, and it finally penetrates the distillation column. The temperature of the distillation column is low at the top (approximately 135 °C), which further reduces to 100 °C in overhead lines that come into contact with water. This causes HCl to build a corrosion reaction with iron, as mentioned in equation 3 (*18*).

$$Fe + HCl \rightarrow FeCl_2 + H_2 \quad (3)$$

When H_2S combines with this reaction product, it replicates numerous times and results in more severe corrosion.

$$FeCl_2 + H_2S \rightarrow FeS + 2HCl \quad (4)$$

NaOH, NH_3, and organic corrosion inhibitors are appropriately injected in the overhead system to lessen the corrosion and to maintain the pH levels of water (higher than 5). The temperature of the top section of the crude distillation column should be higher than the temperature of the water dew point (i.e., 100 °C).

Organic Acids Corrosion

At low temperatures, organic acids are not extremely corrosive. However, they are highly corrosive at their boiling temperatures, but the most severe form occurs generally upon condensation. The rate of corrosion under these circumstances has been observed to be as high as 0.35 inch/year on carbon steel.

CO_2 Corrosion

CO_2 produced from bicarbonate decomposition or steam in combination with H_2O cause corrosion. CO_2 corrosion results in severe metal dissolution, causing intergranular changes and creating a stress corrosion cracking atmosphere. Significant amounts of corrosion take place in carbonic acid solutions at 6.0 pH.

Nitrogen Compounds Corrosion

The formation of organic ammonia, or hydrogen cyanide, by the presence of nitrogen compounds is responsible for corrosion since nitrogen compounds present in crude oil do not fully contribute to corrosion, which happens mainly in hydro treating, catalytic cracking, and hydro-cracking operations. Ammonia and hydrogen cyanide react with H_2S and other accelerating components to form a highly corrosive mixture.

Dissolved Water and Oxygen Corrosion

Storage reservoirs contain moisture and oxygen, which may be present within the oil. This oil can be drawn via temperature variations and pumping. The quantity of oxygen and moisture, and, therefore, the rate of corrosion, is directly related to the number of pumping and temperature variations.

Used Chemicals Corrosion

Corrosion regularly caused by utilization of harsh chemicals such as H_2SO_4, ammonia, chlorine and caustic in refining process. x. At lower concentrations of H_2SO_4 (below 85%), which is used in water treatment plants, cause severe damage to carbon steel, forming pits that cause metal loss. Caustic and ammonia are generally utilized for the neutralization process of acid components in oil refinery operations. Overdosing of these compounds causes severe metal loss or stress corrosion cracking to metal (carbon steel and copper-based alloys).

Environmental Corrosion

Carbon steel is the most preferable material used for the construction of tools in oil refineries. It regularly risks corrosion because of the presence of water and oxygen at the surface. This corrosion can be reduced via restricting the contact of water to the metal surface via means of an applying protective barrier or generally coating. Other forms of atmospheric corrosion in refineries are pipeline corrosion, structural corrosion, cooling water corrosion, and corrosion under insulation or fireproofing.

Corrosion Protective Actions for Refinery Units

This section presents the protective actions that can be used to reduce the drastic effects of corrosion in refining operations of crude oil (*19*).

Appropriate Selection of Materials

Most construction materials for oil refineries fall into the following general categories: Carbon-Molybdenum ,Chromium-Molybdenum, carbon steels, alloy steels (ferritic chromium stainless steels, martensitic, and austenitic chromium-nickel stainless steels), copper alloys (bronze, brass, and cupro-nickel), nickel alloys (monel, nickel, incoloy, and inconel), aluminum alloys, hastelloys, and titanium (*18*, *20*). The following sections present a summary of generally utilized materials and protective lining for minimizing corrosion changes because of particular corrosion compounds.

Hydrochloric Acid and Inorganic and Organic Chloride

Monel is usually resistant to corrosion through weak hydrochloric acid and is generally utilized in the condensers, distillate drum liners, the top parts of crude towers, and overhead systems of pipe still. Generally, a protective monel cladding is applied at the topmost parts of crude distillation columns. The characteristic corrosion rates reach about 0.002 inch per year for monel.

Organic Sulfur Compounds and H_2S

Linear chromium steel with 12% chromium and about 0.5–1% molybdenum, for example 5,7,9% Cr provides better resistance to this type of corrosion. Along with this, the austenitic stainless steels—for example, 18-Cr-8nickel steels—have outstanding corrosion resistance.

Al, SS, Carpo, and CO_2-Monel

Nickel is better against the corrosion because of CO_2. A wet CO_2 attack on steel can be defended with the addition of 12% chromium. However, PE 304 Stainless Steel is broadly utilized and was found to be most suitable for protecting against corrosion.

Organic Acids

In presence of organic acids, copper alloys are more resistant to corrosion than steel and 5~9 % chromium steel, but type 316 of the Cr-Ni SS is highly resistive to a naphthalic acid corrosive environment (*17*).

Nitrogen Compounds

Plastic lining or stainless steel may be needed for prevention against corrosion caused by nitrogen compounds.

Desalting

Crude oils are complex mixtures containing high level of impurities—for instance, salts, solids, water, and metals. Therefore, impurities can cause adverse reactions such as corrosion, which causes fouling and even problems within the unit. The key function of using a desalter is to eliminate the salts, usually calcium ,magnesium, and sodium, to decrease corrosion. Among various classes of techniques, desalination is generally used, as it has a stronger reaction compared to other techniques such as centrifugal separation and settling. It is also cheaper and less time consuming. Desalination and dehydration can be attained simultaneously with the elimination of salts (approximately > 90%) from crude oil in less than 0.5 hr. The principle in this separation is extremely easy: the polar molecules orient under a charged electric field. A potential of about 20–70kV is applied to the electrodes through which crude oil travels. In this procedure, water forms an emulsion, which then forms a coalescence and agglomeration into a stream, entrapping all the salts. Crude oil floats above the surface, forming a separate steam, and brine accumulates at the base of the desalter (Figure 3).

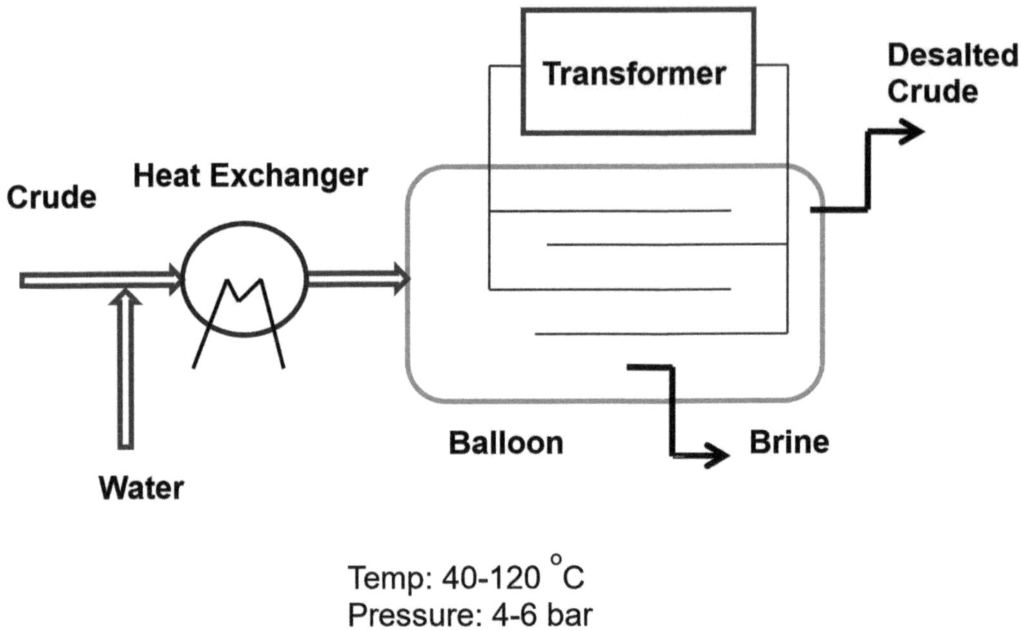

Figure 3. Schematic diagram for electric desalting system. Reproduced with permission from reference (15). Copyright 2019 © IOP Publishing Ltd. Published by IOP Science under a Creative Commons Attribution License 3.0 (CC BY).

The action of the electric desalting system depends on the following main parameters: pressure, properties of crude oil, temperature of desalting, residence time, pH, water wash standard and rate, mixing of wash water, and efficient separation. A pressure of approximately 5–6 kg/cm^2, pH levels of about 5.05–7.0, and a temperature range of about 120–260 °C are needed for successful desalination. Chemical supports such as wetting agents, demulsifiers, and emulsion breakers are very important for this process. The power needed for consumption is quite small, normally about 0.01 KW/hr per barrel. Studies have shown that $Cl-$ content of the desalted crude oil exceeding 20 ppm have the possibility of leading severe corrosion and fouling concern in overhead lines (*21*).

Chemical Dosing

In terms of protection against corrosion, chemical dosing generally depends on the type of crude oil, operating parameters, and tools. The following sections will provide some information on the general methods for chemical dosing.

Demulsifiers or Emulsion Breakers

These chemicals are required to help in the desalting process. They will not only increase the rate of separation, but they will also improve the residence time, enhancing the removal of solids, reducing water carryover or not carrying enough oil, and decreasing the emulsion film. For efficient performance, it should be well dispersed before it reaches the desalting vessel, and it is usually injected into crude charge pump suction. A dose limit of up to about 100 ppm has been suggested. In practice, a dose of Eastern Refinery Limited (ERL) is about 6 ppm (namely, 6kg of separol/1000MT) of crude oil. Overdosing should not be applied because it can risk the reversal of emulsion (*18*).

Caustic Soda Injections

Crude oil includes water and inorganic salts such as NaCl, $CaCl_2$, and $MgCl_2$. Hydrolysis of $MgCl_2$ and $CaCl_2$arises when crude oil is heated in a furnace and preheat exchangers. The hydrolysis of HCl (gaseous) is shown in the following chemical reaction (*13*, *17*):

$$CaCl_2 + H_2O \rightarrow Ca(OH)_2 + 2HCl \rangle 120^oC \quad (5)$$

$$MgCl_2 + H_2O \rightarrow Mg(OH)_2 + 2HCl \rangle 205^oC \quad (6)$$

$$NaCl + H_2O \rightarrow NaOH + HCl \rangle 480^oC \quad (7)$$

The temperature of furnace outlet and crude generally ranges from 360–380 °C in which NaCl is usually stable but most $CaCl_2$ and $MgCl_2$ (about 20–50%) are hydrolyzed. Caustic is generally added to the charging unit of crude oil, the downstream of the desalter, to react with $CaCl_2$ and $MgCl_2$ to prepare more stable NaCl. Caustic also counters with HCl vapor and generates more stable NaCl and collects at the bottoms of distillation column instead of in the overhead system. The accurate quantity of caustic injection is based on the chloride content of the water in the reflux drum of crude oil column overhead. A practical chloride goal is about 10–20 ppm. For a majority of positions, the highest rate of caustic injection is 2.27 kg/1000 bbl crude oil (22). In practice, caustic dose of Eastern Refinery Limited (ERL) is about 5–10 ppm of crude oil (i.e., approximately 0.66–0.32 kg NaOH/1000 bbl of crude oil) (*18*).Overdosing must be avoided because it sustains carbon steel general corrosion, stress corrosion cracking in not post welded heat treated (PWHT) materials such as carbon steel, nickel alloys, and stainless steel.

Neutralizers

Neutralizers are chemicals that are mixed to neutralize acids in condensed water in distillation of refinery units (23–25). Dissolved acidic gas—for example, H_2S and strong HCl—are neutralized to sustain the pH levels of around 5.5 and 7. Corrosion turns very aggressive when the pH values are higher than 7,and fouling occurs at pH values less than 5.5. Furthermore, at pH values greater than 7, amines and ammonia can attack copper alloys—for example, Monel and admiralty brass. Gaseous HCl, produced via the salt hydrolysis, dissolves rapidly in water at the onset of condensation process. The efficient neutralizers to have high solubility in H_2O have been formulated.

As water condenses in the first droplets, it can stop HCl instantly from attacking steel. Neutralizers would include low molecular weight amines—probably one or more to counteract HCl and simultaneously to avoid corrosive salts formation. The paradigms of neutralizers with restricted use are ammonium hydroxide and ammonia gas, which do not have good solubility in water at the onset of water vapor condensation and also produce salts that have undesirable effects on the process unit (*26*). The reaction of neutralization is a common acid base reaction characterized via the following reactions:

$$HCl + NaOH \rightarrow NaCl + H_2O \quad (8)$$

$$HCl + NH_3 \rightarrow NH_4^+Cl^- \quad (9)$$

$$HCl + RNH_2 \rightarrow RNH_3^+Cl^- \quad (10)$$

$$2HCl + H_2N - R - NH_2 \rightarrow {}^-Cl^+H_3N - R - NH_3^+Cl^- \quad (11)$$

The general paradigms of neutralizing amines contain methoxypropylamine (CH_3–O–CH_2–CH_2–CH_2–NH_2; MOPA), ethylamine(CH_3–CH_2–NH_2), ethylenediamine (H_2N–CH_2–CH_2–NH_2; EDA), morpholine ($O(CH_2CH_2)_2NH$) and monoethanolamine (HO–CH_2–CH_2–NH_2; MEA)[26]. Polyamines—for example, EDA—have a better control of pH levels than morpholine or ammonia.

Ammonia is used in acidic component neutralization in the refining industry in such areas as overhead streams. The condensation of HCl in the overhead systems can lead to a strong acid attack if it is not neutralized. pH values of dew point of water as low as 1.5 have been examined in instances in which neutralization was interrupted.

Water Dosing

Water is mixed with the crude oil in the desalter to properly separate salt. Demineralized water is also dosed in the overhead systems to rub NH_4Cl salt and increase the pH levels at the dew point. Additional concerns should be taken to assure that there are separator vessels to support better hydrocarbon and water separation in order to protect water reflux, including salt, back to the crude oil column. Extreme doses of water may affect the grooved lines and erosion problems. Insufficient doses of water can contribute to acid making. Water includes highly dissolved oxygen, which can noticeably increase the corrosion speed. The perfect rate of water injection is approximately 5–10% of the overhead stream. An efficient, trustworthy water washing system should have the following features: plenty of water with 25–50% excess at all times, injection in a position that will distribute the water into all of the bundles, and a maximum pressure drop atomizing spray nozzle to provide better interactions with the liquid and vapor.

Corrosion Inhibitors

Most of the corrosion control programs contain the injection of proprietary film forming organic inhibitors, generally known as "filmers."Film-forming inhibitors are usually organic compounds that have a polar head group joined to an alkyl chain as a tail. The head group generally includes nitrogen and oxygen, and sometimes sulfur or phosphorus. This group connects to the metal's surface, whereas the framework of hydrocarbon forms a film that is hydrophobic in nature that keeps water molecules away and later retards the reactions of corrosion. The term filmers comes from a thin film (which is hydrophobic), in which some molecules are thick. This film thus supports an "oil-wet" surface of the metal. These inhibitors create an incessantly thin replenished film, which acts as a protective barrier for the acid-containing system and other surfaces of metals underneath the film. For the best results, appropriate pH control is necessary, and it has to be maintained in a systematic manner. Usually, it is dosed to the crude oil distillation column overhead immediately prior to the aerocondenser to avoid corrosion of its tubes.

The rate of injection is about 3–5 ppm for the process (*20*). During unit upset or other startup issues in the units, the results of injection may be temporarily raised to stages , for example, 12 ppm to assist in reestablishing the protective film. In the practice of ERL, this dose is about 5–15 ppm of total gasoline(i.e., 0.60~1.7 kg of TG/ 1000 bbl of crude oil) (*18*). Inhibitors are noncorrosive to tools at the dilutions of action dosage, but a concentration of near 100%, they may be corrosive to injection tools. Some examples of amines that form films include alkylquaternary amines, imidazolines, alkylamines, amides, and aminoamides (*19*, *27*).

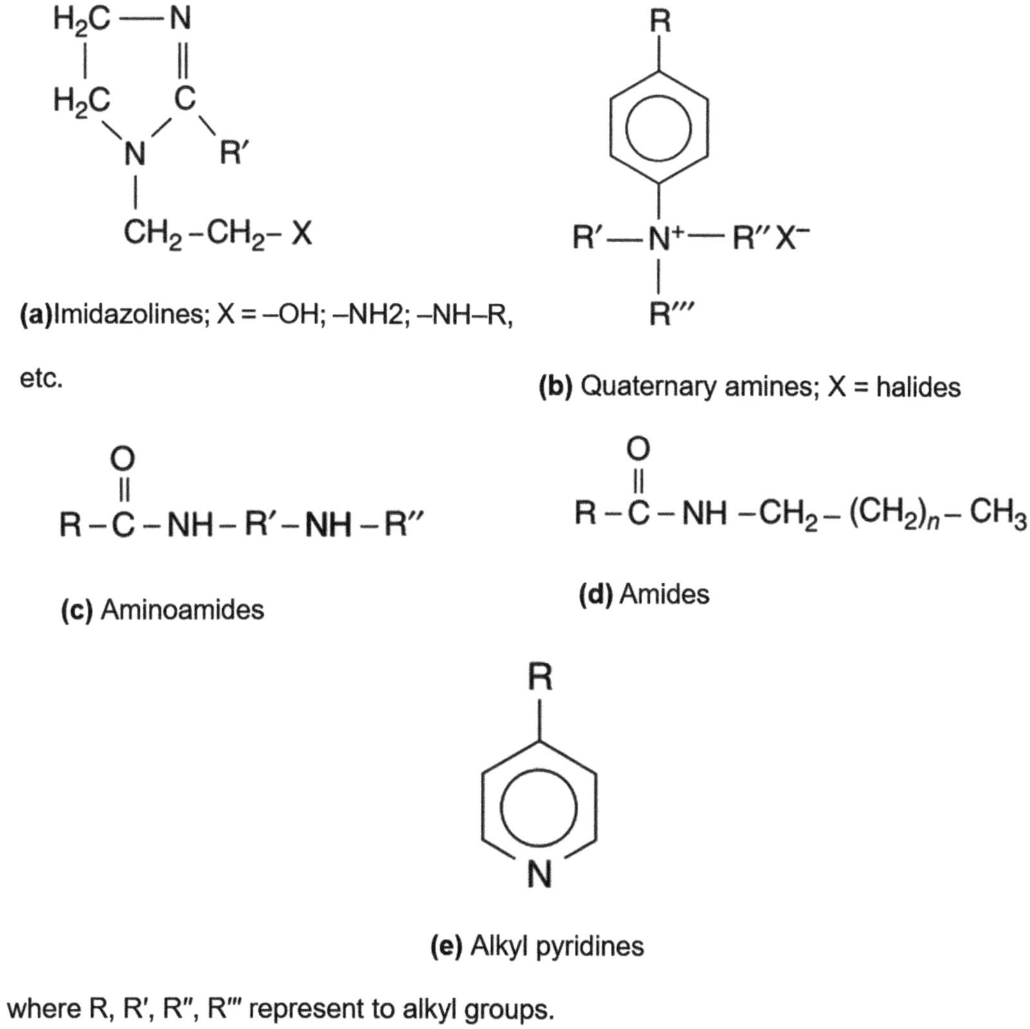

Scheme 1. Chemical name and structures of some common filming amines.

Imidazolines can be formed by the reaction of naphthenic acid with amines (such as dipropylenetriamine). The package of inhibitors usually consists of 20% imidazoline surfactant and 80% kerosene (i.e., formed via the addition of one equivalent part of dipropylenetriamine with two equivalent parts of commercially obtained naphthenic acids). Commercial naphthenic acid has substitute naphthenic ring configurations that can be obtained from crude oil. The applications of these inhibitors are usually in the refinery overheads distillation columns, in other units of refinery, and in the unit of fluid catalytic cracking (FCC) fractionators overhead, having distillation systems. The film-forming amine and neutralizing corrosion inhibitors are injected at the top of the overhead lines to mitigate corrosion. At relatively low pH levels, the film-forming amine turns into a protonated form and is simply desorbed via the water from the metal surface. To facilitate the establishment of a protective film quickly, the inhibitor is primarily attached at 2 to 4 times the suggested rate of dosage. Too much of a dosage of such chemicals creates problem of "gunking" or "fouling" on the surface of the metal. It could also create problem in overhead fluids or processing units by forming tight emulsion. Usually, a package of amines would comprise the corrosion inhibitor, one or more solvents with a combination of a small percent of wetting agents or emulsion breakers. The solution of inhibitors is generally hydrocarbon based, whereas the water-based preparations are sometimes applied. The film-forming amines are also applied in different systems of water washers (for example, in the units of hydroprocessing). Further applications consist of sour water strippers, overhead units of cokers, and amine units. To ensure the efficiency of these chemicals, corrosion rate should be examined constantly using coupons weight loss analysis , electric probes resistance analysis, and water analysis methods.

Hydrogen induced cracking (HIC), blistering, sulphide stress cracking (SSC), and stress oriented hydrogen induced cracking (SOHIC) harm can take place in the refinery where low alloys and carbon steels and wetH_2S atmospheres are present. Water-soluble, amine-based film-forming inhibitors can be applied to prevent the corrosion and blistering of hydrogen in the vapor lines of overhead. These corrosion inhibitors could become extremely corrosive in the hot vapor stream if they are injected while they are undiluted. To examine the efficiency of these types of inhibitors, activity probes of hydrogen are utilized, along with residual corrosion inhibitors analysis in water condensation. Film-persistent and thermally stable inhibitors are applied to inhibit naphthenic acid corrosion (NAC) (*28*). NAC is a high-temperature corrosion form that takes place mostly in vacuums ,crude oil units, and the downstream units that process certain fractions or cuts that contain naphthenic acids. Additionally, because the temperatures are quite high, NAC is susceptible to the effects of velocity, which causes shear stress on high walls. High-temperature NAC inhibitors were applied with moderate achievement; however, the effects of potential detriments on the catalysts' activity downstream must be considered. The effectiveness of corrosion inhibitor under these conditions should be observed carefully. In refineries and petrochemical units, caustic is utilized for neutralizing the acid content. Corrosion by caustic could happen in crude distillation unit (CDU)s because the insertion of 40 wt% of caustic into hot, desalted transfer lines of crude oil. Amine corrosion takes place in units (gas-treating or recovery units) of sulfur, where corrosion is really caused via the dissolved CO_2 or H_2S and via degradation amine products. The system consisting of alkanolamine can be rated from higher to lower in the order of their aggressiveness as follows: monoethanolamine> diglycolamine > diisopropylamine > diethanolamine > methyl diethanolamine Localized or general corrosion, often pitting, usually happens under amine salt or ammonium chloride deposits—frequently when there is no free phase of water. MOPA and MEA chloride salts are liquids at general operating situations and are thus chosen over amines that generate solid deposits (*18*). The hygroscopic salts of ammonium chloride and a little amount of water can notably aggravate the corrosion (>2.5 mm/yr). Carbon steel is the material of choice for the construction of amine units. While corrosion inhibitors of film-forming amine are not efficient in these cases, for this exclusive oxidizing corrosion, inhibitors are applied. If the scale or deposit control or magnetite protection treatment plan in boilers does not decrease attacks by CO_2, then amine inhibitors might be required in the return system of condensate. Pitting corrosion occurs via the H2CO3 and general corrosion results from dissolved CO_2 in water. Inhibitors can be used to decrease the corrosion in the systems of steam condensation. Furthermore, vapor phase inhibitors might be needed to prevent against condensing vapors. Stress corrosion cracking by carbonate refers to cracks or surface breaking that takes place near by to metal welds under the joint tensile stress action and destruction of carbonate-containing systems caused by corrosion. In warm carbonate systems in CO2 elimination units, an inhibitor comprised of meta vanadate can be applied to avoid the cracking.

Some other locations in which corrosion inhibitors could be applied to inhibit the corrosion are in: (i) rotating tools to protect corrosion fatigue, (ii) surroundings(including carbolic acid or phenol)where corrosiveness appears

Monitoring and Measuring Corrosion

In refining units, most severe corrosion arises during the startup phase of the crude oil refining plant and unstable crisis blackout phase. Around 90% of crude oil unit overhead corrosion happens during just 10% of the operating time. This unstable operation phase may take place during the switching of crude oil tanks, processing of slop oil, opportunity crudes processing, or other disruptions to regular processes. The following ways are applied to ensure effectiveness and efficiency of crude unit corrosion control plans:

(i) Water examination: for hardness, chloride, pH, and metals.
(ii) Hydrocarbon examination: residual, inhibitors, and metals.
(iii) The measurement of corrosion rate: via coupons loss in weight, linear polarization resistance probes, and electrical resistance probes.
(iv) On stream nondestructive analysis: via ultrasonic thickness gauging and radiography tests.

Recommended Protective Action

To reduce corrosion influences in crude oil refineries, the following successful, efficient protective actions can be utilized:

(i) Determine the nature of crude oil (mainly sulfur content, chlorine salts, and number of acids) for the purpose of expanding an accurate strategy of corrosion management.
(ii) Maintain proper operating process parameters such as pressure, temperature, pH level, and water content, which helps in controlling corrosion without breaking the key process system.
(iii) Apply chemical dosing entirely without cut off. During unit issues or startup, injection of corrosion inhibitors may be provisionally raised to higher levels than normal.
(iv) Ensure efficient crude oil desaling and suitable selection of materials.
(v) Efficient corrosion analysis techniques should be installed in the correct location, and corrosion trends should be examined to enforce essential protective action.

Conclusion

Determining the rate and form of corrosion and then taking action based on advice is insufficient. Without correct, regular, and appropriate testing, corrosion incidents may be totally neglected or revealed only after severe damage has taken place. Thus, protective actions should be regular. To ensure maximum effectiveness, the efficiency and consistency of corrosion management program should be maintained in an economical manner. This chapter has a vast scope. It explains how corrosion experts can achieve enhanced flexibility in corrosion control customization with minimum effort. This chapter covers the integral fractions of all processes, their performance, and the difficulties that can occur, as these are significant matters.

References

1. Speight, J. G.; Ozum, B., Eds. *Petroleum Refining Processes*; CRC Press, 2001.
2. Hsu, C. S.; Robinson, P. R., Eds. *Practical Advances in Petroleum Processing*; Springer-Verlag New York, 2007.
3. Gary, J. H.; Handwerk, J. H.; Kaiser, M. J.; Geddes, D., Eds. *Petroleum Refining: Technology and Economics*; CRC Press, 2007.
4. Speight, J. G., Ed. *The Chemistry and Technology of Petroleum, 5th ed*; CRC Press, 2014.
5. Speight, J. G., Corrosion in Refinery Units. In *Oil and Gas Corrosion Prevention*, Speight, J. G., Ed.; Gulf Publishing Professional Publishing, 2014, pp 39–66.
6. Kane, R. Corrosion in Petroleum Refining and Petrochemical Operations. *Corrosion: Environments and Industries* **2006**, *13*, 967–1014.
7. Dettman, D.; Li, N.; Wickramasinghe, D.; Xu, Z.; Chen, X. N.; Elliott, G. R.; Luo, J. *The Influence of Naphthenic Acid and Sulfur Compound Structure on Global Crude Corrosively Under Vacuum Distillation Conditions Heather, CORROSION 2012, OnePetro*, 2012, paper no. NACE-2012-1326.
8. Tebbal, S.; Kane, R. *Methodologies for Assessment of Crude Oil Corrosivity in Petroleum Refining, Industrial Corrosion and Corrosion Control Technology*, 1996, pp 225–232.
9. Tebbal, S.; Kane, R. D. *Assessment of Crude Oil Corrosivity, CORROSION 98, OnePetro*, 1998, paper no. NACE-98578.
10. Tebbal, S.; Kane, R. D.; Yamada, K. *Assessment of the Corrosivity of Crude Fractions from Varying Feedstock, Corrosion97, OnePetro*, 1997, paper no. 97498.
11. Tebbal, S. *Critical Review of Naphthenic Acid Corrosion, CORROSION 99, OnePetro*, 1999, paper no. NACE-99380.
12. Ayello, F.; Robbins, W.; Richter, S.; Nešic, S. *Crude Oil Chemistry Effects on Inhibition of Corrosion and Phase Wetting, CORROSION 2011, OnePetro*, 2011, pp 1–19, paper no. 3149.
13. *Condition Causing Deterioration of Failures, Guide for Inspection of Refinery Equipment*; Washington, DC, 1973, pp 8–15.
14. Bhowmik, P. K.; Hossain, M.; Shamim, J. A. *Corrosion and Its Control in Crude Oil Refining Process, 6th International Mechanical Engineering & 14th Conference Annual Paper Meet (6IMEC&14APM)*; Dhaka, Bangladesh, 2012, pp 1–5.
15. Akpanyung, K. V.; Loto, R. T.; Fajobi, M. A. An Overview of Ammonium Chloride (NH4Cl) Corrosion in the Refining Unit. *J. Phys. Conf. S.* **2019**, *1378*, 022089DOI:10.1088/1742-6596/1378/2/022089.
16. Gutzeit, J.; Merrick, R.; Scharfstein, L. Corrosion in Petroleum Refining and Petrochemical Operations. *ASM Handbook* **1987**, *13*, 1262–1287.
17. Nugent, M. J.; Dobis, J. D. *Experience with Naphthenic Acid Corrosion in Low TAN Crudes, CORROSION 98, OnePetro*, 1998, paper no. NACE-98577.
18. *Operation Guidelines and Process Operation Data of Eastern Refinery Limited, Chittagong, Bangladesh.*
19. Al-Janabi, Y. T. *Corrosion Inhibitors for Refinery Operations, Corrosion Inhibitors in the Oil and Gas Industry*, 2020, pp 255–269.

20. Bagdasarian, A.; Feather, J.; Hull, B.; Stephenson, R.; Strong, R. *Crude Unit Corrosion and Corrosion Control, National Association of Corrosion Engineers Annual Corrosion Conference*, NACE, 1996, pp 1–14.
21. *Damage Mechanisms Affecting Fixed Equipment in the Refining Industry, Recommended Practice571*, 1st ed. API Publishing Services; Washington, DC, 2003, pp. 5–65.
22. Al-Omari, A. S.; Al-Zahrani, A. M.; Lobley, G. R.; Tems, R. D.; Dias, O. C. *Refinery Caustic Injection Systems: Design, Operation, and Case Studies, CORROSION 2008, OnePetro*, 2008, paper no. 08551.
23. *N.P. 34109, Crude Distillation Unit—Distillation TowerOverhead System Corrosion*, 2009, pp. 1–85.
24. Fajobi, M.; Loto, R.; Oluwole, O. Corrosion in Crude Distillation Overhead System: A Review. *J. Bio Tribo-Corros.* **2019**, 5, 1–9.
25. Wills, K. A.; Sarpong, K. O. Survey on Crude Unit Overhead Corrosion Control Practices, NACE International Corrosion Conference. *Proceedings*; NACE International 2019, 1–16.
26. Petersen, P. R.; De Jong, A.; Sigmon, J. L.; Minyard, W. F. *Impact of Ammonium Chloride Salt Deposition on Refinery Operations, CORROSION 2001, OnePetro*, 2001, paper no. 01540.
27. Petersen, P. R. *The Use of Corrosion Inhibitors in the Refining Industry, CORROSION 96, OnePetro*, 1996, paper no. NACE-96594.
28. Saji, V. S. A Review on Recent Patents in Corrosion Inhibitors, Recent Patents on. *Corros. Sci.* **2010**, 2 (1), 6–12.

Chapter 11

Corrosion Inhibitors for High Temperature in Oil and Gas Industries

Saman Zehra,[1] Mohammad Mobin,[1,*] and Ruby Aslam[1]

[1]Corrosion Research Laboratory, Department of Applied Chemistry, Faculty of Engineering and Technology, Aligarh Muslim University, Aligarh, Uttar Pradesh-202002, India

[*]Email: drmmobin@hotmail.com.

As corrosion concerns have become more prevalent, particularly in the oil and gas sectors, corrosion scientists have devised various solutions, including the applicability of corrosion inhibitors. Field operators, pipeline engineers, and designers must be corrosion-aware to maintain a smooth and continuous flow of oil and gas to end customers. High-temperature corrosion can be seen in many refining units, but it is most severe in atmospheric and vacuum distillation units, with temperatures ranging from 175 to 400 °C. Pressure and aggressive gases, such as CO_2 and H_2S, and organic acids accelerate high-temperature corrosion here. High temperatures can speed up the corrosion process and cause inhibitors to desorb, diminishing inhibition effectiveness. To control high-temperature corrosion in oil and gas, the properties of the inhibitors must be selected and upgraded to match the demanding environment. This chapter offers a comprehensive assessment of the risks and hazards related to high-temperature corrosion in the oil and gas sectors and their mitigation through the use of corrosion inhibitors.

Introduction

Surface science, metallurgy and materials science, electrochemistry, thermodynamics and kinetics, mechanics, hydrodynamics, and chemistry are all needed to understand corrosion, which is a stochastic, probabilistic phenomenon. Because of corrosion, the oil and gas sectors lose tens of billions of dollars as lost income and treatment expenses every year. Corrosion experts have devised numerous strategies to alleviate the problem as it has become more prevalent, particularly in the oil and gas sectors. Corrosion inhibitors are used substantially in the oil and gas industry. Corrosion issues in this domain are influenced by many sources, including H_2S, CO_2, brines, O_2, naphthenic acid, and high-temperature activities. Corrosion inhibition in this area is more intricate than in any other sector, necessitating the use of specific inhibitors based on the location of application, such as pipelines, well heads, refineries, and recovery units (*1*). High-temperature corrosion inhibitors have been among the most complicated to develop because of the complex composition of the field and the high rate of reactions that occur at high temperatures.

Furthermore, because of the complex assessment techniques, selecting relevant inhibitors and laboratory simulations for assessing the intrinsic inhibitor function at high temperatures is quite challenging (*2*). Thus, despite the significant need for inhibitors in this industry, attempts toward evaluating corrosion inhibitors at high temperatures have been restricted. Moreover, complications in understanding the mechanics of high-temperature corrosion in order to match it with proper inhibitors do not assist the issue (*3, 4*). To address the inherent difficulties in assessment, substantial research has already been accomplished in the last several decades in understanding the mechanisms by which metals corrode at high temperatures, as well as in developing a variety of inhibitors for the abatement of the problems pertaining to high-temperature Corrosion (*5*). Several successful high-temperature corrosion inhibitors have been patented (*6–9*).

In the production of oil, the corrosion of steel and other metal well equipment is caused by the action of certain types of S-bearing waters, dissolved CO_2, and natural brines. Many methods have been employed to keep such equipment from corroding. In a well extending into a producing formation, it is standard practice to introduce an inhibitor into the annular region between the casing and the tubing at the well head. Liquid inhibitors that have been used are expected to flow to a producing zone in the well to mix with the corrosive liquids flowing from that area and then flow up through the tubing to the surface. In the past, the temperature in the well where the inhibitor was used has reached as high as 121 °C. However, because of the exhaustion of main wells and other natural processes, a large number of wells now operate at temperatures well above 121 °C , ranging from 148 to 287.7 °C. These higher temperatures have caused the breakdown of corrosion inhibitors. Therefore, the present compositions of corrosion inhibitors and the methods for their use are designed to be effective in the high-temperature operation of gas and oil wells.

The primary emphasis of this chapter will be on the varied operating circumstances, mechanisms, and improvements in the formulation of high-temperature corrosion inhibitors in the oil and gas sectors. High-temperature corrosion occurs in many refining units, although it is most intense in atmospheric and vacuum distillation units where temperatures range from 175 to 400 °C (*10*). Pressure and aggressive gases, such as CO_2 and H_2S, and organic acids exacerbate high-temperature corrosion.

High-Temperature Corrosion and Mechanistic Understanding

Chemical Corrosion driven by gases, solid or molten salts, or molten metals at temperatures exceeding 400 °C is referred to as high-temperature corrosion. In a hot corrosion environment, the main chemical reactions that deteriorate the protective scales on the metal surface can be categorized on the basis of the nature of the stable products of corrosion formed in the presence of hot gases on the surface of the metal. These chemical reactions include oxidation, sulfidation, carburization, chlorination, and nitridation.

When the stable product is an oxide, the mechanism is called oxidation; when the stable product is a sulfide, the mechanism is called sulfidation. Carbides are produced by carburization, chlorides by chlorination, nitrides by nitridation, and, when a combination of sulfides and oxides is produced, the process is called sulfidation–oxidation. Other methods include ash production, deposits, molten salt formation, H embrittlement, and decarburization (*11*). These mechanisms are explained in the following text to provide clear insight into the selection and formulation of inhibitors for corrosion at high temperatures.

High-Temperature Oxidation

Oxidation often takes place in the presence of oxidizing gases when metals are exposed to temperatures exceeding 300 °C. Common ways to confirm oxidation are either by X-ray study using diffraction of the surface area or by gas composition analysis. High-temperature oxidation is a type of degradation caused by the exposure of metal structures to oxidative high-temperature environments with hot gases at greater than 300 °C and which leads to the development of oxides on the metal surface. Films are considered thin layers of oxide (often thinner than 3000 Å). Scales are thick oxide layers (thicker than 3000 Å) (1.0 Å = 10^{-10} m). The generated oxide layer tends to protect the base metal against further oxidation. However, in a high-temperature environment, the problem is always worsened by an unequal thermal expansion between the metal and its oxide film, resulting in film cracking.

If the partial pressure of O_2 (pO_2) is much less than the oxide disintegration pressure of a given metal, no oxide layer develops. Such circumstances are typical of noble metal oxide films (e.g., Au, Pt, Ag), which are unstable at temperatures below 932 °F (500 °C). The oxides of some metals (such as Mo and Os) are volatile, and the oxide molecules evaporate from the oxide layer surface in tandem with the oxidation process. At a certain thickness of the oxide layer, the rates of the two processes (volatilization and oxidation) equalize. Many layers of various oxides can make up the oxide scale. At temperatures greater than 1050 °F (566 °C), Fe scale has three layers: FeO (ferrous oxide; the closest layer to the Fe), Fe_3O_4 (magnetite; the intermediate layer), and Fe_2O_3 (ferrite; the outermost or surface layer). The mechanics of high-temperature oxidation can be represented by the equations below, using an Fe surface as an example. When exposed to atmospheric O_2 at temperatures between 700 and 900 °C, Fe oxidizes to FeO, Fe_3O_4, and Fe_2O_3 (*12*):

$$O_2 + 2Fe \rightarrow 2\,FeO \qquad (1)$$

$$O_2 + 2FeO \rightarrow 2Fe_3O_4 \qquad (2)$$

$$\tfrac{1}{2}\,O_2 + 2\,Fe_3O_4 \rightarrow 3Fe_2O_3 \qquad (3)$$

Further oxidation can occur if O_2 penetrates the coating and reaches the metal surface. The rapid pace of reaction caused by the high temperature facilitates and complicates the reaction of O_2 with the metal surface.

High-Temperature Sulfidation

Sulfidation (sulfidic) corrosion occurs when a metal reacts with S compounds in high-temperature settings, resulting in the formation of a surface sulfide scale, with S penetrating slightly beyond the initial thickness. In the corrosion literature, sulfidation is also known as sulfidic corrosion. Sulfidation is sometimes referred to as H_2–H_2S Corrosion when H_2/H_2S is present. Wherever S compounds are present in a hydrocarbon stream and the temperature surpasses roughly 230 °C, sulfidation corrosion can occur. In the absence of hydrocarbon, H_2–H_2S Corrosion can develop. Sulfidation corrosion is caused by two distinct and discrete processes. One happens in various refining processes, such as hydrotreating and hydrocracking, when H_2 is present in addition to the sulfidation-causing S species. The other arises when there is no H_2 (i.e., the system is H-free) in processing units that do not need H_2 as part of the operation. Both are diffusion-based nonaqueous corrosion processes that occur at high temperatures. The precise temperature threshold for H-free sulfidation is a source of controversy in the industry; the API 571 utilized a temperature threshold of 260 °C for H-free services. For H-free sulfidation services without the impact of naphthenic acid corrosion, experience has shown that working temperatures below 260 °C cause no significant corrosion. The temperature barrier for H_2-containing services is established at 230 °C.

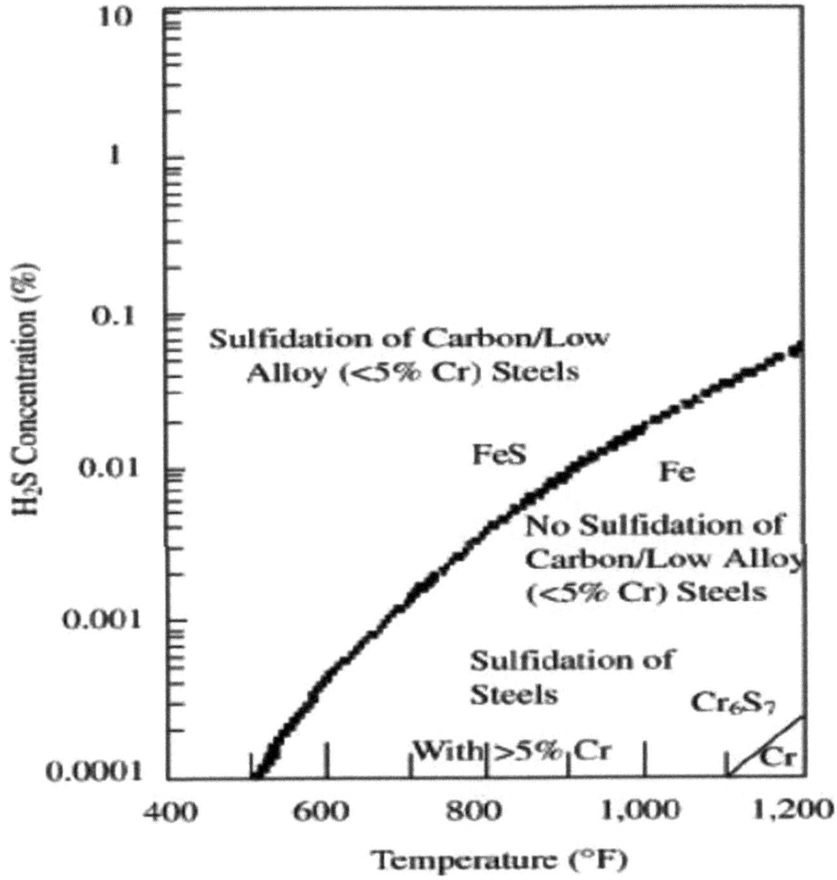

Figure 1. H_2S concentrations in H_2/H_2S gases and temperatures above the FeS/Fe line for C/low-alloy steels and above the Cr6S7/Cr line for alloys with >5% Cr. 204–648 °C. Reproduced with permission from reference (13). Copyright 2004 SciELO Brazil.

Because there are so few "sweet" refineries left in operation, high-temperature sulfidation is perhaps the most common high-temperature corrosion problem in the refining sector. Sulfidation corrosion is an issue in "sour" oil services, commencing at temperatures of around 260 °C. Except in the most sour (oils containing H_2S or other active S compounds) situations, C steels are frequently suitable up to that point. Although sulfidation corrosion rates are quite predictable and broad, corrosion rates are significantly less predictable in sour oil boiling and catalytic cracking activities, and they occur at higher rates and are more localized when sulfidation occurs in hydroprocess operations. When there is no H present, alloying with Cr improves resistance to sulfidation, nearly in direct proportion to the Cr level. Lower Cr alloys are less effective in resisting sulfidation when H is present.

When H is not present, the 300 series austenitic stainless steels with greater than 18% Cr exhibit significant sulfidation resistance and are frequently the material of choice under the more severe sulfidation conditions, corroding at only 1–2 mils penetration per year (mpy) even up to 750°F (400 °C). Austenitic stainless steels, on the other hand, are not a solution for sour oil process equipment owing to their vulnerability to chloride cracking at lower temperatures when moisture is present.

Total penetration measurement for a typical corrosion product, depicted graphically for steels with the morphological changes progressing from lower left to upper right, can be seen in Figure 1. During sulfidation in the presence of H_2/H_2S gases, this line represents the limiting conditions for the formation of FeS, the primary corrosion product of C/low-alloy steels (less than 5% Cr). H_2S concentrations along the line encourage the formation of FeS, which is necessary for the sulfidation of Fe and low-alloy steels. It should be emphasized that implying that alloys with substantial Cr concentrations cannot corrode at circumstances corresponding to lower H_2S concentrations and higher temperatures, as shown by this line, is erroneous. CrS, which develops at lower H_2S concentrations and higher temperatures than FeS on low-alloy steels, can develop in alloys with Cr. The limitations of H_2S concentration and temperature related to Cr corrosion products are shown in Figure 1. In settings where low-alloy steels cannot corrode, alloys having more than 5% Cr can corrode. Corrosion rates may be moderate (e.g., 0.025–0.25 mm per year), but they are substantial in environments where steels have historically been assumed to be resistant.

High-Temperature Carburization

High-temperature carburization occurs when metals are exposed to temperatures exceeding 700 °C in the presence of gases such as CO and CH_4, as well as hydrocarbons and solid carbons. This results in the creation of carbides that grow in size as the temperature rises. In order to carry out the reactions of carburization, at least a source of C, such as atmospheric CO is needed, or we can say it is the type of high-temperature corrosion occurring in the C-rich environment, as shown below:

$$CO + H_2 \rightarrow H_2O + C \text{ (in equilibrium with } H_2\text{–}CO\text{–}H_2O \text{ mixture)} \quad (4)$$

$$C + 3Fe \rightarrow Fe_3C \quad (5)$$

$$Fe_3C \rightarrow 3Fe + C \quad (6)$$

It is worth noting that when the concentration of H_2S rises, the generation of carbides or carburization decreases. Sulfidation will occur instead of carburization as the concentration of H_2S rises to a crucial level for the synthesis of sulfides. At 300-ppm H_2S, sulfidization occurs instead of carburization in an ethylene furnace containing hydrocarbons at temperatures ranging from 980 to 1090 °C. Other variables—such as partial pressures of H_2, CH_4, and H_2S; the exposure period; and the alloy type—are likely to affect the rate of carburization, in addition to temperature (*4*, *13*, *14*). With rising temperature, gas penetration into alloys is significantly faster than in pure metals. Carburization of certain alloys exposed to C and 2×10^{-4} atmospheric (atm) H_2S after 1 y could be seen in Figure 2.

Carburization frequently has a parabolic time dependency. Carburization rates measured after periods of only hundreds of hours (as is often the case for available data) may not be useful in estimating carburization corrosion rates for long-term service, implying that data collected after only hundreds of hours may not be useful in estimating carburization corrosion rates for long-term service.

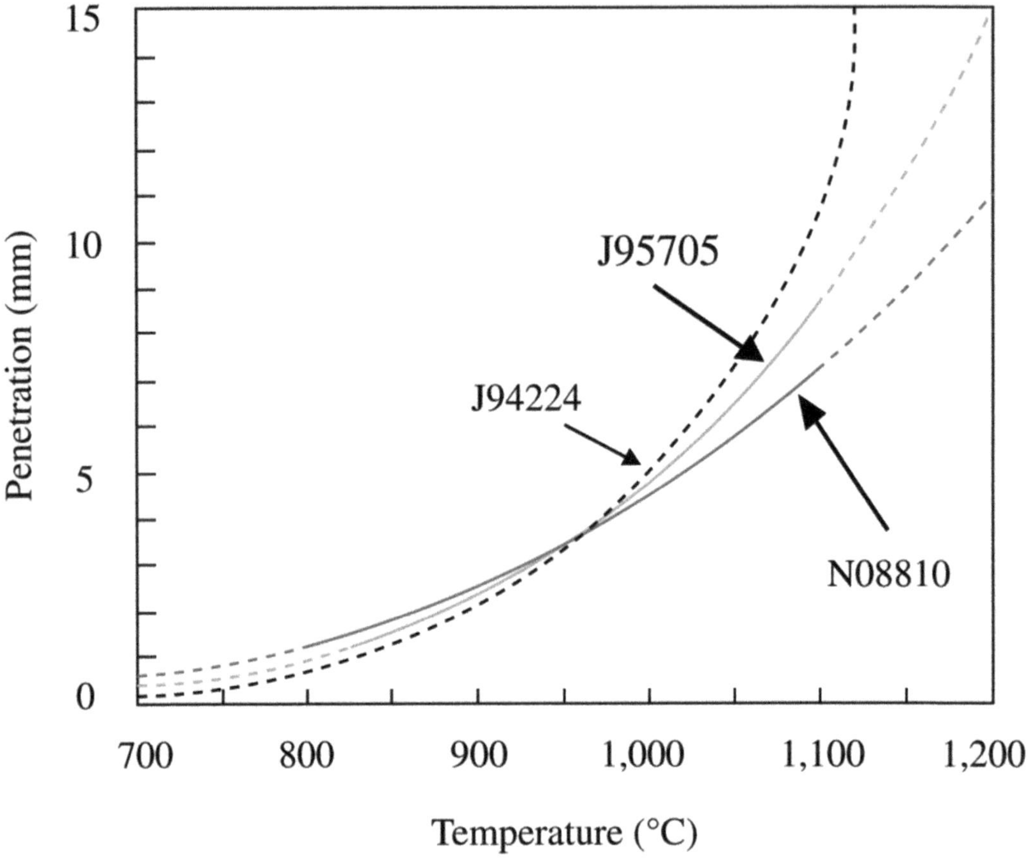

Figure 2. After 1 y, the effect of temperature on carburization of many alloys exposed to solid carbon and 2×10^{-4} atm H_2S. Reproduced with permission from reference (13). Copyright 2004 SciELO Brazil.

High-Temperature Chlorination, Nitridation, and Sulfidation–Oxidation

Chloride-induced high-temperature corrosion that results in the development of surface metal chlorides, as well as chlorine-containing impurities, is known as high-temperature chlorination. In the oil and gas industry, HCl and Ch^- play a key role in corrosion, which is exacerbated by the presence of NaCl in seawater. The rate of corrosion rises when the concentration of HCl is more than 500 ppm at 500 °C (*14*). For the chlorination process, the Ch^- ion source is HCl:

at 676 °C,

$$Cl_2 + Fe \rightarrow FeCl_2 \quad (7)$$

at 303 °C,

$$Cl_2 + 2FeCl_2 \rightarrow 2FeCl_3 \quad (8)$$

When a metal, such as Cr, reacts with N at 950 °C and produces nitrides as a surface corrosion product in the presence of a small quantity of O_2, high-temperature nitridation occurs. Because of the poor reactivity of N, nitridation occurs at a slower rate than oxidation in the presence of a tiny amount of O_2. Porosity of the surface scales as an indication of N penetration has been recognized as a sign of nitridation on Cr alloys. The N in the reactive gases in Cr-based alloys, depending on the pO_2, leads to the synthesis of Cr_2N at high temperatures (*15*):

at 950 °C,

$$1/2\ N_2 + 2Cr \rightarrow Cr_2N \quad (9)$$

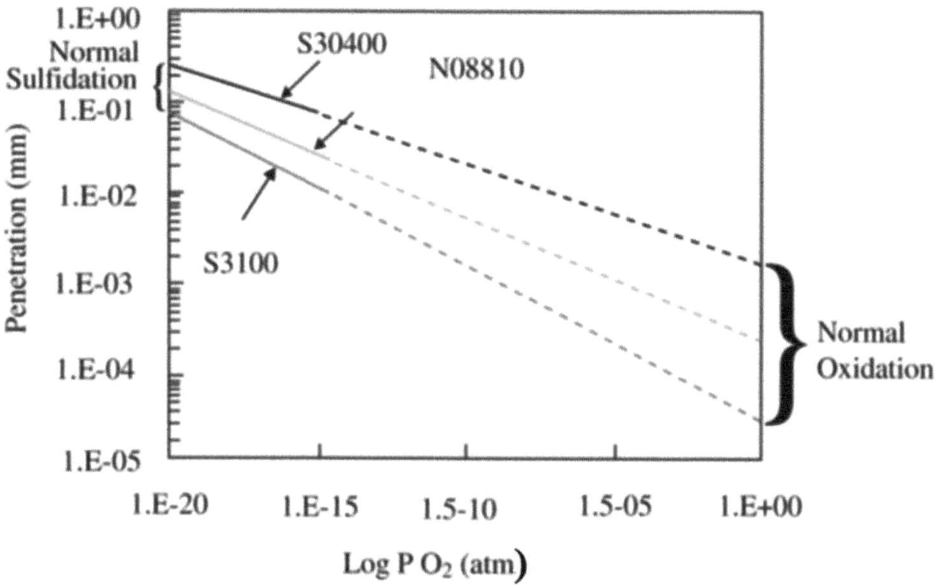

Figure 3. After 1 y, the effect of H_2O or CO_2, as indicated in the PO_2, on the extent of sulfidation/oxidation of a variety of alloys at 700 °C in H_2–H_2S–H_2O gases. Reproduced with permission from reference (13). Copyright 2004 SciELO Brazil.

Sulfidation–oxidation is a type of high-temperature corrosion that arises when aluminum or Cr alloys are exposed to hot gases comprising specific combinations of CO, CO_2, COS, H_2, H_2O, and H_2S gases and contributes to the development of mixed sulfides and oxides on the alloy surface. Oil and gas equipment—such as hydrocrackers, hydrotreaters, coal/coke/oil gasifiers, crude drilling columns, vacuum flashers, flexicokers, and S-removing plants—are particularly susceptible to this form of corrosion (*16*). The presence of oxidizing gases, such as H_2O or CO_2, lowers the sulfidation rate down below what would be predicted if only H_2S/H_2 was present. This is significant because, owing to exposure to liquid water, gases that are supposed to contain exclusively H_2S/H_2 often contain some H_2O. If saturated with water at room temperature, a gas exposed to water at room temperature (such as a water wash in a process) may include up to 2% water in the gas. This slower rate of corrosion can be viewed as a transition between rapid sulfidation and gradual oxidation corrosion. Based on the analysis methods previously reviewed in this chapter, this is shown in Figure

3 for S30400, S3100, and N08810 at 700 °C in a gas H_2S–H_2–H_2O mixture. The right-hand pO_2 represents air, and the left-hand pO_2 represents O_2 depletion (and sulfidation). The rate of oxidation in O_2-containing gases is the lowest, whereas the rate of sulfidation in H_2S/H_2 gases is the highest.

High-Temperature Corrosion Due to the Formation of Ash, Deposits, and Molten Salts

Molten salt corrosion occurs when Na, S, or V oxides, such as Na_2O, SO_3, and V_2O_5, react with the protective surface oxide coating on a metal substrate at high temperatures, usually around 900 °C. Low-grade petroleum in the oil and gas business contains a certain quantity of NaCl, S, and V. Low-grade marine residual fuels, for example, may contain up to 0.2 weight percent (wt %) ash, 5 wt % S, and 0.06 wt % V, whereas low-grade marine distillate fuels may contain up to 0.05 wt % Na, 2 wt % S, and 0.01 wt % V. These elements are oxidized to their corresponding oxides in the oil and gas industry's high-temperature O_2-rich combustion units, and the oxides then condense and interact as corrosive agents with the surface oxides on industrial metallic structures, resulting in a high rate of corrosion and a shortened service life for the structure (*12*).

Corrosion Inhibitors for Oil and Gas

Since the 1980s, an increasing demand for high-temperature corrosion inhibitor materials has been observed (*9, 12*). The reasons include increasing drilling depths along with increasingly hostile drilling environments. Thus, the corrosion inhibitor systems have to become more complex and more efficient. At present, only a very few corrosion inhibitor formulations for high-temperature applications are available on the market to serve this demand, and most products start to fail at temperatures higher than 250 °C. The most effective type of corrosion inhibitors for these applications are film-forming amines and their salts. A wide variety of inhibitor formulations are available. However, most of these inhibitors are produced from only a few basic types of starting molecules. The employment of some inhibitors (patented and non-patented) in various high-temperature units in the oil and gas sectors, where distinct mechanisms of high-temperature corrosion occur, is reviewed next.

Corrosion Inhibitors for High-Temperature Acidization or Well Acidization

It is common knowledge that wellbore acidizing is a difficult task in the oil and gas industry. The use of hot acids to clear the scales and rocks so that the oil and gas from the wells may be easily accessible is a widespread procedure. HCl acid alone, or in combination with HF (i.e., HF + HCl), is often used in the provided method. These particular acids are used because of their high acid concentration, which allows them to dissolve rocks and other obstacles quickly. However, this method causes fast corrosion of various components, including casing, tubing, and downhole equipment. This situation is exacerbated in high-temperature oil wells. As a result, acidification is a less-than-successful procedure that can result in serious issues. Deep wells can operate at temperatures of up to 200 °C or more owing to the depletion of main wells.

When the applied temperature is 100 °C or higher, organic inhibitors that are both environmentally acceptable and thermally stable can be used at high concentrations. However, they have difficulties operating at temperatures of up to 260 °C because of breakdown, even at high concentrations. Organic inhibitors, despite this, have been effectively employed in a number of applications. Although they are not totally environmentally benign, acetylenic alcohol, heterocyclic N compounds, aldehydes (aromatic), condensation products (amines and carbonyls), quaternary salts, and other inhibitors and formulations are efficient in the acidification process. As a result, researchers are seeking more environmentally friendly organic acidification inhibitors that are also biodegradable and thermally stable.

Abdel-Ghany et al. (*17*) evaluated two inhibitor formulations—I (made mostly of coal-tar distillation products) and II (consisting primarily of coal-tar distillation products)—at 65 and 90 °C in order to inhibit mild steel corrosion in 15 and 20% HCl solutions (derived from an aqueous extract of ginger). They used a variety of ways to assess their performance. The weight loss after 6 h of immersion in acid solution indicated a drop in corrosion rate and an increase in inhibition efficiency as the concentration of inhibitors I and II increased. In a 15% HCl solution, inhibitor I displayed the lowest corrosion rate values of 1.867 $g/m^2/h$ (at 65 °C) and 22.75 $g/m^2/h$ (at 90°C). At 65 and 90 °C, the values for the comparable blank were 136.03 and 233.667 $g/m^2/h$, respectively. At 65 °C, acquired inhibitor efficiency reached 99.3%, whereas at 90 °C, it reached 96%. There was no significant difference in efficiency between the two formulations when the HCl concentration was raised to 20%. When the findings were compared with those obtained with currently available commercial inhibitors, it was discovered that the two formulations were more powerful.

Fingar and Jackson (*18*) evaluated the efficacy of 3-phenyl-2-propyl-1-ol as a possible inhibitor against API J55 oilfield tubing corrosion caused by the HCl acidization process. The tested inhibitor (3-phenyl-2-propyl-1-ol) has the unusual characteristic of dehydrating onto the steel surface and forming 4-hydroxypropiophenone. This dehydrated product is adsorbed onto the surface of the steel and converted into phenyl vinyl ketones by losing additional water molecules. The generated phenyl vinyl ketones oligimerize to create an adherent coating on the steel's surface.

The corrosion prevention effectiveness of synthetic Fe_3O_4 and ferrous gluconate for oil-well steel (N-80) in a range of HF concentrations at temperatures of 66, 135, and 177 °C was reported by Amosa et al. (*19*). The ferrous complex, which had a 99.2% inhibition effectiveness at 177°C, was shown to be an excellent corrosion inhibitor compared with synthetic Fe_3O_4, which had a 75.1% inhibition efficiency. For acidizing oil and gas fields at relatively high temperatures, a number of benign organic corrosion inhibitors have been investigated (*20*, *21*).

Abayarathna et al. (*22*) published a study in which they assessed several imidazoline-based corrosion inhibitor formulations for C1018 steel in a 3.3% NaCl solution using a schematic experiment. The solution was saturated with 150 pounds per square inch (psi) pCO_2 at 135 °C. To make the system more aggressive, it was purposely contaminated with 10 g/L elemental S or 30 psi H_2S. Based on weight loss experiments at 2000 revolutions per minute (rpm) in a rotating cylinder autoclave, a corrosion inhibition effectiveness of 500 ppm concentration was determined. The corrosion rate decreased from 256 mpy in the absence of an inhibitor to 27 mpy in the presence of 500 ppm of inhibitor, despite the formulation's sufficient inhibitory effectiveness in the presence of CO_2 and H_2S. The formulation performed poorly in the presence of elemental S, with a corrosion rate of 153 mpy and significant localized corrosion.

Schauhoff and Kissel (*23*) patented the use of trimer acid salts of primary amines as an inhibitor formulation. The formulations included alkyl sulfonate and a mixture of dispersants, including lesser alkanols and highly aromatic solvents (e.g., benzene, toluene, xylene), as well as a small amount of antifoam or water in an oil emulsifier. This mixture was shown to be especially effective in avoiding corrosion in sweet oil wells (those that do not contain mercaptans or H_2S). Also effective were alkylbenzene sulfonic acid salts of partial amide of organic polyamines and carboxylic acids. This can be prepared as a combination of polyalkylene polyamine (an alkylbenzene sulfonic acid) plus dimer and trimer fatty acids.

Quraishi and Jamal (*24*) investigated the anticorrosive characteristics of three long-chain fatty acid oxadiazoles: 2-undecan-5-mercapto-1-oxa-3,4-diazole (A); 2-heptadecene-5-mercapto-1-oxa-3,4-diazole (B); and 2-decene-5-mercapto-1-oxa-3,4-diazole (C). Under the same conditions, inhibition experiments were performed on N80 steel in 15% HCl with 5000 ppm A. These findings demonstrated that oxadiazole derivatives are effective corrosion inhibitors, with A being the most effective.

Chen et al. (*25*) looked at imidazoline and its precursor amide as a potentially useful inhibitor in the oil and gas industry. Their high-temperature corrosion inhibition effectiveness was evaluated and compared. The results reveal that at 65.5°C imidazoline and amide have a greater than 95% inhibitory effectiveness. The temperature determines the lowest inhibitor concentration necessary to provide maximum protection for imidazoline and amide. For imidazoline and amide, the concentration required at a high temperature of 148.8 °C is roughly 40 times that required at 65.5 °C. At 65.5 °C, a minimum of 25 ppm of an inhibitor concentration is required for the maximum protection of 95%, at 148.8 °C/3000 pounds per square inch gauge, and 1000 ppm is required for the maximum protection of 90%. Amide, on the other hand, has a somewhat higher performance at high temperatures, implying greater long-term thermal stability.

Quraishi and Jamal (*26*) investigated the action of 500 ppm of 3-undecane-4-aryl-5-mercapto-1,2,4-triazole (A), 3(heptadeca-8-ene)-4-aryl-5-mercapto-1,2,4-triazole (B), and 3(deca-9-ene)-4-aryl-5-mercapto-1,2-4 triazole (C)—three fatty acid triazoles—as corrosion inhibitors for the oil-well steel N80 in 15% HCl at temperatures of above 105 °C. These inhibitors showed a significant decrease in corrosion current. All three inhibitors showed significant corrosion inhibition at 500 ppm following the order: C (96.2%) > B (90.5%) > A (60.2%). The investigated inhibitor C had the best inhibition performance. It is possible that this is due to the presence of a double bond at the terminal position. These triazoles cause no substantial changes in corrosion potential values, indicating that they are mixed-type inhibitors.

Quraishi et al. (*27*) investigated the effects of hydrazides, such as undecenoic acid hydrazide (a), oleic acid hydrazide (b), and lauric acid hydrazide (c), and semithiocarbazides, such as 1-decene-4-phenyl-thiosemicarbazide (A) > 1-heptadecene-4-phenyl-thiose. The order of percentage inhibition efficiency at 5000 ppm for the chemicals evaluated as inhibitors for mild steel corrosion in boiling 15% HCl is as follows: for hydrazides, a (90.4%) > b (84.4%) > c (71.3%), and for semithiocarbazides, A (96.0%) > B (86.1%) > C (77.8%).

High-Temperature Naphthenic Acid Corrosion Inhibitors

The presence of naphthenic acids (NAP), which are present in crude oils, causes high-temperature corrosion. Temperatures between 200 and 400 °C, are most likely to cause NAP-induced Corrosion. The use of inhibitors is a common strategy for avoiding NAP-induced corrosion. Both mono- and dibasic carboxylic acids are referred to as NAP. NAP accounts for about half of the total acidic components in crude oil and can be written as follows:

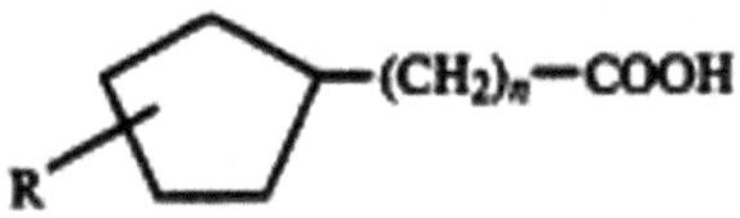

Figure 4. General structure of NAP.

where R represents an alkyl or cycloalkyl group and n represents a number (which ranges generally from 2 to 10).

The corrosion produced by NAP can be reduced with the use of phosphorus-based compounds, which are employed as inhibitors. However, certain consumers have imposed phosphorous content limitations owing to severe environmental laws and regulations, requiring the adoption of phosphorus-free alternative formulations. Some components that impact corrosion include the presence of S, the velocity of the flow regime, and the position in the unit, such as at the liquid–vapor phase boundary. When crude oil is processed, the amount of corrosion caused by NAP in the distillation unit is reduced at lower than 204 °C. The synergistic corrosivity of high temperature and NAP is most noticeable at 204 °C. As a result, corrosion inhibitors based on N are useless at this high temperature. The process varies from that of the liquid form given that the corrosion rate climbs slightly because NAP is present in vapor form at high temperatures. This might get considerably worse if S-containing compounds, such as H_2S, are present.

At the same temperature range, NAP causes sulfidation corrosion, which is caused by reactive S compounds found in crude oils. Blending with lower-acid crudes, neutralizing or eliminating NAP, and using corrosion inhibitors are all methods for minimizing NAP corrosion of existing equipment caused by high total acid number oil. This finding indicates that developing inhibitors for high-temperature NAP-induced corrosion in the oil and gas sectors is a critical and difficult task. It may be accomplished in a variety of ways, one of which is by the use of different combinations. Babaian-Kibal et al. (*9*) documented one such combination in the literature. They investigated the performance of an inhibitor formulation using an alkaline earth metal phosphonate–phenate sulfide and trialkyl phosphate combination (in the ratio of 1:1, up to 5:1).

For high-temperature NAP corrosion, several mercaptan and S compounds, such as simple alkyl mercaptans, dialkyl polysulfide, and di-*t*-nonylpentasulfide, have previously been discovered. Only di-*t*-nonylpentasulfide comes close to the mercaptotriazine compounds' high-temperature corrosion inhibition in the presence of H_2S, elemental S, and mercaptans that are present in crude oil and distillates. S-substituted mercaptotriazine compounds having one or two mercapto hydrogens replaced with alkyl, aryl, cycloalkyl, or other substituents are similarly likely to be effective inhibitors. Another possibility is to use phosphite-based compounds with at least one aryl group as an inhibitor. The operating conditions of temperature range, flow regime, and inhibitor concentration have a significant impact on the efficiency of all of these compounds. As a result, the inhibitor should be begun at a rather high concentration, in the range of 2000–5000 ppm. The subsequent dose will be reduced to a range of 100–1500 ppm once the inhibitor has created a corrosion-resistant layer on the metal surface. High-temperature corrosion-inhibiting phosphite compounds include diphenyl phosphite, isooctyl diphenyl phosphite, and triphenyl phosphite (*28*).

Sulfiding agents might potentially be employed in the oil and gas sectors to increase the efficacy of phosphorus in mitigating high-temperature corrosion. A phosphate ester and organic polysulfide were shown to be a great and effective combination in this circumstance. The phosphate ester can have phosphate, diphosphate, triphosphate, thiophosphate, dithiophosphate, and trithiophosphate as phosphate groups in various combinations. Other phosphates include methyl phosphate, ethyl phosphate, *n*-propyl phosphate, isopropyl phosphate, butyl phosphate, and others, with favored sulfide groups being terpene polysulfides or olefin polysulfides. The polysulfide can contain alkyl or cycloalkyl groups (aromatics with 6–30 C atoms), as well as different stoichiometric ratios of sulfide. The S content of the polysulfide can range from 10 to 60%, although the optimal range for good performance is 25 to 50%. Because polysulfide is soluble in oil, it may be readily introduced to the petroleum stream at concentrations ranging from 30 to 80% or 25 to 2000 ppm (*29*).

Organic polysufides have been demonstrated to be excellent corrosion inhibitors for refinery distillation units when it comes to NAP-induced Corrosion. Petersen et al. (*30*) published their findings in a patent. They claimed that the corrosion inhibitor may be introduced to the oil upstream of the furnaces to protect the furnace tubes and distillation equipment. A reflux recycling stream might also get the inhibitor (which is returned to the atmospheric or vacuum distillation tower above the NAP-corroding section). The treated liquid will then descend into the tower, protecting all metal surfaces from NAP-induced Corrosion. The concentration of corrosion inhibitor in the oil should be sufficient to protect the oil from the acid's corrosive effects. However, economics dictates that, with sufficient amounts of therapy, the percentage protection should be greater than 40%, ideally 50–80%. The corrosion inhibitor concentration will typically range from 10 to 5000 ppm, ideally between 25 and 2000 ppm, and most preferably between 100 and 1500 ppm, depending on the weight of the input stream. Organic polysufides are very effective at treating crude oil that contains corrosive quantities of NAP and H_2S.

Calcium Carbonate Scale and Ash Deposit Inhibitors

Equipment made of superalloys based on Fe, Ni, and Co is frequently used in steam boilers and in the harsh environments of gas turbines. Heat corrosion degradation is a problem for these metals. Many commercially available corrosion inhibitors, such as Mn and Mg based compounds, can help to minimize the severity of oil ash corrosion. MnO_2, CaO, ZnO, Al_2O_3, PbO, BaO, BaO, SiO_2, $Ba(OH)_2$, $CaCO_3$, $Ca(OH)_2$, and oil-soluble Al, Ni, Fe, and other compounds are also suitable for hot corrosion in various oil and gas units (*31*).

In the presence of Na_2SO_4 and V_2O_5, Goya et al. studied the effect of a superficially applied a ZrO_2 inhibitor on the high-temperature corrosion performance of various Fe-, Ni-, and Co-based superalloys as a corrosion inhibitor for ash deposits (*32*). The corrosion rate was lowering after the weight reduction experiment, and it followed the Fe-, Ni-, and Co-based configurations pattern.

Singh et al. (33) investigated the high-temperature corrosion behavior of certain Fe-, Ni-, and Co-based superalloys in the presence of Y_2O_3 as a salt or ash deposit inhibitor in the presence of Na_2SO_4 and V_2O_5. The rate of corrosion for the Co-based alloy was determined to be the highest, whereas the rate of corrosion for the Ni-based alloy was the lowest. Surface-applied Y_2O_3 was shown to be effective in reducing the corrosion of high-temperature metals.

Jones (*34*) patented SnO_2 as a high-temperature corrosion inhibitor via liquid O20S5V2–10 deposits and gaseous SO_3 on engines and other high-temperature equipment, primarily in the petroleum sector, where components such as S, V, and Na are burned. SnO_2 was used as a stabilizer for Zr as a thermal barrier coating in this application, resulting in a Zr-stabilized SnO_2 that is highly resistant to molten $O_{20}S_5V_2{}^{-10}$ hot corrosion in the presence of SO_3 gas.

Du et al. (*1*) investigated the efficacy of an HZG inhibitor. The major components of HZG include imidazoline derivatives, organic amines, and fluorocarbon imidazoline, which has a high corrosion-resistant strength and an average corrosion-resistant percentage of 90% at 130°C. The HZG has a high corrosion-resistant effect and can address the corrosion issues of most oil wells, according to the field application.

Corrosion Inhibitors for High-Temperature Sweet Corrosion

In oil and gas production wells, sweet corrosion appears to be the most frequent kind of corrosion. In the industry, it is also recognized as a primary source of serious corrosion-related material failure. This is what happens when steel-based structural materials come into contact with water solutions containing dissolved CO_2 gas. Farquhar et al. (*34*) described the first evident application of a corrosion inhibitor for high-temperature sweet corrosion of steel. In a synthetic brine solution saturated with 0.54 megapascal (MPa) CO_2 gas at 150 °C, various polyamines and alkyl amides were used as corrosion inhibitors for X65 steel. The most-often reported high-temperature sweet corrosion inhibitors appear to be imidazoline-based compounds. Imidazoline is a good film former, and its derivatives are key starting materials in the synthesis of several sweet corrosion inhibitors. This is due to the steel surface's efficient adsorption characteristics (*35*, *36*). During sweet corrosion, they operate as mixed-type corrosion inhibitors, preventing both anodic and cathodic half-reactions. Jovancicevic et al. (*37*) are credited with being the first to describe an imidazoline derivative as a high-temperature sweet corrosion inhibitor. The impact of hydrocarbon chain length on the inhibitory behavior of an imidazoline derivative was studied by the authors. Fifty parts per million of imidazoline may reduce the corrosion rate of C1018 steel to 2 mpy at 150°C, 15 psi pCO_2, and 6000 rpm rotation speed. The imidazolines' inhibitory effectiveness improved with the length of the alkyl chain. The longer alkyl chain allowed for the development of bigger, more stable lamellar micelles with stronger contacts with the steel surface. Several publications (see Table 1) have since documented the use of imidazoline derivatives as high-temperature sweet corrosion inhibitors for various types of steel. Table 1 shows some of the information found in the literature.

Table 1. Some of the Data Available in the Literature on Corrosion Inhibitors for Sweet Environment

S. no.	*Inhibitor*	*Material*	*Experimental conditions*	*Performance*	*Ref.*
1	Carboxylic acid	MS	Test solution: 0.5 M NaCl solution Concentration used: 200 ppm Temperature: 120 °C Pressure: 1450 psi Technique: Autoclave	IE: 80%	(*38*)
2	Carboxylic acid	MS	Test solution: 0.5 M NaCl solution Concentration used: 200 ppm Temperature: 150 °C Pressure: 479 psi Technique: Autoclave	IE: 80%	(*39*)
3	Gemini imidazoline	C1018 CS	Test solution: Synthetic brine solution Concentration used: 500 ppm Temperature: 100 °C Pressure: 15 psi Technique: Sparged beaker test	CR: 2 mpy	(*40*)
4	Imidazoline	MS	Test solution: 0.5% NaCl Solution Concentration used: 200 ppm Temperature: 120 °C Pressure: 145 psi Technique: Autoclave	IE: 97%	(*41*)

Table 1. (Continued). Some of the Data Available in the Literature on Corrosion Inhibitors for Sweet Environment

S. no.	*Inhibitor*	*Material*	*Experimental conditions*	*Performance*	*Ref.*
5	Imidazoline	MS	Test solution: 0.5% NaCl solution Concentration used: 200 ppm Temperature: 120 °C Pressure: 145 psi Technique: Autoclave	IE: 90%	(42)
6	Oleic imidazoline	X65 Steel	Test solution: 3 g/L NaCl solution Concentration used: 200 ppm Temperature: 150 °C Pressure: 15 psi Technique: Jet impingement	CR: 0.4 mpy	(43)
7	Aldehyde	MS	Test solution: 0.5 M NaCl solution Concentration used: 200 ppm Temperature: 120 °C Pressure: 1450 psi Technique: Autoclave	IE: 88.3%	(44)
8	Imidazoline	X65 Steel	Test solution: 3 g/L NaCl solution Concentration used: 200 ppm Temperature: 150 °C Pressure: 15 psi Technique: Jet impingement	CR: 0.4 mpy	(45)
9	Oleic imidazoline	X65 Steel	Test solution: 3 g/L NaCl solution Concentration used: 200 ppm Temperature: 150 °C Pressure: 15 psi Technique: Jet impingement	CR = 2.4 – 15.7 mpy	(43)
10	Imidazoline	X65 CS	Test solution: Synthetic brine solution Concentration used: 250 ppm Temperature: 93 °C Pressure: 29 psi Technique: Autoclave	IE: 84%	(46)
11	Polyamine	C1018 Steel	Test solution: Synthetic brine solution Concentration used: 400 ppm Temperature: 121 °C Pressure: 100 psi Technique: Autoclave	CR: 59 mpy	(47)

Table 1. (Continued). Some of the Data Available in the Literature on Corrosion Inhibitors for Sweet Environment

S. no.	*Inhibitor*	*Material*	*Experimental conditions*	*Performance*	*Ref.*
12	Quaternary amine	13 Cr Steel	Test solution: Synthetic brine solution Concentration used: 100 ppm Temperature: 120 °C Pressure: 15 psi Technique: Autoclave	CR: 1.6 mpy IE: 98.27%	(*48*)
13	Amidoamine	C1018 steel	Test solution: Synthetic brine solution Concentration used: 200 ppm Temperature: 116 °C Pressure: 100 psi Technique: Autoclave	CR: 20 mpy	(*49*)
14	Amidoamine	C1018 steel	Test solution: Synthetic brine solution Concentration used: 200 ppm Temperature: 204 °C Pressure: 797 psi Technique: Autoclave	CR: 5 mpy	(*50*)
15	Amidoimidazoline	C1018 steel	Test solution: 5% NaCl solution Concentration used: 110 ppm Temperature: 93 °C Pressure: 80 psi Technique: Jet impingement test	CR: 10 mpy	(*50*)
16	Imidazoline	MS	Test solution: 80% NaCl solution Concentration used: 1000 ppm Temperature: 232 °C Pressure: 450 psi Technique: Autoclave	IE: 90%	(*51*)
17	Amine	MS	Test solution: Synthetic brine solution Concentration used: 1000 ppm Temperature: 232 °C Pressure: 450 psi Technique: Autoclave	CR: 16.1 mpy IE: 96%	(*51*)
18	Amine	C1018 Steel	Test solution: Synthetic brine solution Concentration used: 1000 ppm Temperature: 110 °C Pressure: 4351 psi Technique: Autoclave	CR: 2.3 mpy	(*52*)

Table 1. (Continued). Some of the Data Available in the Literature on Corrosion Inhibitors for Sweet Environment

S. no.	*Inhibitor*	*Material*	*Experimental conditions*	*Performance*	*Ref.*
19	Imidazoline	C1018 Steel	Test solution: Synthetic brine solution Concentration used: 200 ppm Temperature: 135 °C Pressure: 149 psi Technique: Jet impingement test	CR: 153 mpy	(22)
20	Quaternary amine	13 Cr Steel	Test solution: Synthetic brine solution Concentration used: 1000 ppm Temperature: 135 °C Pressure: 149 psi Technique: Autoclave	CR: 54.7 mpy	(22)
21	Amine	X65	Test solution: 1% NaCl solution Concentration used: 30 ppm Temperature: 120 °C Pressure: 15 psi Technique: Electrochemistry in specially designed loop	CR: 10 mpy	(53)
22	Imidazoline	CS	Test solution: Synthetic brine solution Concentration used: 100 ppm Temperature: 130 °C Pressure: 116 psi Technique: Autoclave	IE: 96% CR: 2 mpy	(54)
23	Imidazoline	MS	Test solution: 5% NaCl solution Concentration used: 1000 ppm Temperature: 149 °C Pressure: 3000 psi Technique: Autoclave	IE: 90% CR: 2.5 mpy	(55)
24	Amine	API 13 Cr Steel	Test solution: 5% NaCl solution Concentration used: 1000 ppm Temperature: 150 °C Pressure: 580 psi Technique: Electrochemistry in specially designed loop	IE: 95%	(56)
25	Amidoamine	CS	Test solution: Synthetic brine solution Concentration used: 100 ppm Temperature: 130 °C Pressure: 116 psi Technique: Multiphase flow loop	IE: 96%	(54)

[a] Note: MS, mild steel; Ref., reference; CR, corrosion rate; IE, inhibition efficiency.

Corrosion Inhibitors for High-Temperature Sour Corrosion

Sour corrosion has been a concern in the oil and gas industry since the 1940s when it was first discovered in Louisiana and Texas oil wells (*57, 58*). Sour corrosion is defined as metal corrosion in an aqueous environment with acid gases (H_2S/CO_2), but with H_2S concentrations sufficiently higher than CO_2 to dominate corrosion behavior and products. The pH_2S should be greater than 0.30 kilopascal (kPa) (0.05 psi), according to NACE (*59*).

The following test conditions were used by Stewart et al. (*60*): pressure, 830 kPa; temperature, 60 °C (140 °F); time, 72 h; pH_2S, 620 kPa; pCO_2, 210 kPa; brine, synthetic; brine volume, 270 mL; hydrocarbon volume, 30 mL; and stir rate, 120 rpm. The oil and gas sectors have frequently utilized inhibitors A and E; inhibitors B, C, and D were novel potential products; C and D included amines and quaternary amines, respectively; and B had a unique amine-based chemistry. The new candidate inhibitors were shown to be more chemically and thermally stable than commercial inhibitors, with more inhibitive strength against C steel pitting corrosion. Kashkovskiy et al. [612] identified tributylamine as a promising volatile inhibitor for sour steel corrosion. Imidazolines and its derivatives are effective corrosion inhibitors, and most commercial inhibitors for oilfield applications—such as Solvay's Marcor 310, 550, 900, and Ps9—are based on imidazoline chemistry (*62*). One of the controversies among corrosion experts is the mechanism of inhibition by imidazoline in a corrosive environment.

According to Edwards et al. (*63*), corrosion inhibition by imidazoline molecules is primarily due to the imidazoline ring's bonding to the metal surface in a planar orientation. According to the authors, the hydrophobic portion of the imidazoline ring plays a critical role in inhibition performance, whereas the pendant group contributes less. Although the orientation was not examined, Gusmano et al. (*64*) seemed to agree with Edwards et al. on the strong contact between the imidazoline ring and the substrate surface, but they varied on the importance of the pendant group. The pendant group, according to Gusmano et al., might increase inhibitory efficacy by increasing molecule adsorption. Zhao et al. (*65*) proved this by designing and analyzing an imidazoline inhibitor with three phenyl groups. With 50 mg/L of this inhibitor, a C steel inhibition effectiveness of 94% was reached in a H_2S/CO_2 environment. Szyprowski (*66*) synthesized seven homogeneous imidazoline inhibitors and tested their effectiveness on Corrosion of St3S C steel and 1H18N9T austenitic stainless steel in a 2 wt % NaCl-containing hydrocarbon phase saturated with H_2S gas. The inhibitors with the smallest hydrophobic region exhibited the most inhibition.

Table 2. Some of the Surfactant Based Inhibitors for Low-C Steel in a Sour Environment

S. no.	*Inhibitors*	*Experimental conditions*	*Performance*	*Ref.*
1	20–30% Mannich base quaternary ammonium propargyl alcohol, 15–20% quaternary ammonium salt 8-hydroxyquinoline, 10–17% fatty alcohol polyoxyethylene ether phosphate potassium salt, 5–12% 2-ethyamino ethanol, 1–2% emulsifier, and 30–35% solvent	Temperature: 60 °C Test duration: 168 h H_2S: 2000 mg/L pCO_2: 0.5 MPa Pressure: 10 MPa Inhibitor concentration: 200 mg/L	IE: 98.5%	(73)
2	Quaternary ammonium salt prepared by a reaction of a polyepihalohydrin with tertiary amine	NACE brine and LVT 200 (90:10) solution made sour by bubbling H_2S through the solution. Inhibitor concentration: 15 and 25 ppm	IE: 86.0–90.0%	(74)
3	Quaternary pyridinium salt plus lower alkylene polyamine/corresponding polyalkylene polyamine	Test solution: 30% ethanolamine saturated with CO_2 and containing the equivalent of 100 ppm, 300 ppm, and 500 ppm H_2S as $(NH_4)_2S$ Pressure: 40 psi Temperature: 125 °C Test duration: 16-18 h Concentration: 100 and 500 ppm	IE: 76.5–95.2%	(75)
4	1-Benzyl-quinolinium chloride	Concentration: 100 mg/L Temperature: 60 °C pH: 4.5 Test duration: 72 h	IE: 95.2%	(67)
5	Cocodimethyl ammonium chloride and alkyl pyridinium ammonium chloride	Water cut: 100% pCO_2 saturated aqueous phase ASTM method D-1141-52 Temperature: 25 °C S: 1.0 g/L and 10 g/L; Test duration: 6 h Optimum concentration:1000 ppm	CR: 0.1 mpy (0.0025 mm/y)	(70)

[a] Note: Ref., reference; CR, corrosion rate; IE, inhibition efficiency.

To explain the corrosion inhibition mechanism of imidazoline, Ramachandran et al. (*67*) utilized atomistic simulations (quantum mechanics and molecular dynamics) and came to the conclusion that a self-assembled monolayer form on the metal's native oxide surface. By providing a hydrophobic barrier against the entrance of tiny ions, this layer acts as a corrosion inhibitor. For monolayer construction to occur, the hydrocarbon tail must be sufficiently long, with roughly 12 or more C atoms, and the hydrophilic head must result in an octanol/water partition coefficient (log P) below a threshold value (*68*). The description by Blair et. al (*68*) of the most efficient imidazoline compounds appears to support this hypothesis. Blair et al. characterized the most potent imidazoline inhibitors as having a lengthy hydrocarbon chain linked to one of the imidazoline ring's N atoms or a relatively tiny organic radical attached to one of the ring's N atoms in US Patent 2,466,517. Cocodimethyl ammonium chloride (*69*), alkyl pyridinium ammonium chloride, and 1,3-bis (dodecyl dimethyl ammonium chloride)-2-propanol (*70*) are some of the Gemini surfactants that have been evaluated as low-C steel inhibitors in a sour environment. 1,3-Bis(myristyl dimethyl ammonium chloride)-2-propanol (*71*), 1,3-bis(hexadecyl dimethyl ammonium chloride)-2-propanol (*71*), and 1,3-bis(octadecyl dimethyl ammonium chloride)-2-propanol (*72*) are some of the others. There have also been several claims for sour corrosion inhibitions using this family of inhibitors, some of which are described in Table 2.

Summary and Future Considerations

This chapter concentrated on the inhibitors that could be used for high-temperature corrosion management and control in the oil and gas industry. The focus was on gaining a better knowledge of the mechanisms, identifying vulnerable units, and selecting optimal inhibitors based on the various contributing elements that cause high-temperature corrosion. As a result of this knowledge, inhibitor formulations can be developed by creating a novel chemical or a combination of compounds with the necessary qualities to endure such conditions and effectively stop corrosion. Structure and electron density are key prerequisites for choosing inhibitors for high-temperature corrosion mitigation according to early testing detailed in this chapter. To withstand the effects of high-temperature degradation, inhibitors should have a high molecular weight and a large volume electron cloud to permit good adsorption. In formulations that combine two or more inhibitors to improve temperature resistance, the combining ratio is critical to achieving optimal efficiency. Because the metal elements of each structural unit and the presence of other corrosive agents—such as acids, CO_2, H_2S, and other hot gases—tend to worsen the rate of high-temperature corrosion, the area or units of operation is also highly essential in the inhibitor selection and application. As a result, inhibitors like ZrO_2 have been found to be particularly effective in preventing oil ash corrosion in nickel- and iron-based alloys. Some future directions in the development of high-temperature sweet corrosion inhibitors may be suggested based on the recognized state (efficiencies and gaps) of corrosion inhibitors revealed for high-temperature sweet corrosion of steel. When several corrosion inhibitor chemistries with good performance can be combined into a single formulation to create a high level of synergism, some surprising outcomes will undoubtedly be obtained. One of the ways could be by employing the corrosion inhibitors that have already been used in other high-temperature environments, such as in the geothermal field. Polyethylene imines, metal salts of glycol phosphate esters, and amine sulfurized borate are some examples.

Acknowledgments

The authors acknowledge financial support from the Council of Scientific and Industrial Research, New Delhi, India, through the major research project file no. **22(0832)/20/EMR-II.**

References

1. Du, Q.; Xie, G.; Yan, X.; Zhang, L.; Pei, Y.; Shang, F.; Wang, Y.; Li, H.; Chen, S. Corrosion inhibitor for high temperature oil well. *J. Chem. Pharm. Res.* **2015**, 7, 1062–1068.
2. Fiaud, C.; Harch, A.; Mallouh, D.; Tzinmann, M. The inhibition of iron corrosion by acetylenic alcohols in acid solutions at electrochemical testing. *Corros. Sci.* **1993**, 35, 1437–1444.
3. Gao, W.; Li, Z. Developments in high temperature corrosion and protection of materials. *Woodhead publishing series in metals and surface engineering* **2008**, Pages 1–5.
4. Prabha, S. S.; Rathish, R. J.; Dorothy, R.; Brindha, G.; Pandiarajan, M.; Al-Hashem, A.; Rajendran, S. Corrosion problems in petroleum industry and their solution. *Eur. Chem. Bull.* **2014**, 3, 300–307.
5. Hackerman, N. Use of inhibitors in corrosion control. *Corros.* **1948**, 4, 45–60.
6. Hong, T.; Jepson, W. P. Corrosion inhibitor studies in large flow loop at high temperature and high pressure. *Corros. Sci.* **2001**, 43, 1839–1849.
7. Edmondson J. G. (1996) *High temperature corrosion inhibitor*. US Patent 5,500,107, filed 15 March 1994 and issued 19 March 1996.
8. Zetlmeisl M. J., Outlaw B. T. (1990) *Corrosion inhibitors for use in hot hydrocarbons*, US Patent 4,941,994, filed 18 July 1989 and issued 17 July 1990.
9. Babaian-Kibal E., Walker J. S., Hyatt J. G., Ferguson S. (1996) *High temperature corrosion inhibitors simulator*. US Patent 5,503,006, filed 23 October 1980 and issued 04 February 1986.
10. Saji, V. S. A review on recent patents in corrosion inhibitors. *Rec. Pat. Corros. Sci.* **2010**, 2, 6–12.
11. Yepez, O. J. (2018). *On the Mechanism of High Temperature Corrosion*. The Woodlands, TX: Clariant Oil Services.
12. Deal, B. E.; Grove, A. S. Mechanisms of oxidation and Corrosion. *J. Appl. Phys.* **1999**, 85, 7646.
13. John, R. C.; Pelton, A. D.; Young, A. L.; Thompson, W. T.; Wright, I. G.; Besmann, T. M. Assessing Corrosion in oil refining and petrochemical processing. *Mater. Res.* **2004**, 7, 163–173.
14. May, W. R.; Zetlmeisl, M. J.; Annand, R. R. High-temperature corrosion in gas turbines and steam boilers by fuel impurities—Part VII: evaluation of magnesium-aluminum-silicon combinations as corrosion inhibitors. *J. Eng. Power.* **2010**, 98, 511.
15. Hurst, R. C.; Norton, J. F. The Effect of High Temperature Carburization upon the Ambient Temperature Ductility of Alloy 800H. *High Temperature Technology* **1983**, *1*, 319–325.
16. Stellwag, B. The mechanism of oxide film formation on austenitic stainless steels in high temperature water. *Corros. Sci.* **1998**, *40*, 337–370.
17. Abdel-Ghany, N. A.; Shehata, M. F.; Saleh, R. M.; El Hosary, A. A. Novel corrosion inhibitors for acidizing oil wells. *Mat. Corros.* **2016**, *68*. https://doi.org/10.1002/maco.201609146.

18. Finšgar, M.; Jackson, J. Application of corrosion inhibitors for steels in acidic media for the oil and gas industry: a review. *Corros. Sci.* **2014**, *86*, 17–41.
19. Amosa, M. K.; Mohammed, I. A.; Yaro, S. A.; Arinkoola, A. O.; Ogunleye, O. O. Corrosion inhibition of oil-well steel (N-80) in simulated hydrogen sulphide environment by ferrous gluconate and synthetic magnetite. *NAFTA* **2010**, *61* (5), 239–246. https://hrcak.srce.hr/index.php?show=clanak&id_clanak_jezik=82178.
20. Popoola, L. T.; Grema, A. S.; Latinwo, G. K. Corrosion problems during oil and gas production and its mitigation. *Intl. J. Indus. Chem.* **2013**, *4*, 1–35.
21. Rajeev, P.; Surendranathan, A. O; Murthy, S. N. Corrosion mitigation of the oil well steels using organic inhibitors—a review. *J. Mater. Environ. Sci.* **2012**, *3*, 856–869.
22. Abayarathna, D.; Naraghi, A.; Obeyesekere, N. (2003) Inhibition of Corrosion of carbon steel in the presence of CO_2, H_2S and S. In: *NACE International Conference*, 16–20 March. National Association of Corrosion Engineers, San Diego, California, USA.
23. Schauhoff, S.; Kissel, C. L. New corrosion inhibitors for high temperature applications. *Mater. Corros.* **2002**, *51*, 141–146.
24. Quraishi, M. A.; Jamal, D. Corrosion inhibition by fatty acid oxadiazoles for oil well steel (N-80) and mild steel. *Materials chemistry and physics* **2001**, *71*, 202–205.
25. Chen, H. J. *High temperature corrosion inhibition performance of imidazoline and amide, corrosion*, 2000, paper no. 00035.
26. Quraishi, M. A.; Jamal, D. Fatty acid triazoles: novel corrosion inhibitors for oil well steel (N-80) and mild steel. *J. Am. Oil Chem. Soc.* **2000**, *77*, 1107–1111.
27. Quraishi, M. A.; Jamal, D.; Saeed, M. T. Fatty Acid Derivatives as Corrosion Inhibitors for Mild Steel and Oil-Well Tubular Steel in 15% Boiling Hydrochloric Acid. *J. Am. Oil Chem. Soc.* **2000**, *77*, 265–268.
28. Robertson, I. D.; Dean, L. M.; Rudebusch, G. E.; Sottos, N. R.; White, S. R.; Moore, J. S. Alkyl phosphite inhibitors for frontal ring-opening metathesis polymerization greatly increase pot life. *ACS Macro Lett.* **2017**, *6*, 609–612.
29. Greyson, E.; Manna, J.; Mehta, S. C. (2012). *Scale and corrosion inhibitors for high temperature and pressure conditions*. US Patent 8,158,561 B2, filed 13 August 2010 and issued 17 April 2012.
30. Petersen, P. R.; Robbins, F. P.; Winston, W. G. (1993). *Naphthenic acid corrosion inhibitors*. US patent, 5,182,013, Date of Patent: Jan. 26, 1993.
31. Paul, L. D.; Seeley, R. R. Oil Ash Corrosion—A Review of Utility Boiler Experience. *Corrosion* **1991**, *47* (2), 152.
32. Goya, G.; Singh, H.; Prakash, S. Effect of superficially applied ZrO_2 inhibitor on the high temperature corrosion performance of some Fe-, Co- and Ni-base superalloys. *Appl. Surf. Sci.* **2008**, *254*, 6653–6661.
33. Singh, H.; Singh, S.; Prakash, S. High temperature corrosion behaviour of some Fe-, Co- and Ni-base superalloys in the presence of Y_2O_3 as inhibitor. *Appl. Surf. Sci.* **2009**, *255*, 7062–7069.
34. Jones, R. L. (1993). *Corrosion inhibition in high temperature environment*, NC 75, 053, 22217-5660.
35. Farquhar, G. B.; Kane, R. D.; Abayarathna, D. 1997. Prediction of corrosion inhibition performance using simulated CO2/H2S environment autoclave and flow loop tests. In: *NACE*

International Conference, 9–14 March. National Association of Corrosion Engineers, New Orleans, Louisiana, USA.

36. Ortega-Toledo, D. M.; Gonzalez-Rodriguez, J. G.; Casales, M.; Martinez, L.; Martinez-Villafane, A. CO_2 corrosion inhibition of X-120 pipeline steel by a modified imidazoline under flow conditions. *Corros. Sci.* **2011**, *53*, 3780–3787.
37. Jovancicevic, V.; Ramachandran, S.; Philippe, P. Inhibition of CO2 Corrosion of Mild Steel by Imidazolines and Their Precursors. Paper presented at the *CORROSION 98*, San Diego, California, March 1998.
38. Ali, S. A.; Jafar, M. A.; Al-Muallem, H. A. 2017. *Mild steel corrosion inhibitor compounds*. US 9,758,878 B2.
39. Tomoe, Y.; Shimizu, M.; Ito, T.; Sunaba T. 2013. Inhibitor evaluation for 13%Cr steel in high temperature, sweet corrosions, and the effects of co-existence of liquid hydrocarbons. In: *NACE International Conference* 17–21 March. National Association of Corrosion Engineers, Orlando, Florida, USA.
40. Yang, J.; Jovancicevic, V. 2009. *Corrosion inhibitors for oilfield applications*. US2009/0181867 A1.
41. Mazumder, M. A. J.; Al-Muallem, H. A.; Ali, S. A. The effects of N-pendants and electron-rich amidine motifs in 2-(p-alkoxyphenyl)-2-imidazolines on mild steel corrosion in CO2-saturated 0.5 M NaCl. *Corros. Sci.* **2015**, *90*, 54–68.
42. Mazumder, A. J.; Ali, S. A.; Al-Muallem, H. A. 2018. *Corrosion inhibition with alkoxy aromatic imidazolined*. US 9,868,894 B1.
43. Palenscar, A.; Gulbrandsen, E.; Kosoru, K. 2013. High temperature testing of corrosion inhibitor performance. In: *NACE International Conference* 17–21 March, National Association of Corrosion Engineers, Orlando, Florida, USA.
44. Mazumder, M. A. J.; Al-Muallem, H. A.; Faiz, M.; Ali, S. A. Design and synthesis of a novel class of inhibitors for mild steel corrosion in acidic and carbon dioxide-saturated saline media. *Corros. Sci.* **2014**, *87*, 187–198.
45. Palenscar, A.; Gulbrandsen, E.; Kosoru, K. 2013. Corrosion inhibition under $FeCO_3$-forming conditions at elevated temperatures. In: *SPE International Oilfield Corrosion Conference and Exhibition*, 12–13 May. Society of Petroleum Engineers, Aberdeen, Scotland.
46. Hassani, Sh.; Roberts, K. P.; Shirazi, S. A; Rybicki, E. F., Shadley, J. R.; Joia, C., 2012. Characterization and prediction of chemical inhibition performance for erosion-corrosion conditions in sweet oil and gas production. In: *NACE International Conference* 11–15 March. National Association of Corrosion Engineers, Salt Lake City, Utah, USA.
47. Ramachandran, S.; Menendez, C.; Jovancicevic, V.; Long, J. Film persistency of new high-temperature water-based batch corrosion inhibitors for oil and gas wells. *J. Petrol. Explor. Prod. Technol.* **2012**, *2*, 125–131.
48. Jenkins, A. 2011. Performance of high-temperature, biodegradable corrosion inhibitors. In: *NACE International Conference*, 13–17 March. National Association of Corrosion Engineers, Houston, Texas, USA.
49. Ramachandran, S.; Jovancicevic, V.; Jeff, L. 2009. Development of a new water soluble high temperature corrosion inhibitor. In: *NACE International Conference*, 22–26 March, National Association of Corrosion Engineers, Atlanta Georgia, USA.

50. Meng, Q.; Jovancicevic, V. 2008. Electrochemical evaluation of CO2 corrosion inhibitors in high turbulence multiphase fluid flow. In: *NACE International Conference* 16–20 March, National Association of Corrosion Engineers, New Orleans, Louisiana, USA.
51. Ramachandran, S.; Soo Ahn, Y.; Greaves, M.; Jovanciecevic, V.; Basset J. 2006. Development of high temperature, high pressure corrosion inhibitor. In: *NACE International Conference* 12–16 March. National Association of Corrosion Engineers, San Diego, California, USA.
52. Gregg, M.; John L. Non Foaming Corrosion Inhibitor Developments Providing Mitigating Benefits against CO2 Corrosion for Application in High Volume Gas Wells. Paper presented at the *CORROSION 2004*, New Orleans, Louisiana, March 2004.
53. Kvarekval, J.; Gulbrandsen, E. 2001. High temperature and high flow velocity tests for CO2 corrosion inhibitors. In: *NACE International Conference* 11–16 March. National Association of Corrosion Engineers, Houston, Texas, USA.
54. Swidzinski, M.; Fu, B.; Kelly, A.; Jepson, W. P. Inhibitor tests achieve 97% corrosion-protection level. *P&GJ* **2000**, 71–74.
55. Chen, H. J.; Hong; T.; Jepson, W. P. 2000. High temperature corrosion inhibition performance of imidazoline and amide. In: *NACE International Conference*, 26–31 March. National Association of Corrosion Engineers, Orlando, Florida, USA.
56. Tomoe, Y.; Shimizu, M.; Hausier R. H. 2000. Evaluation of commercial corrosion inhibitors for CO2 Corrosion of API-13Cr tubing. In: *NACE International Conference* 26–31 March. National Association of Corrosion Engineers, Orlando Florida, USA.
57. El-Sherik, A. M.; Ajwad, H. A. Sour corrosion: a review of current gaps and challenges. *Mater. Perform.* **2018**, *57* (5), 48.
58. Crolet, J. L. Acid Corrosion in Wells (CO2, H2S): Metallurgical Aspects. *J. Pet. Technol.* **1983**, 1553.
59. Bardal, E. Corrosion and protection, *Engineering Materials and Processes*, Springer-Verlag London Limited, 2004.
60. Stewart, S.; Jovancicevic, V.; Menendez, C.; Moloney, J.; Wamburi, W. Evaluation of a new sour gas corrosion inhibitor for field applications via localized corrosion monitoring techniques. Paper No. 10274, *NACE International Corrosion Conference & Expo* (2010).
61. Kashkovskiy, R. V.; Kuznetsov, Yu. I.; Kazansky, L. P. Evaluation of Partial Protective Contributions of Surface Films Under Inhibition of Metal Corrosion: Ways of Development. *Corros. Sci.* **2012**, *64*, 126.
62. *Solvay products*; https://www.solvay.com/en/search/products?s=Ercabac +50. (Accessed 8 November 2018).
63. Edwards, A.; Osborne, C.; Webster, S.; Klenerman, D.; Joseph, M.; Ostovar, P.; Doyle, M. Mechanistic studies of the corrosion inhibitor oleic imidazoline. *Corros. Sci.* **1994**, *36* (2), 315.
64. Gusmano, G.; Labella, P.; Montesperelli, G.; Privitera, A.; Tassinari, S. Study of the inhibition mechanism of imidazolines by electrochemical impedance spectroscopy. *Corrosion, NACE International* **2006**, 576.
65. Zhao, X.; Chen, C.; Yu, H.; Chen, Q. Evaluation and mechanism of corrosion inhibition performance of new corrosion inhibitor. Paper No. 11029, *NACE International Corrosion Conference & Expo* (2018).

66. Zhao, J.; Duan, H.; Jiang, R. Synergistic corrosion inhibition effect of quinoline quaternary ammonium salt and Gemini surfactant in H2S and CO2 saturated brine solution. *Corros. Sci.* **2015**, *91*, 108–119.
67. Ramachandran, S.; Tsai, B.-L.; Blanco, M.; Chen, H.; Tang, Y.; Goddard, W. A. Self-assembled monolayer mechanism for corrosion inhibition of iron by imidazolines. *Langmuir* **1996**, *12*, 6419.
68. Blair, C. M.; Groves, W., Jr.; Gross; W. F. *Processes for preventing corrosion and corrosion inhibitors*. US Patent No.: 2,466,517, April, 1949.
69. Moore, J.; Liu, J. F.; Tang, X. Corrosion inhibitors in the presence of elemental sulfur, Paper No. 09363, Corrosion 2009, *NACE International Corrosion Conference & Expo* (2009).
70. Blair, C. M.; Mako, E. Adsorption and desorption of corrosion inhibitors in anaerobic brine, *Society of Petroleum Engineers Symposium on Handling of Oilfield Water*, 4–5 December 1972, Los Angeles, California, 2019 Document ID: SPE-4222-MS.
71. Zhao, J. M.; Li, J. Corrosion Inhibition Performance of Carbon Steel in Brine Solution Containing H2S and CO2 by Novel Gemini Surfactants. *Acta Phys. Chim. Sin.* **2012**, *28*, 623.
72. Wafts, M. M. Imidazoline hydrolysis in alkaline and acidic media—A review. *J. Am. Oil Chem. Soc.* **1990**, *67* (12), 993.
73. Yuan, W.; Juantao, Z.; Feng, T.; Qiang, L.; Jinheng, L.; Gang, W.; Qiang, B. *Suitable for oil and gas fields containing H2S/CO2 conveying corrosive environment inhibitor and preparation method line*. Chinese Patent CN 105483708B, 2018.
74. Naraghi, A.; Obeyesekere, N. *Polymeric quaternary ammonium salts useful as corrosion inhibitors and biocides*. US Patent US2006/0062753A1, 2006.
75. Asperger, R. G.; Clouse, R. C. *Quaternary salt-polyamine inhibitor for sour gas conditioning solutions*. US Patent 4,100,099, 1978.

Editors' Biographies

Chaudhery Mustansar Hussain, Ph.D.

Chaudhery Mustansar Hussain, Ph.D., is an adjunct professor, academic advisor, and director of chemistry and EVSC Labs in the Department of Chemistry & Environmental Sciences at the New Jersey Institute of Technology (NJIT), Newark, New Jersey, United States. His research is focused on the applications of nanotechnology and advanced materials, environmental management, analytical chemistry, and other various industries. Dr. Hussain is the author of numerous papers in peer-reviewed journals as well as a prolific author and editor of around 100 books, including scientific monographs and handbooks in his research areas. He has published with Elsevier, Royal Society of Chemistry, John Wiley & Sons, CRC Press, and Springer.

Chandrabhan Verma, Ph.D.

Chandrabhan Verma, Ph.D., works at Centre of Research Excellence in Corrosion, Research Institute, King Fahd University of Petroleum and Minerals, Dhahran 31261, Saudi Arabia. He is a member of the American Chemical Society (ACS). His research is mainly focused on the synthesis and design of environmentally friendly corrosion inhibitors useful for industries. Dr. Verma is the author of numerous research and review articles published in peer-reviewed international journals, including platforms of ACS, Elsevier, Royal Society of Chemistry, Wiley, and Springer. Dr. Verma is currently editing various books at ACS, Elsevier, Royal Society of Chemistry and Wiley platforms. He has total citations of more than 5000 with an H-index of 40 and i-10 index of 91. Dr. Verma has received several national and international awards for his academic achievements.

Indexes

Author Index

Subject Index

Printed in the USA/Agawam, MA
October 27, 2022

800369.015